DRYING OF SOLIDS IN THE CHEMICAL INDUSTRY

DRYING OF SOLIDS IN THE CHEMICAL INDUSTRY

G. Nonhebel, M.A., D.Sc., C.Eng., F.Inst.F., F.R.I.C., M.I.Chem.E.

A. A. H. Moss, B.Sc., C.Eng., A.R.I.C., A.M.I.Chem.E.

LONDON
BUTTERWORTHS
1971.

THE BUTTERWORTH GROUP

ENGLAND:

BUTTERWORTH & CO. (PUBLISHERS) LTD.
LONDON: 88 Kingsway WC2B 6AB

AUSTRALIA:

BUTTERWORTH & CO. (AUSTRALIA) LTD.
SYDNEY: 20 Loftus Street
MELBOURNE: 343 Little Collins Street
BRISBANE: 240 Queen Street

CANADA:

BUTTERWORTH & CO. (CANADA) LTD
TORONTO: 14 Curity Avenue, 374

NEW ZEALAND:

BUTTERWORTH & CO. (NEW ZEALAND) LTD.
WELLINGTON: 49/51 Ballance Street
AUCKLAND: 35 High Street

SOUTH AFRICA:

BUTTERWORTH & CO. (SOUTH AFRICA) (PTY) LTD.
DURBAN: 33/35 Beach Grove

First published 1971

ISBN 0 408 70113 7

FILMSET BY PHOTOPRINT PLATES, RAYLEIGH, ESSEX

Printed by photo-lithography and made in Great Britain at the Pitman Press, Bath

Preface

A joint panel of specialists was appointed by the Chemical Industries Association and the British Chemical Plant Manufacturers Association to study gaps in knowledge on the process of drying chemicals in both large and small manufacturing plants. This panel has sponsored research at the University of Birmingham on basic drying characteristics with particular relevance to the various classes of materials handled in the chemical industries.

The panel decided that a book was required on the practical aspects of drying.* Imperial Chemical Industries has greatly assisted by making available to the authors for this purpose a set of internal technical reports on practical aspects of drying of solids.†

Code letters have been given to the types of commercial equipment described in this book. Readers requiring additional information should apply to the Director of the British Chemical Plant Manufacturers Association, 14 Suffolk St., London, S.W.1.

*The members of the C.I.A./B.C.P.M.A. panel when this recommendation was made were: G. Nonhebel (chairman), P. J. Barr, D. J. Breeze, R. R. Clegg, C. Dodson, C. J. Grayston, C. W. Heathcote, R. Hutt, J. Mackenzie, A. A. H. Moss, S. C. M. Salter, A. J. Scott, R. H. Simpson, O. G. Weller (secretary)
†The I.C.I. reports were prepared by a panel comprising: A. A. H. Moss (chairman), A. L. Cude, E. Davies, J. Denman, G. H. Gostling, H. R. Hinckliеff, J. A. Lodge, V. G. Lovelock, J. B. McGowan, J. Q. Picken, H. G. Usher

Contents

Symbols, Units and Conversion Factors

Quantities and associated symbols in this book (except in Appendix 10.1) are expressed in the International System units (SI) and units derived from them. Readers not fully conversant with SI units should note that the unit of force is the Newton (N); 1 kgf = 9·807 N. Time is expressed in seconds; consequently metric quantities such as kg/h are conveniently expressed as g/s, though the strict SI unit is kg/s. The unit of energy for both work and heat is the Joule. Heat rates are expressed as Watts in preference to Joules per second (1W = 1J/s).

In order to avoid cumbersome multipliers of powers of 10, many of the quantities to which the symbols refer are in units derived from strict SI units, e.g. mm and μm instead of metres. Pressures are expressed as bars instead of the strict SI unit of N/m^2 known as the Pascal (Pa). The standard pressure of one atmosphere = 101 325 kg/s m^2 or N/m^2 or Pa or J/m^3. One millibar (mb) = approximately 1 cm water gauge = 100 Pa. Use of partial pressures in atmospheres instead of in bars is within the error of measurement of operations described in this book.

Tables of conversion to SI units from British and kg.m.h units follow this list of symbols. For further information on the use of SI units in chemical engineering see J. W. Mullin, *The Chemical Engineer,* No. 217, CE 176, (1967), published by The Institution of Chemical Engineers, reprinted with correction in *J. Inst. Fuel,* **41,** 233 (1968).

Chapters in which symbols appear	*Symbol*	*Definition*	*Derived SI unit employed*
2, 8, 9, 10	a	A constant	—
2, 5, 7, 8, 9, 11	A	Area of drying surface at which heat and matter transfer takes place	m^2
2, 8, 11	A'	Cross-sectional area of dryer bed	m^2
2	$\bar{A}$	Concentration of component A	kg/kg
5	A_b	Surface of dryer body, including floor	m^2
8	A_c	Area of dryer band for constant-rate drying	m^2
8	A_f	Area of dryer band for falling-rate drying	m^2
2, 6, 7	A_h	Area of heating surface	m^2
2	A_r	Area of radiating surface	m^2
2	A_s	Area of surface receiving radiation	m^2
6	A_t	Total required drying area	m^2
12	A_u	Surface area per unit mass of solid	m^2
8	b	A constant dependent on particle size of material	—
10	B	Mass hold-up in rotary dryer	kg
	c_b	Capacity of conveyor	g/s
2, 5	c_g	Mean specific heat of (wet) drying gas	kJ/kg°C
5	c_h	Specific heat of liquid water	kJ/kg°C
11	c_p	Mean specific heat of dry air at constant pressure, including dry fluidising gas	kJ/kg°C
11, 12	c_s	Specific heat of dry solids	kJ/kg°C
2, 5	c_v	Mean specific heat of vapour	kJ/kg°C
13	c_ρ	Coefficient of air resistance	kg^2/s^4
13	d	Diameter of spray disc	mm
13	d_d	Maximum drop diameter	mm
5, 11, 12	d_m	Mean diameter of particle	Ch 12 μm Ch 5, 11 mm

Chapters in which symbols appear	*Symbol*	*Definition*	*Derived SI unit employed*
12	d_s	Diameter of sphere having the same surface area as particle	m
12	d_v	Diameter of sphere having the same volume as the particle	m
2	D	Diffusion coefficient	m^2/s
10, 13	D_s	Diameter of rotary dryer or spinning atomiser disc in spray dryer	m
8	$f(\ldots)$	A function of (...)	—
2	f	Friction factor	—
15	f^*	Frequency of electro-magnetic vibrations, Hz	s^{-1}
10	f_r	Average number of lifts (flights) per revolution of the shell of a rotary dryer	—
5	F	Void fraction of granular beds	—
2	F_g	Geometrical factor for shape and emissivity of surface	—
11, 12	F_p	Particle shape factor	—
2	F_r	Geometrical factor for shape and emissivity of radiating surface	—
2	F_s	Geometrical factor for shape and emissivity of drying surface	—
12	g	Acceleration due to gravity (=9·81)	m/s^2
2, 5, 8, 11	G	Drying gas mass flow per unit area of bed or band	$g/s\,m^2$
5	G_e	Air leaving dryer	g/s
5	G_f	Air circulated by dryer fans	g/s
8, 10	h	Heat-transfer coefficient	$W/m^2\,^\circ C$
12	h^*	Distance fallen by particle in laboratory pneumatic dryer	m
2, 5,	h_c	Heat-transfer coefficient by convection	$W/m^2\,^\circ C$
2	h_e	Heat-transfer coefficient for evaporation from surface	$W/m^2\,^\circ C$

Chapters in which symbols appear	Symbol	Definition	Derived SI unit employed
2	h_f	Heat-transfer coefficient from wall to material	W/m^2 °C
2, 9	h_h	Heat-transfer coefficient from heating medium to wall	W/m^2 °C
9	h_l	Heat-transfer coefficient through liquid film	W/m^2 °C
2	h_m	Heat-transfer coefficient by conduction through material being dried	W/m^2 °C
2, 5, 6	h_r	Radiation heat-transfer coefficient	W/m^2 °C
2, 9	h_w	Heat-transfer coefficient by conduction through wall	W/m^2 °C
12	(hA_u)	Heat-transfer coefficient per unit mass of dry solid	W/g °C
2, 8	H	Absolute humidity, mass water/dry air	kg/kg
5	H_0	Humidity of ambient air entering dryer	kg/kg
11	H_1	Approach humidity within a dryer zone	kg/kg
11	H_2	Humidity of air leaving a dryer zone	kg/kg
2, 5, 8, 11,	H_g	Humidity of drying gas, including for fluidisation, mass water/dry air	kg/kg
10	H_l	Hold-up per unit length in rotary dryer	kg
2	H_s	Saturation humidity of drying gas at adiabatic saturation temperature T_{sg}	kg/kg
2, 5, 11	H_{ss}	Saturation humidity of drying gas at drying surface temperature, T_s	kg/kg
2, 5, 8	H_{sw}	Saturation humidity of drying gas at wet-bulb temperature T_{wb}	kg/kg
2	H'_{sw}	Saturation humidity of drying gas over evaporating solid or solution at corresponding wet-bulb temperature T'_{wb}	kg/kg

Chapters in which symbols appear	*Symbol*	*Definition*	*Derived SI unit employed*
8	ΔH	Humidity driving force between surface of evaporation and drying air	kg/kg
8, 11	ΔH_{lm}	Log mean humidity driving force between surface of evaporation and drying air or for inlet and outlet conditions in a dryer zone	kg/kg
2, 9	k_l	Thermal conductivity of liquid film	W/m°C
2, 9	k_m	Thermal conductivity of material	W/m°C
2, 9	k_w	Thermal conductivity of dryer wall	W/m °C
5	K	The appropriate constant for the equation	—
11	K_1	Drying-rate coefficient for individual particle in a fluidised bed	kg/kg s = s^{-1}
12	(KA_u)	Mass-transfer drying coefficient referred to unit mass of solid	kg/kg s bar = s^{-1} bar^{-1}
2, 6, 8	K_c	Conduction drying coefficient	W/m^2 °C
2, 5, 8	K_H	Drying coefficient per unit humidity difference (see footnote*)	g/s m^2
8	K'_H	Drying coefficient on a humidity basis referred to area of bed A'	g/s m^2
11	K^*_H	Drying coefficient on a humidity basis referred to area of bed A' and unit mass of solid	kg/kg s = s^{-1}
2	$(K_H)_c$	Constant-rate value of K_H	g/s m^2
2, 8	K_p	Drying coefficient on partial pressure basis	g/s m^2 bar
2	$(K_p)_c$	Constant-rate value of K_p	g/s m^2 bar

*Unit humidity difference expresses the existence of the dimensionless absolute humidity driving force in the same way as the dimensionless partial pressure driving force appears in the heat- and matter-transfer equation

Chapters in which symbols appear	*Symbol*	*Definition*	*Derived SI unit employed*
	ln	Natural logarithm to base e	—
8	L	Loading of dryer band, dry basis	kg/m^2
10, 12, 14	L_d	Effective length of drying tube, conveyor, rotary dryer	m
5	L_m	Mean length of cylindrical particles	mm
2	m_m	Mean molecular weight	kg
2	m_g	Molecular weight of drying gas	kg
2	m_v	Molecular weight of vapour	kg
2, 7, 8, 11, 12	M	Moisture content, mass water/dry solids*	kg/kg
11, 12	M_1, M_2	Moisture content, inlet and exit	kg/kg
2, 5, 7, 8	M_c	Critical moisture content	kg/kg
2, 5, 6, 7, 8	M_f	Final moisture content	kg/kg
11	M_p	Moisture content of an individual particle	kg/kg
2, 5, 6, 8	M_s	Initial moisture content	kg/kg
8	M_x	Intermediate (interzone) moisture content	kg/kg
12	n	Number of passes in a recycling dryer	—
2	N	Molecular rate of diffusion	$kmol/s\ m^2$
	N_v	Molecular rate of diffusion of water vapour	$kmol/s\ m^2$
2	N_A, N_B	Molecular rate of diffusion for components A and B	$kmol/s\ m^2$
10, 13	N_r	Speed of rotation of rotary dryer or of disc in spray dryer	rev/min
	NTP	Normal (standard) conditions of temperature and pressure to which gas volume reduced, namely 0°C and one standard atmosphere pressure (= 101·25 kN/m^2 = 1·013 25 bar)	—

*To find mass of dry solids W from M, divide the mass of wet solids by $1 + M$ — e.g. for $M = 0{\cdot}42$, divide by 1·42

Chapters in which symbols appear	*Symbol*	*Definition*	*Derived SI unit employed*
	OTB	Gas condition reduced to 0°C and one bar pressure. (This abbreviation is special to this book; an internationally agreed abbreviation is not yet available)	—
2, 8, 10	p	Partial pressure of vapour	mb
2	p_A, p_B	Partial pressure of vapour of components A and B	mb
8	Δp	Partial pressure driving force between surface of evaporation and drying air	mb
2, 12	Δp_{lm}	Log mean partial pressure driving force	mb
2	p_s	Saturation partial pressure of vapour at the drying surface temperature T_s	mb
2	p_v	Partial pressure of water vapour in drying gas	mb
2	p_{wb}	Saturation partial pressure of water vapour at the wet-bulb temperature	mb
2	p'_{wb}	Saturation partial pressure of water vapour over solution at corresponding wet-bulb temperature T'_{wb}	mb
2	P	Total pressure, absolute	mb
2	(P_r)	Prandtl number $= c_p\eta/k$	—
11	ΔP	Pressure of fluidising gas for incipient fluidisation of bed	bar
7, 9,	Q	Total heat transferred	kJ/g
12	Q'	Total heat transferred per kg dry solid	kJ/kg
12	Q'_c	Heat of crystallisation of new solid formed per kg of dry solid	kJ/kg
12	Q'_{c1}, Q'_{c2}	Heat of crystallisation of new solid formed at solubilities S_1, S_2	kJ/kg
2, 7	$\mathrm{d}Q/\mathrm{d}\theta$	Instantaneous rate of heat supply	kW

Chapters in which symbols appear	*Symbol*	*Definition*	*Derived SI unit employed*
12	Q_g	Sensible heat gained by wet solid per unit mass dry solid	kJ/kg
5	Q_l	Rate of heat loss from dryer by radiation and convection	kW
15	Q_r^*	Energy flux, radiation, per unit area	W/m^2
5	Q_s	Sensible heat in gas leaving dryer	kW
5	Q_λ	Heat flow to evaporate water	kW
12	Q'_λ	Heat to evaporate water per unit mass of dry solid	kJ/kg
2, 10	R	Universal gas constant (= 8·31)	kJ/kmol K
2, 5, 8	R_c	Constant drying rate	$g/s\ m^2$
8	R'_c	Constant drying rate referred to cross-sectional area of bed A'	$g/s\ m^2$
5	R_d	Rate of drying	$g/s\ m^2$
12	(Re)	Reynolds number, $vd\rho/\eta$	—
2, 8	R_f	Overall falling-drying rate	$g/s\ m^2$
9	R_l	Heat resistance of liquid film	$m^2\ °C/kW$
2, 6, 8	R_m	Overall drying rate	$g/s\ m^2$
11	R_r	Recycle ratio	s
12	R'_r	Recycle ratio	—
9	R_s	Heat resistance of solids film	$m^2\ °C/kW$
9	R_{st}	Heat resistance of condensing steam film	$m^2\ °C/kW$
9	R_w	Heat resistance of metal wall of dryer	$m^2\ °C/kW$
12	s	Solubility	kg/m^3
12	s_1, s_2	Solubilities at temperatures T_1, T_2	kg/m^3
10	s_d	Slope of rotary dryer	m/m
2, 5, 11	S	Humid heat of drying gas*	kJ/kg°C
11	S_1, S_2	Humid heat of fluidising gas entering and leaving bed	kJ/kg °C

*Humid Heat, S, is the heat capacity of 1 kg of dry air and the moisture it contains. For most drying calculations, $S = 1{\cdot}00 + 1{\cdot}88\ H$ kJ/kg°C where 1·00 and 1·88 are the heat capacities of dry air and water vapour respectively, and both are assumed constant.

Chapters in which symbols appear	*Symbol*	*Definition*	*Derived SI unit employed*
2	(Sc)	Schmidt number $\eta/\rho D$	—
5	S_d	Steam consumption of dryer	g/s
8	t	Thickness of preformed rod	mm
9	t_l	Thickness of liquid film	mm
2, 9, 11	t_m	Thickness of bed of drying material, depth of fluidised bed	m
2	t_w	Thickness of dryer wall	m
2, 8	T	Temperature of system	°C
10	T^*	Absolute temperature of system	K
5	T_0	Temperature of ambient air entering dryer	°C
5, 11, 12	T_1, T_2	Temperature of air passing onto or leaving material, all types of dryer	°C
12	T_1', T_2'	Temperature of solid to and from (pneumatic) dryer	°C
2, 6, 8, 9, 11	T_b	Boiling point of liquid in contact with solids being dried	°C
2, 12	T_e	Boiling point elevation of solution	°C
11, 12	T_f	Temperature of wet solids fed to dryer	°C
2, 5, 11	T_g	Temperature of drying gas	°C
2, 6, 8, 9, 11	T_h	Temperature of heating medium	°C
5	T_l	Temperature of air leaving dryer	°C
2, 5	T_r	Temperature of radiating surface	°C
2, 5, 15	T_r^*	Absolute temperature of radiating surface	K
2, 5, 11, 12	T_s	Temperature of drying surface or surface of evaporation or of fluidised bed	°C
2	T_s^*	Absolute temperature of drying surface	K
2	T_{sg}	Adiabatic saturation temperature of drying gas	°C

Chapters in which symbols appear	*Symbol*	*Definition*	*Derived SI unit employed*
2	ΔT_{sr}	Surface temperature elevation due to radiation	°C
2, 5	T_{wb}	Wet-bulb temperature of drying gas	°C
2	T'_{wb}	Wet-bulb temperature over solution = $T_{wb} - T_e$	°C
2, 7, 8	ΔT	Temperature difference at any time, and temperature driving force between drying air and surface of evaporation	°C
2, 7	ΔT_m	Mean temperature difference between heating medium and batch, assuming a batch temperature equivalent to the operating pressure within the dryer	°C
2, 11, 12	ΔT_{lm}	Log mean temperature difference	°C
2	ΔT_{sr}	Surface temperature elevation from radiation	°C
9	ΔT_θ	Average temperature difference over the drying period	°C
13	u	Velocity of droplet (spray drying)	m/s
2, 5, 9, 11	U	Overall heat-transfer coefficient by conduction and in fluidised bed	W/m² °C
7	U_c	Mean heat-transfer coefficient for complete drying period	W/m² °C
8	U_f	Falling drying rate	W/m² °C
2	U_m	Mean overall heat-transfer coefficient by conduction	W/m² °C
7	U_t	Overall heat-transfer coefficient for drying by evaporation	W/m² °C
5, 10	v	Linear velocity of air passing over material	m/s
12	v^*	Velocity of particle after falling distance h^*	m/s
9	v_d	Peripheral drum velocity	m/s
2	v_l	Fluid velocity at the edge of the laminar layer	m/s

Chapters in which symbols appear	*Symbol*	*Definition*	*Derived SI unit employed*
2, 10	v_m	Mean velocity of fluid	m/s
10	v_r	Peripheral speed of carrying rollers	m/s
12	v_s	Terminal velocity of falling particles by Stokes Law	m/s
12	v_t	Terminal velocity of particles of size within the 'intermediate region' not governed by the Stokes and Newton equations	m/s
10	V	Volume rate of drying gas in rotary dryer	m^3/h
15	V^*	Peak radio frequency potential	V/m
2, 5, 6, 8, 11, 12	W	Weight of dry solid in dryer; see note under the symbol *M*	kg
8	W_c	Constant drying rate	g/s
8, 10, 11, 14	W_d	Rate at which solids (dry basis) are fed to dryer	g/s
5	W_e	Water in gas leaving dryer	g/s
8	W_f	Falling drying rate	g/s
8	W_l	Weight of water	kg
9	W_o	Initial water content; mass water/dry solid	kg/kg
9	W_q	Equilibrium moisture content, mass water/mass dry solid	kg/kg
10	W_t	Total weight of rotating parts of rotary dryer	kg
9	W_θ	Moisture content at time θ: mass water/dry solid	kg/kg
2, 6, 8	$dW/d\theta$	Instantaneous rate of drying	g/s
5	x	Ratio of recirculated to exit air rate	kg/kg
12	x_s	Ratio of product to total flow of solid in a drying tube	kg/kg
12	Y	A number	—
2	z	Thickness of laminar layer	m

Chapters in which symbols appear	*Symbol*	*Definition*	*Derived SI unit employed*
9, 13	γ	Surface tension	10^{-3} J/m^2
9	γ_a	Adhesion tension	10^{-3} J/m^2
2, 5	ε	Emissivity of drying surface	—
15	ε^*	Dielectric constant	—
2	ε_r	Emissivity of radiating surface	—
2, 5	ε_s	Emissivity of surface receiving radiation	—
2, 11, 12	η	Viscosity (dynamic)	10^{-3} Ns/m^2
11	η_t	Theoretical thermal efficiency	—
2, 6, 7, 8, 10, 11, 12	θ	Drying time–average residence time in drying tube (Minutes in 10)	s
2, 5, 8	θ_c	Constant-rate drying time	s
2, 8	θ_f	Falling-rate drying time	s
9, 11, 12	θ_h	Drying time derived from heat transfer	s
11, 12	θ_m	Drying time derived from mass-transfer equation	s
6	θ_o	Outage time for charging and discharging	s
2, 5, 6	θ_t	Total drying time	s
2, 5, 6, 8, 10, 11, 12	λ	Latent heat of evaporation at drying surface at stated temperature	kJ/kg
15	λ^*_{max}	Peak wave length of infra-red radiant heat	m
5	λ_{st}	Latent heat of steam in heating coil	kJ/kg
2, 5, 9, 11, 12	ρ	Density of fluid	kg/m^3
11	ρ_b	Bulk density of powder	kg/m^3
13	ρ_g	Density of gas passing over or through material	kg/m^3
11, 12	σ	Density of powder particles	kg/m^3
2, 15	(σ)	Stefan-Boltzmann constant $(= 56{\cdot}7 \times 10^{-9})$	W/m^2 K^4
9	ϕ	Angle of contact of liquid with drum	degrees
15	$\cos \phi$	Power factor	—

Chapters in which symbols appear	*Symbol*	*Definition*	*Derived SI unit employed*

Other subscripts
$(\ldots)_{in}$ refers to inlet conditions
$(\ldots)_{out}$ refers to outlet conditions
$(\ldots)_{c}$ refers to constant-rate value
$(\ldots)_{corr}$ refers to corrected value
$(\ldots)_{m}$ refers to mean or average
$(\ldots)_{lm}$ refers to log mean

CONVERSION FACTORS TO STRICT S.I. UNITS

The following table was prepared by F. P. Lees, M.A., B.Sc., A.M.I.Chem.E., of the Department of Chemical Engineering, University of Technology, Loughborough and is reproduced by permission from *The Chemical Engineer*, page CE 341–344 (October 1968).

Acceleration	1 cm/s^2	:	$1{\cdot}000\,0 \times 10^{-2}$ m/s^2
	1 m/h^2	:	$7{\cdot}716\,0 \times 10^{-8}$ m/s^2
	1 ft/s^2	:	$3{\cdot}048\,0 \times 10^{-1}$ m/s^2
	1 ft/h^2	:	$2{\cdot}351\,9 \times 10^{-8}$ m/s^2
Area	1 cm^2	:	$1{\cdot}000\,0 \times 10^{-4}$ m^2
	1 ft^2	:	$9{\cdot}290\,3 \times 10^{-2}$ m^2
	1 in^2	:	$6{\cdot}451\,6 \times 10^{-4}$ m^2
	1 yd^2	:	$8{\cdot}361\,3 \times 10^{-1}$ m^2
	1 acre	:	$4{\cdot}046\,9 \times 10^{3}$ m^2
	1 mile^2	:	$2{\cdot}590\,0 \times 10^{6}$ m^2
Calorific value (volumetric)	1 cal/cm^3	:	$4{\cdot}186\,8 \times 10^{6}$ J/m^3
	1 kcal/m^3	:	$4{\cdot}186\,8 \times 10^{3}$ J/m^3
	1 Btu/ft^3	:	$3{\cdot}726\,0 \times 10^{4}$ J/m^3
	1 Chu/ft^3	:	$6{\cdot}706\,7 \times 10^{4}$ J/m^3
	1 therm/ft^3	:	$3{\cdot}726\,0 \times 10^{9}$ J/m^3
	1 kcal/ft^3	:	$1{\cdot}478\,6 \times 10^{5}$ J/m^3
Coefficient of expansion	1 g/cm^3 °C	:	$1{\cdot}000\,0 \times 10^{3}$ kg/m^3 °K
(volumetric)	1 lb/ft^3 °F	:	$2{\cdot}883\,3 \times 10$ kg/m^3 °K
	1 lb/ft^3 °C	:	$1{\cdot}601\,8 \times 10$ kg/m^3 °K
Density	1 g/cm^3	:	$1{\cdot}000\,0 \times 10^{3}$ kg/m^3
	1 lb/ft^3	:	$1{\cdot}601\,8 \times 10$ kg/m^3
	1 lb/U.K.gal	:	$9{\cdot}977\,9 \times 10$ kg/m^3
	1 lb/U.S.gal	:	$1{\cdot}198\,3 \times 10^{2}$ kg/m^3
	1 kg/ft^3	:	$3{\cdot}531\,5 \times 10$ kg/m^3
Diffusion coefficient—see Viscosity, kinematic			
Energy	1 cal	:	$4{\cdot}186\,8$ J
	1 kcal	:	$4{\cdot}186\,8 \times 10^{3}$ J
	1 Btu	:	$1{\cdot}055\,1 \times 10^{3}$ J
	1 erg	:	$1{\cdot}000\,0 \times 10^{-7}$ J
	1 hp h (metric)	:	$2{\cdot}647\,7 \times 10^{6}$ J
	1 kW h		$3{\cdot}600\,0 \times 10^{6}$ J
	1 ft pdl	:	$4{\cdot}213\,9 \times 10^{-2}$ J
	1 ft lbf	:	$1{\cdot}355\,8$ J
	1 Chu	:	$1{\cdot}899\,1 \times 10^{3}$ J
	1 hp h (British)	:	$2{\cdot}684\,5 \times 10^{6}$ J
	1 therm	:	$1{\cdot}055\,1 \times 10^{8}$ J
	1 thermie	:	$4{\cdot}185\,5 \times 10^{6}$ J
	1 ft kgf	:	$2{\cdot}989\,1$ J
Power	1 cal/s	:	$4{\cdot}186\,8$ W
	1 kcal/h	:	$1{\cdot}163\,0$ W
	1 Btu/s	:	$1{\cdot}055\,1 \times 10^{3}$ W
	1 erg/s	:	$1{\cdot}000\,0 \times 10^{-7}$ W
	1 tonne cal/h	:	$1{\cdot}163\,0 \times 10^{3}$ W
	1 hp (metric)	:	$7{\cdot}354\,8 \times 10^{2}$ W
	1 ft pdl/s	:	$4{\cdot}213\,9 \times 10^{-2}$ W
	1 ft lbf/s	:	$1{\cdot}355\,8$ W
	1 Btu/h	:	$2{\cdot}930\,8 \times 10^{-1}$ W
	1 Chu/h	:	$5{\cdot}275\,4 \times 10^{-1}$ W
	1 hp (British)	:	$7{\cdot}457\,0 \times 10^{2}$ W
	1 ton refrigeration	:	$3{\cdot}516\,9 \times 10^{3}$ W

Pressure	1 dyn/cm^2	: 1·000 0 × 10^{-1} N/m^2
	1 kgf/m^2	: 9·806 7 N/m^2
	1 pdl/ft^2	: 1·488 1 N/m^2
	1 standard atmosphere	: 1·013 3 × 10^{5} N/m^2
	1 at (1 kgf/cm^2)	: 9·806 7 × 10^{4} N/m^2
	1 bar	: 1·000 0 × 10^{5} N/m^2
	1 lbf/ft^2	: 4·788 0 × 10 N/m^2
	1 lbf/in^2	: 6·894 8 × 10^{3} N/m^2
	1 tonf/in^2	: 1·544 4 × 10^{7} N/m^2
	1 in water	: 2·490 9 × 10^{2} N/m^2
	1 ft water	: 2·989 1 × 10^{3} N/m^2
	1 mm Hg	: 1·333 3 × 10^{2} N/m^2
	1 in Hg	: 3·386 6 × 10^{3} N/m^2
Reaction rate – see Mass release rate		
Shear stress – see Pressure		
Specific enthalpy	1 cal/g	: 4·186 8 × 10^{3} J/kg
	1 Btu/lb	: 2·326 0 × 10^{3} J/kg
	1 Chu/lb	: 4·186 8 × 10^{3} J/kg
Specific heat	1 cal/g °C 1 Btu/lb °F	: 4·186 8 × 10^{3} J/kg K
Specific volume	1 cm^3/g	: 1·000 0 × 10^{-3} m^3/kg
	1 ft^3/lb	: 6·242 8 × 10^{-2} m^3/kg
	1 ft^3/kg	: 2·831 7 × 10^{-2} m^3/kg
Surface per unit mass	1 cm^2/g	: 1·000 0 × 10^{-1} m^2/kg
	1 ft^2/lb	: 2·048 2 × 10^{-1} m^2/kg
	1 m^2/g	: 1·000 0 × 10^{3} m^2/kg
	1 ft^2/kg	: 9·290 3 × 10^{-}– m^2/kg
Surface per unit volume	1 cm^2/cm^3	: 1·000 0 × 10^{2} m^2/m^3
	1 ft^2/ft^3	: 3·280 8 m^2/m^3
Surface tension	1 dyn/cm	: 1·000 0 × 10^{-3} N/m
Temperature difference	1 deg F (deg R)	: 5/9 deg C (deg K)
Thermal conductivity	1 cal/s cm^2 (°C/cm)	: 4·186 8 × 10^{2} W/m^2 (°C/m)
	1 kcal/h m^2 (°C/m)	: 1·163 0 W/m^2 (°C/m)
	1 Btu/h ft^2 (°F/ft)	: 1·730 8 W/m^2 (°C/m)
	1 Btu/h ft^2 (°F/in)	: 1·442 3 × 10^{-1} W/m^2 (°C/m)
	1 kcal/h ft^2 (°C/ft)	: 3·815 6 W/m^2 (°C/m)
Time	1 h	: 3·600 0 × 10^{3} s
	1 min	: 6·000 0 × 10 s
	1 day	: 8·640 0 × 10^{4} s
	1 year	: 3·155 8 × 10^{7} s
Force	1 dyn	: 1·000 0 × 10^{-5} N
	1 kgf	: 9·806 7 N
	1 pdl	: 1·382 5 × 10^{-1} N
	1 lbf	: 4·448 2 N
	1 tonf	: 9·964 0 × 10^{3} N
Heat – see Energy		
Heat of combustion, formation, *etc.* – see Specific enthalpy		
Heat capacity – see Specific heat		
Heat flow – see Power		
Heat flux	1 cal/s cm^2	: 4·186 8 × 10^{4} W/m^2
	1 kcal/h m^2	: 1·163 0 W/m^2
	1 Btu/h ft^2	: 3·154 6 W/m^2
	1 Chu/h ft^2	: 5·678 4 W/m^2
	1 kcal/h ft^2	: 1·251 8 × 10 W/m^2

Heat release rate (mass)	1 cal/s g	: $4{\cdot}186\,8 \times 10^3$ W/kg
	1 kcal/h kg	: $1{\cdot}163\,0$ W/kg
	1 Btu/h lb	: $6{\cdot}461\,2 \times 10^{-1}$ W/kg
Heat release rate (volumetric)	1 cal/s cm^3	: $4{\cdot}186\,8 \times 10^6$ W/m^3
	1 kcal/h m^3	: $1{\cdot}163\,0$ W/m^3
	1 Btu/h ft^3	: $1{\cdot}035\,0 \times 10$ W/m^3
	1 Chu/h ft^3	: $1{\cdot}863\,0 \times 10$ W/m^3
	1 kcal/h ft^3	: $4{\cdot}107\,1 \times 10$ W/m^3
Heat-transfer coefficient	1 cal/s cm^2 °C	: $4{\cdot}186\,8 \times 10^4$ W/m^2 °C
	1 kcal/h m^2 °C	: $1{\cdot}163\,0$ W/m^2 °C
	1 Btu/h ft^2 °F 1 Chu/h ft^2 °C	: $5{\cdot}678\,4$ W/m^2 °C
	1 kcal/h ft^2 °C	: $1{\cdot}251\,8 \times 10$ W/m^2 °C
Henry's law constant	1 atm/(g/cm^3)	: $1{\cdot}013\,3 \times 10^2$ (N/m^2)/(kg/m^3)
	1 atm/(kg/m^3)	: $1{\cdot}013\,3 \times 10^5$ (N/m^2)/(kg/m^3)
	1 atm/(lb/ft^3)	: $6{\cdot}325\,8 \times 10^3$ (N/m^2)/(kg/m^3)
	1 atm/(kg/ft^3)	: $2{\cdot}869\,3 \times 10^3$ (N/m^2)/(kg/m^3)
Latent heat – see Specific enthalpy		
Length	1 cm	: $1{\cdot}000\,0 \times 10^{-2}$ m
	1 ft	: $3{\cdot}048\,0 \times 10^{-1}$ m
	1 Ångstrom	: $1{\cdot}000\,0 \times 10^{-10}$ m
	1 micron	: $1{\cdot}000\,0 \times 10^{-6}$ m
	1 in	: $2{\cdot}540\,0 \times 10^{-2}$ m
	1 yd	: $9{\cdot}144\,0 \times 10^{-1}$ m
	1 mile	: $1{\cdot}609\,3 \times 10^3$ m
Mass	1 g	: $1{\cdot}000\,0 \times 10^{-3}$ kg
	1 lb	: $4{\cdot}535\,923\,7 \times 10^{-1}$ kg
	1 tonne	: $1{\cdot}000\,0 \times 10^3$ kg
	1 grain	: $6{\cdot}480\,0 \times 10^{-5}$ kg
	1 oz	: $2{\cdot}835\,0 \times 10^{-2}$ kg
	1 cwt	: $5{\cdot}080\,2 \times 10$ kg
	1 ton	: $1{\cdot}016\,0 \times 10^3$ kg
Mass per unit area	1 g/cm^2	: $1{\cdot}000\,0 \times 10$ kg/m^2
	1 lb/ft^2	: $4{\cdot}882\,4$ kg/m^2
	1 lb/in^2	: $7{\cdot}030\,7 \times 10^2$ kg/m^2
	1 ton/$mile^2$	: $3{\cdot}923\,0 \times 10^{-4}$ kg/m^2
	1 kg/ft^2	: $1{\cdot}076\,4 \times 10$ kg/m^2
Mass flow	1 g/s	: $1{\cdot}000\,0 \times 10^{-3}$ kg/s
	1 kg/h	: $2{\cdot}777\,8 \times 10^{-4}$ kg/s
	1 lb/s	: $4{\cdot}535\,9 \times 10^{-1}$ kg/s
	1 tonne/h	: $2{\cdot}777\,8 \times 10^{-1}$ kg/s
	1 lb/h	: $1{\cdot}260\,0 \times 10^{-4}$ kg/s
	1 ton/h	: $2{\cdot}822\,4 \times 10^{-1}$ kg/s
Mass flux	1 g/s cm^2	: $1{\cdot}000\,0 \times 10$ kg/s m^2
	1 kg/h m^2	: $2{\cdot}777\,8 \times 10^{-4}$ kg/s m^2
	1 lb/s ft^2	: $4{\cdot}882$ kg/s m^2
	1 lb/h ft^2	: $1{\cdot}356\,2 \times 10^{-3}$ kg/s m^2
	1 kg/h ft^2	: $2{\cdot}990\,0 \times 10^{-3}$ kg/s m^2
Mass release rate (volumetric)	1 g/s cm^3	: $1{\cdot}000\,0 \times 10^3$ kg/s m^3
	1 kg/h m^3	: $2{\cdot}777\,8 \times 10^{-4}$ kg/s m^3
	1 lb/s ft^3	: $1{\cdot}601\,8 \times 10$ kg/s m^3
	1 lb/h ft^3	: $4{\cdot}449\,6 \times 10^{-3}$ kg/s m^3
	1 kg/h ft^3	: $9{\cdot}809\,6 \times 10^{-3}$ kg/s m^3

Mass-transfer coefficient, concentration driving force – see Velocity
Mass-transfer coefficient, dimensionless driving force – see Mass flux

Mass-transfer coefficient, pressure driving force	1 g/s cm^2 atm	$9{\cdot}868\,7 \times 10^{-5}$ kg/s m^2 (N/m^2)
	1 kg/h m^2 atm	$2{\cdot}741\,3 \times 10^{-9}$ kg/s m^2 (N/m^2)
	1 lb/h ft^2 atm	$1{\cdot}338\,4 \times 10^{-8}$ kg/s m^2 (N/m^2)
	1 kg/h ft^2 atm	$2{\cdot}950\,7 \times 10^{-8}$ kg/s m^2 (N/m^2)
Momentum, angular	1 g cm^2/s	$1{\cdot}000\,0 \times 10^{-7}$ kg m^2/s
	1 lb ft^2/s	$4{\cdot}214\,0 \times 10^{-2}$ kg m^2/s
	1 lb ft^2/h	$1{\cdot}170\,6 \times 10^{-5}$ kg m^2/s
Momentum, linear	1 g cm/s	$1{\cdot}000\,0 \times 10^{-5}$ kg m/s
	1 lb ft/s	$1{\cdot}382\,5 \times 10^{-1}$ kg m/s
	1 lb ft/h	$3{\cdot}840\,4 \times 10^{-5}$ kg m/s
Moment of inertia	1 g cm^2	$1{\cdot}000\,0 \times 10^{-7}$ kg m^2
	1 lb ft^2	$4{\cdot}214\,0 \times 10^{-2}$ kg m^2
Torque – see Energy		
Velocity	1 cm/s	$1{\cdot}000\,0 \times 10^{-2}$ m/s
	1 m/h	$2{\cdot}777\,8 \times 10^{-4}$ m/s
	1 ft/s	$3{\cdot}048\,0 \times 10^{-1}$ m/s
	1 ft/h	$8{\cdot}466\,7 \times 10^{-5}$ m/s
	1 mile/h	$4{\cdot}470\,4 \times 10^{-1}$ m/s
Viscosity, absolute (or dynamic)	1 g/cm s (1 poise)	$1{\cdot}000\,0 \times 10^{-1}$ kg/m s (or N s/m^2)
	1 kg/m h	$2{\cdot}777\,8 \times 10^{-4}$ kg/m s
	1 lb/ft s	$1{\cdot}488\,2$ kg/m s
	1 lb/ft h	$4{\cdot}133\,8 \times 10^{-4}$ kg/m s
	1 kg/ft h	$9{\cdot}113\,4 \times 10^{-4}$ kg/m s
Viscosity, kinematic	1 cm^2/s (1 stokes)	$1{\cdot}000\,0 \times 10^{-4}$ m^2/s
	1 m^2/h	$2{\cdot}777\,8 \times 10^{-4}$ m^2/s
	1 ft^2/s	$9{\cdot}290\,3 \times 10^{-2}$ m^2/s
	1 ft^2/h	$2{\cdot}580\,6 \times 10^{-5}$ m^2/s
Volume	1 cm^3	$1{\cdot}000\,0 \times 10^{-6}$ m^3
	1 ft^3	$2{\cdot}831\,7 \times 10^{-2}$ m^3
	1 litre	$1{\cdot}000\,0 \times 10^{-3}$ m^3
	1 in^3	$1{\cdot}638\,7 \times 10^{-5}$ m^3
	1 yd^3	$7{\cdot}645\,5 \times 10^{-1}$ m^3
	1 U.K.gal	$4{\cdot}546\,0 \times 10^{-3}$ m^3
	1 U.S.gal	$3{\cdot}785\,3 \times 10^{-3}$ m^3
Volumetric flow	1 cm^3/s	$1{\cdot}000\,0 \times 10^{-6}$ m^3/s
	1 m^3/h	$2{\cdot}777\,8 \times 10^{-4}$ m^3/s
	1 ft^3/s	$2{\cdot}831\,7 \times 10^{-2}$ m^3/s
	1 cm^3/min	$1{\cdot}666\,7 \times 10^{-8}$ m^3/s
	1 litre/min	$1{\cdot}666\,7 \times 10^{-5}$ m^3/s
	1 ft^3/min	$4{\cdot}719\,5 \times 10^{-4}$ m^3/s
	1 ft^3/h	$7{\cdot}865\,8 \times 10^{-6}$ m^3/s
	1 U.K. gal/min	$7{\cdot}576\,6 \times 10^{-5}$ m^3/s
	1 U.S. gal/min	$6{\cdot}308\,9 \times 10^{-5}$ m^3/s
	1 U.K. gal/h	$1{\cdot}262\,8 \times 10^{-6}$ m^3/s
	1 U.S. gal/h	$1{\cdot}051\,5 \times 10^{-6}$ m^3/s
Wetting rate (mass) – see Viscosity, absolute		
Wetting rate (volumetric)	1 litre/h in	$1{\cdot}093\,6 \times 10^{-5}$ m^3/s m
	– see also Viscosity, kinematic	
Work – see Energy		

CHAPTER ONE

The importance of correct design

1.1 DEFINITION OF DRYING PROCESSES

The essential feature of drying a solid is that it involves the removal of a liquid by converting it to a vapour which can then be separated readily from the solid. In normal practice the energy to vaporise the liquid is supplied as heat. Other forms of energy such as a radio frequency field, mechanical work or chemical reaction are used on occasions; these are, however, specialist cases which need not be considered for the general argument.

The heat transfer is predominantly by convection or by conduction, for though it is true that nearly all dryers transfer some heat by radiation it is rarely the predominating mechanism. This is because the provision of means for convection or conduction heating intrinsically involves means for removal of the vapour; radiation heat transfer requires no medium and does not in itself involve means for removal of the vapour. Since it is a secondary effect we can treat any heat transfer by radiation as a detailed correction to the convection or conduction mechanism.

Drying by conduction is superficially different from drying by convection. In the conduction case the wet solid is placed in a vessel which is heated from the outside and provided with a vent through which the vapour can be removed; the vessel is frequently maintained under reduced pressure to increase the temperature driving force. In the convection case hot gas is blown over the surface of the wet solid and this provides both the source of the heat and the means for removal of the vapour.

As drying proceeds in a conduction dryer, the rate of drying usually decreases steadily. It can be shown that this effect can be generally accounted for by a decrease in the heat transfer from the wall to the drying solid. The operation can thus be regarded as controlled by the heat transfer to the material, and can be treated in terms of an overall heat-transfer coefficient from the heating medium to the drying solid and the temperature of the latter can be taken as the boiling point of the liquid (allowance can be made for the case where the liquid is a solution). At the end of the drying this mechanism does not hold and the rate of removal of water from the material will begin to control the drying, the effect being evinced by a rise in temperature above the boiling point of the saturated solution. This is frequently a minor effect in conduction dryers and can be written into the overall heat-transfer coefficient (perhaps better renamed overall conduction-drying coefficient on this account). It is thus true to a large extent that the drying rate in a conduction dryer is independent of the mass transfer from the material. This is unfortunately not the same as saying that the drying rate is independent of the properties of the material because the heat-transfer coefficient will depend to some extent on the material (e.g. on its rheological properties in agitated dryers).

The convection case is different. The heat-transfer coefficient is generally good and does not vary greatly as drying proceeds. Limitation of drying rate is now largely a matter of the matter-transfer coefficient. This effect shows up as two markedly different drying regimes well known as the constant- and falling-rate drying periods. In the constant-rate period operation is limited by the heat transfer to the material and is predictable and more or less independent of the nature of the material. In the falling-rate period it is limited by the matter transfer which is dependent on the properties of the drying material and is not predictable. In practice the falling-rate period accounts for a substantial part (and frequently all) of the drying time in a convection dryer. This falling-rate period is the same as the period of elevation of temperature which usually occurs at the end of conduction drying.

The change from constant- to falling-rate drying, often called the critical moisture content, is a function both of the constant-drying rate as dictated by the external conditions and geometry of the dryer, and of the matter transfer which is dependent on the material itself. The critical moisture content is thus far removed from the popular concept of a physical constant.

To put the matter in another way, the key operation in drying is the mass transfer from the solid. On a theoretical basis we have no quantitative knowledge of the mechanism of matter transfer from a drying solid. It is presumably a function of the size and shape and disposition of the component particles of the solid material and of the ease with which the liquid and vapour can migrate from the pores and capillaries within the solid to its bulk external surface. That is about as far as one can go. Now in some dryers, particularly conduction dryers, and in some stages (usually the initial ones) in most dryers, the drying rate is limited by the heat transfer to the material rather than by the matter transfer from the material. Under such conditions the rate of drying can be assessed from well understood heat-transfer principles and is to some extent independent of the properties of the material being dried. However, in the general case the rate depends upon the matter transfer from the drying solid and this is not well understood and cannot be predicted. Further, the more the heat transfer efficiency of dryers is improved, the more one becomes dependent upon the unpredictable matter transfer.[1]

The practical effects of this dichotomy and ignorance are two-fold:

(a) The drying rate of a material can only be determined by experiment; we cannot predict it.

(b) The experiments must be related to the type of dryer it is proposed to use.

These points are dealt with in Chapter 3.

1.2 THE PLACE OF DRYING IN CHEMICAL ENGINEERING TECHNOLOGY

The process of drying solids is of considerable practical and economic importance in all industries. The operation varies tremendously with the material to be dried – from a few seconds to several days. Most of the practice is still more of an art than a science, though explicable by scientific principles. Consequently it is to be expected that future advances will come from a stronger partnership between 'inventors' and applied scientists.

A rough estimate of the amount of water removed in dryers in the chemical food and process industries in Britain is ten million tons per year and the corresponding fuel usage may be as much as five million tons per year of coal equivalent. The largest single class of dried material sold by the chemical industry is synthetic nitrogenous fertilizers, including granulated mixtures with phosphorus and potassium compounds; in 1968 the British output of these materials was 3·3 million tons.

1.3 CATEGORIES OF MATERIALS TO DRYERS

In the chemical industry proper we are concerned mainly with the following categories of materials which are solid in the required dry state:

(a) Crystals, either organic or inorganic and granules (usually of mixtures) usually larger than one millimetre.
(b) Porous solids such as synthetic rubber crumbs.
(c) Pastes and sludges—such as those from filters and centrifuges.
(d) Free flowing powders, usually, smaller than 200 μm (microns) which may even be relatively free flowing when wet and awkwardly dusty when dry—such as many pigments and some rapid settling precipitates after centrifuging.
(e) Slurries which are required to be dried without previous separation of wet solids.
(f) Liquids such as extracts, emulsions and a few straight solutions.

The liquid to be removed on drying is not always water and may be any volatile solvent, sometimes a highly inflammable or toxic material. This book is not concerned with the drying of liquids or gases.[2]

1.4 EFFECTS OF INCORRECT DESIGN

Although the capital and running costs of dryers form only a small part of the total cost of manufacture of chemicals, the performance of a dryer can assume overriding importance when it affects the specification of the finished product in terms of marketing—for example free flowing characteristics, size distribution, freedom from caking or sweating in storage. Moreover, whilst too small a dryer can reduce the output of the whole chemical plant, occasions arise when a wrong choice of type of dryer can result in an inferior product either difficult to process further, e.g. to blend with other solids, or unacceptable to customers.

1.5 THE PROCESS DESIGN ENGINEER'S PROBLEM

There are many types of dryer on the market, most but not all of which will enable more than one of the above general classes of wet material to be dried. The process design engineer concerned with the manufacture of a product from given raw materials or semi-finished intermediates has an unenviable task in choosing the most economic form of dryer capable of giving the required output of dried solids to the often difficult specifications required by modern industry. Once the correct dryer has been installed and run in with a single or related series of feed materials, and suitable monitoring methods evolved, little trouble should occur with normal operation unless the characteristics of the feed are altered either deliberately or by unwitting changes in the prior stages of manufacture.

This book is based on the practical experience of chemical engineers concerned with drying problems in many chemical factories. Whilst the classification and selection of dryer types is discussed primarily in Chapter 3, all the chapters on individual dryers are directed towards the practising process design engineer who requires a rapid means of assessing the performance characteristics of the dryers on the market. The engineer is thus put in a better position to discuss his requirements with equipment manufacturers and both parties should be able to ask the right questions of each other when discussing tenders.

When reviews have been made of drying equipment which has fallen short of process requirements, it has been found that all the numerous facts concerned with the drying problem have not been presented to the dryer manufacturers. This is usually through lack of appreciation by the process design engineer of all the factors concerned with choice of drying equipment and not through misplaced secretiveness. It is indeed a common complaint at technical meetings of professional institutions concerned with many technologies that the equipment manufacturers are not taken sufficiently into the confidence of the process engineer when invited to tender.

1.6 ANCILLARY EQUIPMENT

Whilst this book is concerned solely with dryers, it must be realised that unsatisfactory performance of a process can be due to malfunctioning of the ancillary equipment associated with the drying section of a plant. The following items should, therefore, be examined by the engineer in association with the dryer proper when relevant to the rate of production of dried material:

(a) Mechanical de-waterers ahead of the drying section for reasonable uniformity of liquid content in feed to the dryer.

(b) Mixers and preformers when, for example, wet granules are to be dried.

(c) Bunker emptying devices* and all forms of conveyor to and from the dryer. There is a vast range of feedstocks which are easy to dry provided they can be fed uniformly into and out of continuously operating dryers. Hand-loading of most chemicals, even in small quantities, say over 25 kg/h is to be discouraged on account of toxic risks.

(d) Heating equipment: equipment which is too small can wreck a production schedule whilst equipment which is too large, besides being wasteful of capital, may give uneven or fluctuating heating and be difficult to control. With steam heaters, badly placed air vents or malfunctioning of steam water traps are well known to reduce performance substantially; hence in a large works it is always worth while calling for specialist examination of drawings.

(e) Dust collection from the effluent gases:[3] this is a difficult problem with fine powders when the effluent gas is nearly saturated with vapour because any condensation of liquid causes rapid build-up of dust in ducts and dust collecting equipment such as cyclones or fabric filters. Effective lagging and easy methods of warming-up the whole of the dedusting system before dusty effluent gas is sent to it are, therefore, essential; unfortunately the process manager regularly has to rediscover this anew.

Whilst clean abstraction of the collected dust from a dry-process unit for cleaning the effluent gases is not always easy, it is not infrequently found that wet methods of collection give rise to serious problems with fume as well as with effluent liquid. Fume, which can be roughly described in the context of drying as sub-micron particles of volatilised salts, is a cause of substantial difficulty whenever large tonnages of materials containing volatile ammonium and similar salts are dried because some volatile salts, e.g. ammonium chloride, condense on the water particles in the 'steamy vapour' leaving the vents or stacks, thereby stabilising the fume; this mist disperses much less rapidly than uncontaminated steam and gives rise to strong complaints by the public. The seriousness of the problem and its alleviation by use of high chimneys for the discharge of washed dryer effluent is the subject of comment by the Chief Alkali Inspector.[3] As a general guide it has been found that the fine particulate content of the effluent should be reduced below 0·12 g/m^3 NTP in order to produce an emission which is only just above the threshold of visibility. An approximate relationship between visibility of chimney plumes of dusty flue gas and particulate size has been given by Jarman.[4]

(f) Many organic compounds are sufficiently volatile to impart an odour, often highly penetrating and disagreeable, to the gases exhausted from a dryer. The destruction of this odour before the gases are discharged to atmosphere can be a difficult problem particularly with the

*For information on bunker design reference should be made to recent annual reports by Warren Spring Laboratory, Ministry of Technology (published by H.M.S.O.)

increased public antagonism to all air pollution from industrial establishments. Methods of odour destruction are described by Nonhebel.[2]

(g) Noise emission – this is often difficult to predict and seldom easy to cure when resonance occurs in conjunction with large fans. Specialist advice can be obtained from the National Physical Laboratory, Teddington, who have accumulated considerable experience on the solution of noise problems – see also Reference 5.

CHAPTER TWO

Simplified theory of the drying of solids

2.1 PARAMETERS AFFECTING DESIGN AND PERFORMANCE OF SOLIDS

Chemical engineers concerned with choice of drying plant, or design of special plant to meet unprecedented conditions, will need to consider the effect which the 34 factors given at the end of this section may have on the ability of a drying plant to give a rigidly specified product. These factors will vary in importance with each problem. Many of them interact during the drying process. Some will be important during the initial constant-rate period and others during one or more stages of the falling-rate periods. Yet further factors may be found to require consideration after studies of drying of complex materials in small scale dryers.

An examination of the list will at once show the impractibility of even semi-quantitative prediction of the performance of a dryer from the fundamental physical data available or readily obtainable for any system likely to be met in a chemical works. Prediction with wet clean quartz would be accurate but unknowns would immediately become apparent with dirty sand. It is obvious also that, even if quantitative formulae could be developed to cover the numerous parameters, it would take far longer, and be more costly, to determine the physical data required for insertion into the formulae than to carry out correctly chosen laboratory and pilot scale work to determine the drying characteristics of a material, and its likely performance in two or three types of dryer carefully selected by an experienced 'drying technologist'. This conclusion must not be taken as implying that further research should not be directed towards a better understanding of the mechanism of drying and the interactions within a drying solid. Such research would certainly be capable of application towards reducing the operating cost or improving the product. A start on these lines has been sponsored at a British University by a joint panel of the Chemical Industries Association and the British Chemical Plant Manufacturers Association but much more work is desirable both in academic and industrial establishments.

The basic theoretical principles which can be applied quantitatively to design are described in later sections of this chapter. This basic theory has been extended in succeeding chapters on individual types of dryer when, and only when, it aids the understanding of and further development of the design, for example, for the purpose of scaling-up from pilot plant operation.

2.1.1 Parameters involved in drying

A. Heat Transfer

1. Heat transfer from heating medium to liquid surface.
2. Heat transfer at 'bond film' (see 19).
3. Heat transfer from solid directly to the liquid.
4. Heat transfer from solid to the liquid through the surface film within the interstices – pores – of the solid mass.
5. Heat transfer from heating medium to nearly dry solid.
6. Thermal conductivity of the liquid.
7. Thermal conductivity of the wet solid.
8. Thermal conductivity of the nearly dry solid.
9. Latent heat of the liquid.
10. Heat of hydration when combined water is to be removed at the end of the drying process.
11. Relation between temperature and liquefaction (solution) of the wet material; some dryer feeds 'melt' before all the liquid has been evaporated.
12. Effect of electrolytes present in the liquid on the drying characteristics of the materials, example – clays.

B. Drying Atmosphere

13. Pressure and temperature of the drying atmosphere.
14. Composition, of the drying atmosphere, including changes in the partial pressure of the liquid throughout the drying process.
15. Relative velocity of the drying atmosphere past the drying surface.
16. Effective vapour pressure of the liquid, taking into account changes in elevation of the boiling point during the drying process; example NaOH.
17. Degree of saturation of the drying atmosphere within a bed of solid.

C. General Physical Properties of Solid – Liquid System

18. Interfacial tension between solid and liquid.
19. Thickness of 'bond film' between solid and liquid.
20. Ratio of surface area to liquid volume within pores.
21. Diffusion coefficient of vapour between pores.
22. Capillary suction of liquid within pores.
23. Concentration gradients of liquid within pores.

24. Presence of fibrous materials or sacs within the bulk solid.
25. Size of liquid molecule – relevant with some organic liquids.
26. Maximum content of soluble impurities permissible in residual liquor present in wet material, example – imperfectly washed filter cake; important when the impurities bond the particles of the dried material.

D. Properties of the Solid

27. Particle size.
28. Effective area of solid (see 24).
29. Porosity, i.e. size and length of pores and the extent to which they are branched.
30. Solubility of solid in the liquid where the liquid is part of the structure of the wet solid.
31. Surface hardening of sludges as they dry, giving a nearly non-porous surface, and relation of this effect with the rate of drying.
32. General caking of powders during drying; strength of agglomerates.
33. Resistance of dried material to attrition.
34. Maximum content of liquid permissible in the dried product, e.g. to prevent caking in storage under prescribed conditions, or to ensure free flowing when transported in bulk tankers.

2.2 SCOPE OF THEORETICAL TREATMENT

It is assumed that the reader is acquainted with the general theory of drying given in standard works such as Cremer and Watkins,[1] Coulson and Richardson[1] and Perry[2] etc.

The theoretical approach to the drying of solids depends upon whether the mechanism of heat transfer to the solid is by convection, conduction or radiation. In practice the predominant mechanism is usually convection in direct dryers in which the material is dried by means of a hot gas stream passing over or through it and conduction in indirect dryers in which the material is heated via a wall. There may be significant heat transfer by radiation in either case and some degree of conduction in convection dryers and vice versa. It is rare for a dryer to be operated with pure radiation alone.

It is thus convenient to treat the theory of drying in terms of pure convection and pure conduction and then to apply corrections for the effects of subsidiary modes of heat transfer.

Rigorous theoretical treatment of the drying of wet solids requires quantitative knowledge of the factors affecting the movement of both liquid and vapour under given thermal conditions. This amounts to a description of the internal structure of the solid in terms which can be used to calculate the rates of movement of both liquid and vapour from their physical properties and the interior surface properties of the solid. As shown in Section 2.1 the system is complex (especially if the structure of the solid changes during drying, as is likely) and the best that can be expected in the present state of knowledge is an approximate working theory with many reservations.

On the other hand the simpler relationships between the external drying conditions and the rate of drying of a given material can be specified. The theoretical treatment presented here is therefore limited to this overall basis. It should be noted that this limitation of treatment means that any calculation of drying rate (for example) requires experimental knowledge of the drying rate of the particular material under known conditions and further that the translation of data from one set of conditions to another can only be done in suitable cases.

It should also be noted that practical behaviour in dryers is often controlled largely by the nature of the machine and physical properties of the material which may be different from those which determine its drying behaviour. As an example the behaviour of a dryer as a feeding or conveying device may dictate the design rather than the drying operation itself.

The theoretical treatment presented here is thus useful in giving some understanding of the process but the design of dryers must remain largely dependent upon experiment with the material and system involved.

Descriptions are given in some later chapters of tests which should be made with small-scale dryers to assess the drying characteristics of a wet material in the particular types of dryers selected from Chapter 4 as possible candidates for the required duty.

2.3 HEAT TRANSFER BY CONVECTION

2.3.1 General

In this case the wet solid is dried by passing a stream of heated gas across or through it. The hot gas serves to transfer heat to the solid by convection and to remove the evaporated vapour. If the hot gas is supplied to the system at a constant temperature and humidity it is observed that the drying process occurs in two distinct stages. Initially the rate of drying is constant and then at some moisture content it begins to diminish and continues to do so progressively until it is zero when the material is completely dry.

The moisture content at which the drying rate begins to diminish is known as the critical moisture content but the change generally tends to occur gradually over a range of moisture contents as indicated in the typical drying curves of Fig. 2.1.

In some cases the initial moisture content may be below the critical value and the drying will then be entirely falling-rate with no constant rate. The falling-rate curves themselves may be concave or convex or may approximate to a straight line; they are inflected when a change of physical form occurs—for example when shrinking and cracking occur or when a skin forms on the surface of partly dried material.

The constant-rate drying period corresponds to the situation when the surface of the solid is wet with the liquid and in a given system the rate of drying is controlled entirely by the drying conditions; which in the case of pure convection drying are simply the velocity, temperature and humidity of the drying gas. Thus if these are constant the rate of drying is constant; the rate of migration of liquid from inside the solid to the surface at which evaporation occurs is such that it does not in any way limit the process. In the falling-rate period the rate of migration of liquid to the surface has decreased so that it controls the rate of drying. The surface is now no longer fully wetted and as the rate of migration decreases the effect of the external conditions progressively diminishes and the rate is merely a reflection of lack of uniformity in the rate of migration of liquid to the surface. The critical

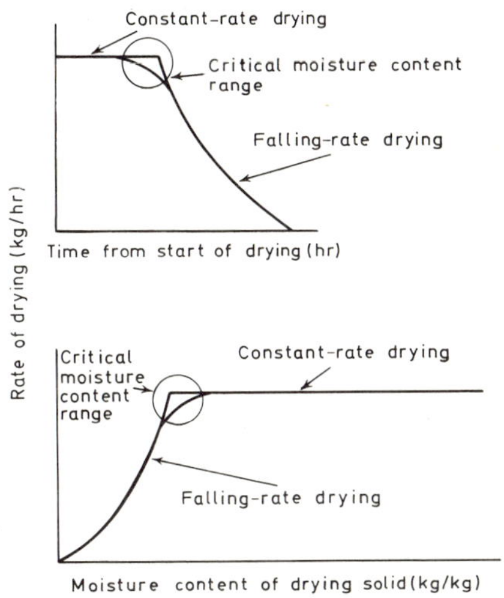

Fig. 2.1. Typical drying rate curves—convection drying—constant external conditions.

moisture content represents the range of moisture content over which the rate of migration to the surface equals the rate of evaporation from the surface and a mean value can be taken for calculation purposes. The critical moisture content will be dependent upon the external drying conditions but in practice the variation is often no greater than the range of uncertainty as to its exact value.

The term *constant-rate drying period* is somewhat misleading as the rate is only constant if the external drying conditions are constant and a better name would be *wet-bulb drying period* as explained in Section 2.3.5. However the term is widely used in the literature[2] and its use is therefore retained.

We can now consider the quantitative relationships involved.

2.3.2 Heat and matter transfer equilibrium

(N.B. Notes on equations are given at the end of the section)

When a wet solid is dried by pure convection heat is supplied solely by the sensible heat of the drying gas and evaporated liquid is removed as vapour in the drying gas stream. There is at any instant (whether in the constant- or falling-rate drying period) a dynamic equilibrium between the rate of heat supply and the rate of liquid removal which can be represented as follows (see also Note a):

$$\frac{dW}{d\theta} = \frac{h_c A(T_g - T_s)}{\lambda} = 10^{-3}\, K_p A(p_s - p_v) \tag{2.1}$$

The numerical term appears here and in later equations because of units derived from strict SI units e.g. use of the more convenient g/s rather than kg/s

Now for perfect gases the partial pressure and humidity are related by Avogadro's law, i.e.

$$H = \frac{m_v}{m_g} \cdot \frac{p}{P - p} \tag{2.2}$$

We can therefore relate the partial pressure driving force to a humidity driving force as

$$(H_{ss} - H_g) = \frac{m_v}{m_g} \cdot \left(\frac{p_s}{P - p_s} - \frac{p_v}{P - p_v}\right) \tag{2.3}$$

If the partial pressure is small compared with the total pressure we can say approximately that

$$P - p_s = P - p_v = (P - p)_{\text{lm}} \tag{2.4}$$

where $(P - p)_{\text{lm}}$ = log mean of $(P - p_s)$ and $(P - p_v)$.

The instantaneous drying rate can thus be specified in terms of a humidity driving force by substitution in equation 2.1, i.e.

$$\frac{dW}{d\theta} = 10^{-3}\, K_p A(p_s - p_v) = 10^{-3}\, K_p A \,.\, \frac{m_g}{m_v} \,.\, (P - p)_{\text{lm}}(H_{ss} - H_g)$$

$$= K_H A(H_{ss} - H_g) \tag{2.5}$$

Equations 2.1 and 2.5 can now be combined to give a complete heat and matter transfer equilibrium in terms of the temperature, partial pressure and humidity driving forces, i.e.

$$\frac{dW}{d\theta} = \frac{h_c A(T_g - T_s)}{\lambda} = 10^{-3}\, K_p A(p_s - p_v) = K_H A(H_{ss} - H_g) \tag{2.6}$$

The use of a matter-transfer or drying coefficient based on humidity, K_H, is not only for convenience in using data in the form in which it is available; it also has a slightly wider applicability than the partial pressure based coefficient, K_p. This is because the transfer of vapour from a region of high partial pressure in the solid, p_s, to a region of lower partial pressure in the gas, p_v, induces a stream of vapour known as drift which enhances the rate of transfer. This effect becomes significant at high temperatures when p_s is high (e.g. above 80°C for the air/water system) and it is necessary to apply a correction in order to obtain the correct value of the partial pressure driving force (see Note b). No such correction is necessary for the humidity driving force provided that the ratio p_s/p_v is not too high.

Equation 2.6. assumes that both the drying gas and vapour behave as perfect gases and that radiation and conduction to and from the evaporating surface are negligible. These assumptions are only likely to hold when the conditions in the drying gas are not very widely different from those in the solid being dried. It should also be noted that the coefficients will vary with the absolute pressure of the system. The limitations are reasonable for direct dryers which are usually operated at about atmospheric pressure but they can be seriously in error at high temperatures and pressures at high humidities. Caution is therefore necessary when considering systems working under extreme conditions.

Notes on Section 2.3.2

Note a. The significance of the matter-transfer coefficient can be shown on a more fundamental basis by deriving it from Fick's law for molecular diffusion across a laminar layer.[2] For a two component system this can be stated as

$$N_A = \frac{D\partial \bar{A}}{\partial z} + (N_A + N_B) \cdot \frac{p_A}{P} \tag{2.7}$$

For ideal gases this simplifies to

$$\frac{N_A p_B}{RT^*} - \frac{N_B p_A}{RT^*} = -\frac{PD}{RT^*} \cdot \frac{1}{RT^*} \cdot \frac{\partial p_A}{\partial z} \tag{2.8}$$

In the case of drying, only one component is diffusing and $N_B = 0$. Further $p_A + p_B = P$, and in the symbols used above for drying p_A is p_v and p_B is p_g. The expression 2.8 can therefore be integrated between the limits p_s and p_v across the stagnant gas layer to give

$$N_v (= N_A) = \frac{DP}{RTz(P-p)_{\mathrm{lm}}} \cdot (p_s - p_v) \tag{2.9}$$

The drying rate per unit area is thus:

$$\frac{\mathrm{d}W}{\mathrm{d}\theta} \cdot \frac{1}{\bar{A}} = N_v m_v = K_p (p_s - p_v) \tag{2.10}$$

where

$$K_p = \frac{m_v DP}{RTz(P-p)_{\mathrm{lm}}} = \text{matter-transfer coefficient in g/s m}^2 \text{ bar} \tag{2.11}$$

Note b. The significance of drift can be shown by examining the fundamental expressions for the matter-transfer coefficient. Thus from equations 2.5 and 2.11

$$K_H = K_p \cdot \frac{m_g}{m_v} \cdot (P-p)_{\mathrm{lm}} = \frac{m_g DP}{RTz} \tag{2.12}$$

The partial pressure based coefficient is thus dependent upon the value of $(P-p)_{\text{lm}}$ whereas the humidity expression is not. On the other hand the derivation of the humidity expression assumes that (equation 2.4)

$$P-p_s = P-p_v = (P-p)_{\text{lm}}$$

The net effect is that the partial pressure expression assumes that p_s and p_v are both similar and small relative to P, whereas the humidity expression only requires p_s and p_v to be similar in magnitude.

The drift factor correction can only be derived from the fundamental expression,[3] for the matter-transfer coefficient K_p. Thus if $(P-p)_{\text{lm}}$ cannot be taken as constant, the drying-rate equation 2.1 becomes

$$\frac{\mathrm{d}W}{\mathrm{d}\theta} = AK'_p \cdot \frac{(p_s-p_v)}{(P-p)_{\text{lm}}} = AK'_p(p_s-p_v)_{\text{corr}} \qquad (2.13)$$

where

$$K'_p = K_p(P-p)_{\text{lm}} = \frac{m_v DP}{RTz}$$

The corrected value of the partial pressure driving force is thus

$$(p_s-p_v)_{\text{corr}} = \frac{(p_s-p_v}{(P-p)_{\text{lm}}} = \frac{(p-p_v)\log_e (P-p_v)/(P-p_s)}{(P-p_v)-(P-p_s)}$$

$$= \log_e \frac{(P-p_v)}{(P-p_s)} \qquad (2.14)$$

And at atmospheric pressure ($P = 1$), we have

$$(p_s-p_v)_{\text{corr}} = \log_e \frac{(1-p_v)}{(1-p_s)} = \log_e \frac{1}{(1-p_s)} - \log_e \frac{1}{(1-p_v)} \qquad (2.15)$$

2.3.2 Driving force ratios

The drying coefficients at any instant can be derived in terms of the ratio of the driving forces by rearrangement of equation 2.6.

Thus

$$K_p = \frac{10^3\, h_c}{\lambda} \frac{(T_g-T_s)}{(p_s-p_v)} \qquad (2.16)$$

and

$$K_H = \frac{h_c}{\lambda} \frac{(T_g-T_s)}{(H_{ss}-H_g)} \qquad (2.17)$$

This form is convenient for graphical integration of the equilibrium equation 2.6 as explained below (Sections 2.3.6 and 2.3.8). Before proceeding to derive specific equations for the constant- and falling-rate drying periods it is necessary to consider the psychrometric properties of gas/liquid mixtures.

2.3.4 Psychrometric considerations

Wet-Bulb Temperature

The wet-bulb temperature is the equilibrium temperature which a liquid will attain when heat is transferred to it by contact with an infinitely large quantity of gas at a given temperature and humidity. This is the condition where the mass transfer from the liquid by evaporation is exactly balanced by the heat supply from the gas and can be represented by a heat- and matter-transfer equilibrium in the form

$$h_c(T_g - T_{wb}) = K_H\lambda(H_{sw} - H_g) \tag{2.18}$$

where λ = the latent heat of evaporation at T_{wb}.

This transfer balance equation can be rearranged to the form

$$\frac{H_{sw} - H_g}{T_{wb} - T_g} = \frac{-h_c}{K_H\lambda} \tag{2.19}$$

The relationship between the wet-bulb temperature T_{wb} and the temperature and humidity of the gas (T_g and H_g) is shown by plotting the data on the usual psychrometric chart as shown in Fig. 2.2.

The X co-ordinate represents the condition of the gas at temperature (T_g) and humidity (H_g) and the Y co-ordinate on the saturation curve represents the corresponding wet-bulb temperature T_w and saturation humidity

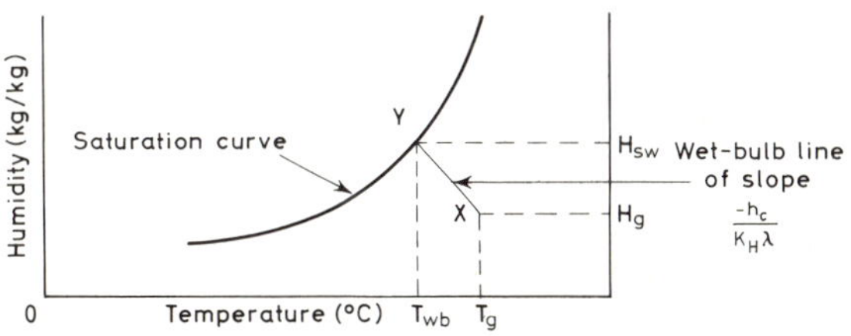

Fig. 2.2. Psychrometric chart showing wet-bulb line

H_w at the wet-bulb temperature. The line XY joining the two co-ordinates has a slope of $-h_c/K_H\lambda$ and is called the wet-bulb line. It should be noted that the wet-bulb line is not actually quite straight because the latent heat λ varies somewhat with temperature.

Adiabatic saturation temperature

The adiabatic saturation temperature is the equilibrium temperature which a mixture of gas and liquid will attain under adiabatic conditions. (There must be an adequate quantity of liquid available at or near the adiabatic saturation temperature for the equilibrium to be attained.)

Since the condition is adiabatic with no heat loss or gain, this equilibrium can be determined by heat balance of the system with the adiabatic saturation temperature as datum. Thus

$$\left\{\begin{array}{l}\text{sensible heat in gas}\\ \text{and vapour at}\\ \text{initial state}\end{array}\right\} + \left\{\begin{array}{l}\text{latent heat in}\\ \text{vapour at initial}\\ \text{state}\end{array}\right\} = \left\{\begin{array}{l}\text{latent heat in}\\ \text{vapour at final}\\ \text{state}\end{array}\right\}$$

or

$$(T_g - T_{sg})(c_g + c_v H_g) + \lambda H_g = \lambda H_s \tag{2.20}$$

Now the specific heat term $(c_g + c_v H_g)$ represents the specific heat of gas of given humidity H_g and is known as the humid heat, S. The equation can therefore be arranged in a similar form to equation 2.19. i.e.

$$\frac{H_s - H_g}{T_{sg} - T_g} = -\frac{S}{\lambda} \tag{2.21}$$

The relationship between the adiabatic saturation temperature and the temperature and humidity of the gas can be represented in exactly the same way as for the wet-bulb temperature in Fig. 2.2 by plotting the data on a saturation curve for the system; this is shown in Fig. 2.3.

In this case the co-ordinate Y on the saturation curve represents the adiabatic temperature T_{sg} and corresponding adiabatic saturation humidity

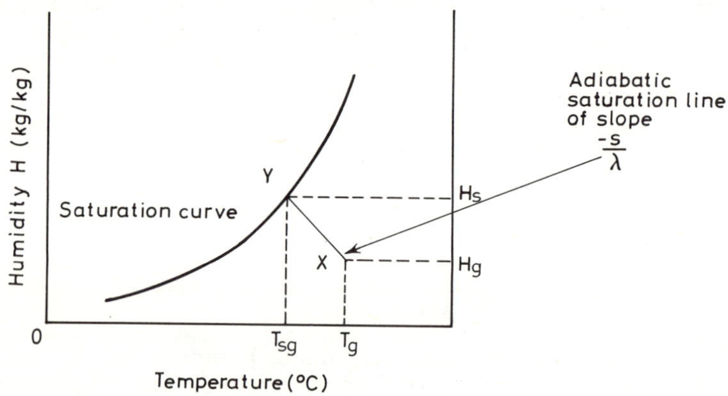

Fig. 2.3. Psychrometric chart showing adiabatic saturation line

H_s, and the line $X-Y$ has a slope of S/λ and is called the adiabatic saturation line. Like the wet-bulb line it is not quite straight because of the variation in latent heat of evaporation and humid heat with temperature. In fact there will strictly speaking be a unique value of the humid heat for each value of humidity as can be seen from equation 2.21 but the term $c_v H_g$ is usually very much smaller than the term c_g at the values of H_g met in dryers; hence it is reasonably true to take the humid heat as being constant at a given adiabatic saturation temperature.

Comparison of wet-bulb and adiabatic saturation temperature

The condition for identity of the wet-bulb and adiabatic saturation temperatures follows clearly from equations 2.19 and 2.21 as

$$\frac{h_c}{K_H} = S = c_g + c_v H_g \tag{2.22}$$

This identity can actually only be completely true for one value of the humidity (which may of course be unreal in being outside the range of the system). If however the humidity H_g is small, or the change in the term $c_v H_g$ is small, then a reasonable identity can occur over a range of humidity and temperature providing the values are suitable.

Now it so happens that equation 2.22 holds approximately in the case of a fluid composed of air and water vapour at relatively dilute concentration of water vapour, (see Note c below). Thus under the conditions usually met with in dryers the wet-bulb and adiabatic saturation temperatures are substantially identical for the air/water system; an illustration of the range of identity is shown in Fig. 2.5.

This identity is peculiar to the air/water system and has no physical significance. It is nevertheless of considerable convenience in dryer calculations as will be discussed in Section 2.3.8. In the case of systems other than air/water the identity of equation 2.22 no longer applies, and for the air/toluene system we have

$$\frac{h_c}{K_H} = 7{\cdot}41S \tag{2.23}$$

Examples for other systems are given by Perry[2] and Sherwood.[4]

Note on Section 2.3.3

Note c. The nature of the relation between h_c/K_H and S can be shown more fundamentally as follows:

The Prandtl–Taylor analogy for heat transfer[4] states that

$$\frac{h_c}{\rho v_m c_g} = \tfrac{1}{2}f \cdot \frac{10^3}{1 + v_l/v_m\{(Pr) - 1\}} \tag{2.24}$$

Similarly the Colburn analogy for mass transfer states that

$$\frac{K_p(P-p)_{\text{lm}}(m_g/m_v)}{\rho v_m m_m} = \tfrac{1}{2}f \cdot \frac{10^6}{1 + v_l/v_m\{(Sc) - 1\}} \tag{2.25}$$

In a fluid composed of air and water vapour at relatively low concentrations of water vapour, the Prandtl and Schmidt numbers are approximately equal, the mean molecular weight of the fluid is approximately equal to that of air ($m_m/m_g \approx 1$) and the specific heat of the fluid is the humid-heat of the air/water vapour ($c_g = S$). Thus for the air/water system only we can write

$$\frac{10^3 h_c}{K_p(P-p)_{\text{lm}}} = \frac{h_c}{K_H} = S \tag{2.26}$$

2.3.5 Constant-rate drying or wet-bulb drying

The mechanism for constant-rate drying is that of evaporation from a liquid surface and the presence of the solid does not affect the mechanism. The solid may, however, affect the properties of the liquid surface so that the rate of evaporation is somewhat different from that obtained with a pure liquid. This solid effect is relatively small and usually corresponds to a reduction in evaporation rate of not more than 20%.

The rate of drying is determined by the rate of diffusion of the vapour through the gas film at the drying surface into the main body of the gas stream and, at equilibrium in presence of a relatively large amount of drying gas, the conditions are those already defined for the wet-bulb temperature (Section 2.3.4.). The surface at any point will thus tend to assume the wet-bulb temperature corresponding to the temperature, humidity and quantity of the drying gas. If the quantity and condition of the drying gas at the point remains constant, the surface or wet-bulb temperature will also be constant.

Correspondingly, the partial pressure and humidity at the surface will be the saturation partial pressure and the saturation humidity at the wet-bulb temperature.

The instantaneous drying rate equation 2.6 can therefore be very simply integrated over the constant-rate period between the limits of initial and critical moisture content in the form

$$R_c = \frac{10^3(M_s - M_c)W}{A\theta_c} = \frac{h_c(T_g - T_{wb})}{\lambda} = 10^{-3}\,K_p(p_{wb} - p_v) = K_H(H_{sw} - H_g)_{\text{lm}} \tag{2.27}$$

This is a key equation.

The velocity of the drying gas is the only other variable affecting the drying rate and it will influence the values of the coefficients h_c, K_p and K_H. It would be more correct to refer to the process as wet-bulb drying rather than constant-rate drying but the latter name will be used in this book because it is widely used in practice.

It should be noted that the quantity or velocity of the gas stream passing over the surface is not always sufficient to ensure that true equilibrium at the wet-bulb temperature is achieved (e.g. a minimum velocity of 3 m/s is recommended in wet-bulb hygrometry for the air/water system) but in practical drying problems it can be assumed that the actual surface will approximate to the wet-bulb temperature for constant-rate drying with heat transfer by pure convection.

2.3.6 Falling-rate drying

During this period drying is controlled by the rate of migration of liquid to the surface at which evaporation occurs. Heat is thus transferred to the surface at a progressively decreasing rate to balance the decreasing rate of matter transfer within the material and, under constant ambient conditions, the surface temperature keeps in step and increases from the wet-bulb temperature T_W at the critical moisture content to the drying gas temperature T_A when drying is complete; i.e. the temperature driving force $(T_A - T_S)$ diminishes as drying proceeds. Measurements in I.C.I. laboratories on coal, coke and pumice have confirmed that the heat-transfer coefficient remains substantially constant. The reduced rate of heat transfer can thus be quantitatively expressed in terms of the rise in surface temperature of the solid alone. Integration of the heat-transfer part of equation 2.6 then takes the form

$$R_f = \frac{10^3 (M_c - M_f)W}{A\theta_f} = \frac{h_c}{\lambda} \int_{\to 0}^{T_g - T_{wb}} \mathrm{d}(T_g - T_s) \qquad (2.28)$$

The process in respect of matter transfer is different since we are now concerned with the overall coefficient rather than a limiting coefficient at the surface. The postulated mechanism for the falling-rate period is that the overall resistance to matter transfer increases with decreasing moisture content, which is the same thing as saying that the matter-transfer or drying coefficient decreases as drying proceeds. Now the partial pressure p_s is the saturation partial pressure corresponding to the surface temperature T_s and will increase as the solid temperature increases from T_{wb} at the critical moisture content to the limiting value T_g when drying is complete. The partial pressure of the vapour in the drying gas, p_v, is, on the other hand, dependent only on the amount of vapour present and will tend to decrease as the drying rate slows down. The driving force $(p_s - p_v)$ will thus increase as drying proceeds. This increase must be less than the decrease in the matter transfer or drying coefficient since the overall effect is for the drying rate to decrease. A similar argument applies to the humidity driving force $(H_{ss} - H_g)$. This effect has been shown experimentally in I.C.I. laboratories in terms of drying resistance or inverse matter-transfer coefficient. Integration of the matter-transfer parts of equation thus takes the form

$$R_f = \frac{10^3 (M_c - M_f)W}{A\theta_f} = 10^3 \int_{\to 0}^{(K_p)_c} \mathrm{d}K_p \int_{\to 0}^{p_{wb} - p_v} \mathrm{d}(p_s - p_v)$$

$$= \int_{\to 0}^{(K_H)_c} \mathrm{d}K_H \int_{\to 0}^{H_{sw} - H_g} \mathrm{d}(H_{ss} - H_g) \qquad (2.29)$$

where $(K_p)_c$ and $(K_H)_c$ are constant-rate values of the drying coefficients as defined in equation 2.27.

Combination of equations 2.28 and 2.29 gives a general equation for falling-rate drying but it is not possible to solve it in absence of knowledge of the functions. In the present state of knowledge this integration is most conveniently carried out step-wise or graphically in terms of an experimentally determined falling-rate drying curve for the particular material. A convenient procedure is to use the relation between the drying coefficients and driving force ratio derived in equations 2.16 and 2.17.

Since the heat-transfer coefficient can be assumed to be constant at a value determined from the constant-rate drying conditions by equation 2.27, equations 2.16 and 2.17 can be used to transform a graph of drying rate against moisture content for specific conditions into a graph of drying coefficient against moisture content which is independent of the drying conditions. The drying-rate curve for other specific conditions can then be derived by the reverse calculations. This procedure is described more explicitly in Section 2.3.9 and the worked examples in Section 2.3.10.

2.3.7 Drying from solutions

If the solid is soluble in the liquid from which it is being dried the solution will have a lower vapour pressure than the pure liquid and the drying rate will in general be reduced. The effect in the constant-rate period is to raise

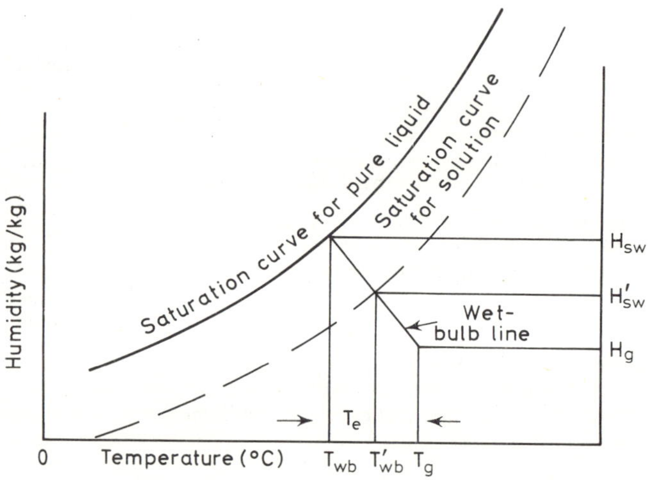

Fig. 2.4. Psychrometric chart for a solution

the surface or wet-bulb temperature by an amount equal to the boiling point elevation of the solution and to correspondingly reduce the surface humidity and partial pressures. Thus equation 2.27 for constant-rate drying becomes

$$R_c = \frac{h_c(T_g - T'_{wb})}{\lambda} = 10^{-3}\, K_p(p'_{wb} - p_v) = K_H(H'_{sw} - H_g) \qquad (2.30)$$

where $T'_{wb} = T_{wb} + T_e$ = wet-bulb temperature over solution.

If we assume that the ratio of heat and matter transfer is unchanged compared with the pure liquid (i.e. the slope of the wet-bulb line is unchanged), then the surface humidity (and partial pressure) can be obtained from a normal psychrometric chart by assuming that the saturation curve for the solution is parallel with that for the pure liquid as shown in Fig. 2.4. It can usually be assumed for drying problems that the solution is saturated, and this simplifies determination of the boiling point elevation T_e. Equation 2.30 does not apply to systems complicated by hydrate formation.

2.3.8 Simplification for the air-water system

It has already been shown in Section 2.3.4 that the wet-bulb and adiabatic saturation temperatures are substantially identical for the air/water system at moderate temperatures and humidities.

Thus over the range 0–100°C we have

Mean specific heat of dry air $c_g = 1{\cdot}05$ kJ/kg°C

Mean specific heat of water vapour $c_v = 1{\cdot}84$ kJ/kg °C

Wet-bulb mass transfer ratio $\dfrac{h_c}{(K_H)_c} = 1{\cdot}09$ kJ/kg °C

(Ref 3)

By substituting these values in equations 2.19, 2.21 and 2.22, the range may be calculated over which the wet-bulb and adiabatic saturation temperatures are substantially identical for the air/water system: this is shown in Fig. 2.5.

The identity of the wet-bulb and adiabatic saturation temperatures in

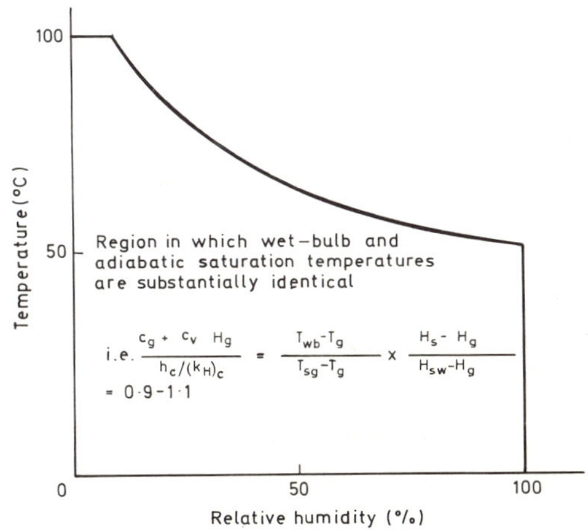

Fig. 2.5. Air/water system; wet-bulb and adiabatic saturation temperatures

the special case of the air/water system has no physical significance but nevertheless leads to a considerable simplification in the treatment of the constant-rate period for this system. Thus consider a bed of material wet with water being dried by the passage over its upper surface of a stream of air of constant temperature and humidity. Heat is supplied solely by this air stream and the process is therefore adiabatic. The material is wetter than its critical moisture content and the air leaving the surface is not saturated. The adiabatic saturation temperature remains constant (within the limits defined in Section 2.3.4) and is dependent solely on the inlet temperature and humidity of the air. For the air/water system the wet-bulb temperature can be taken as equal to the adiabatic saturation temperatures (see Fig. 2.5) and is therefore constant also. Thus, as the air passes over the

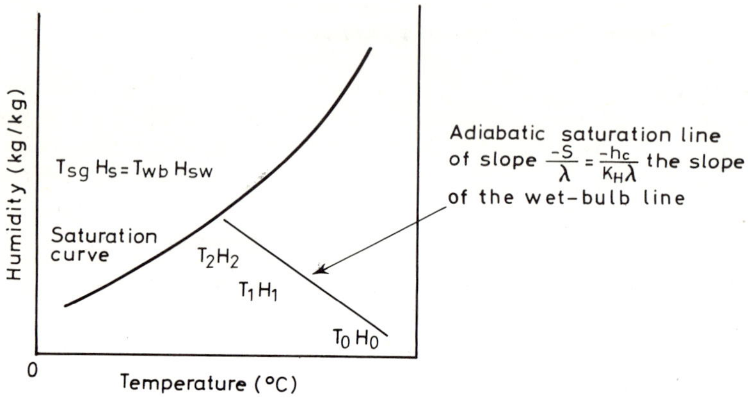

Fig. 2.6. Psychrometric chart for air/toluene system

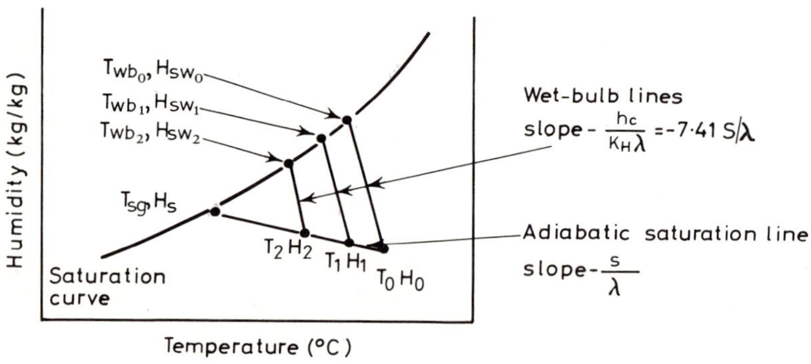

Fig. 2.7. Psychrometric chart for air/toluene system

surface of the material it will cool and humidify along the wet-bulb line, which in this case is the same as the adiabatic saturation line. The surface temperature will be constant at the wet-bulb temperature at all points and the drying rate can be determined from the driving forces derived from a single wet-bulb line as shown in Fig. 2.6.

Consider now a similar case but where the solid is wet with a non-aqueous liquid and the air is 'humid' with the same substance. The adiabatic saturation temperature will still be constant since it is only dependent on the inlet temperature and 'humidity' of the air. The wet-bulb temperature does not now coincide with the adiabatic saturation temperature and will vary at each point of the surface in accordance with the temperature and 'humidity' of the air at that point. The temperature of the surface will thus vary along the length of the air stream. As shown in Fig. 2.7 it is therefore necessary to use a number of different wet-bulb lines at different parts of the surface in order to derive the corresponding driving forces and hence the drying rate.

In general therefore for the air/water system under adiabatic conditions of drying and when the moisture content of the material being dried is greater than the critical moisture content, the surface temperature is equal to the wet-bulb temperature and is a constant dependent only on the inlet air condition. Wet-bulb or constant-rate drying for the air/water system is thus an easily defined process (hence the common use of the term constant-rate drying); in other systems where no identity of the wet-bulb and adiabatic saturation temperature occurs the system is less easily defined. Care must therefore be taken not to use this simplification for systems in which the liquid to be removed by the drying process is not water.

Quantitative values relating the coefficients and for the driving force ratios can be calculated directly for the air/water system. Thus from equations 2.5, 2.6 and 2.27 we have at atmospheric pressure

$$(K_H)_c = (K_p)_c \frac{m_g}{m_v} = 1{\cdot}61(K_p)_c \qquad (2.31)$$

and (above)

$$\frac{h_c}{(K_H)_c} = 1{\cdot}09 \qquad (2.32)$$

The constant-rate relationship between the three coefficients for the air/water system is thus

$$(K_H)_c = 1{\cdot}61(K_p)_c = 0{\cdot}92\, h_c \qquad (2.33)$$

By substitution in equations 2.16 and 2.17 we get the constant-rate values for the driving force ratios are

$$\frac{(T_g - T_{wb})}{(p_{wb} - p_g)} = 0{\cdot}57 \times 10^{-3}\lambda \qquad (2.34)$$

$$\frac{(T_g - T_{wb})}{(H_{sw} - H_g)} = 0{\cdot}92\lambda \qquad (2.35)$$

$$\frac{(p_{wb} - p_g)}{(H_{sw} - H_g)} = 1{\cdot}61 \times 10^{-3}\lambda \qquad (2.36)$$

2.3.9 Analysis and correlation of convection drying data

The principles given above can be used to analyse experimental data in a form which enables the drying rate under different conditions to be predicted.

Thus the heat-transfer coefficient and constant-rate drying coefficients can be obtained from the experimental data by equation 2.27. The heat-transfer coefficient is now taken to be constant throughout the drying cycle allowing the surface temperature at each of a series of moisture contents to be calculated by equation 2.6. From this we obtain the temperature driving force at each moisture content and hence the corresponding partial pressure and humidity driving forces. The drying coefficients can then be obtained from the driving force ratios by equations 2.16 and 2.17 and plotted as drying coefficient against moisture content. The actual drying curve for a different set of ambient conditions can then be obtained by the reverse procedure, though unfortunately trial-and-error calculation is necessary for several steps.

It is not usually necessary in any particular problem to use both these drying coefficients. In general use of the humidity-based coefficient K_H is more convenient and the humidity driving force has the advantage that it can be used over a wider range without being corrected for drift (Section 2.3.2) but the partial pressure coefficient K_P may afford a useful direct comparison against matter-transfer coefficients from other sources.

The basic assumption in this method of calculation is that the drying coefficients are independent of the ambient conditions. Put more specifically this means that the critical moisture content and the shape of the falling-rate curve are assumed to be independent of the drying conditions. Neither assumption is quite correct but the changes are normally relatively small for moderate changes in conditions with thermally-stable materials.

It should also be noted that difficulty is experienced when the drying temperature is much above the boiling point of the liquid as the driving forces lose their meaning. Such cases are best dealt with by regarding the drying gas as a superheated vapour rather than as a humid gas.

2.3.10 Worked examples in convection drying

Example 1. Calculation of drying coefficients

Granules prepared from an organic paste are dried by passage of hot air downwards through a thin bed of the material under conditions such that heat transfer is entirely by convection and heat losses are negligible. The material is thermally stable and shrinkage is negligible. Calculate the drying coefficient – moisture curves. (See Note a.)

The experimental data are:

Granules in the form of rods, 4 mm diameter and 0·6–0·9 mm long, lying in a bed 2·5 cm thick

Bed loading $W/A' = 7\ \mathrm{kg/m^2}$ dry material

Air flow $= G = 4\,880\ \mathrm{kg/m^2\,h} = 1{\cdot}36\ \mathrm{kg/s\,m^2}$

Air condition,	inlet temperature	$(T_g)_{\mathrm{in}}$	$= 70°\mathrm{C}$
	inlet humidity	$(H_g)_{\mathrm{in}}$	$= 0{\cdot}005$ kg/kg
Moisture,	initial	M_s	$= 0{\cdot}890$ kg/kg dry
	critical	M_c	$= 0{\cdot}288$ kg/kg dry (see Note b)
	final	M_f	= negligible

Drying time, constant rate θ_c = 7·9 min = 474 s (see Note b)
total θ_t = 21 min = 1 260 s
Effective exposed area of material A/A' = 10·9 m²/m² bed (see Note c)

(a) Constant Rate
The constant drying rate (eq. 2.27)

$$R_c = \frac{10^3\,(M_s - M_c)W}{A\theta_c} = 0{\cdot}816 \text{ g/s m}^2$$

Wet-bulb temperature from psychrometric tables, T_{wb} = 27·2°C
Latent heat of evaporation at T_{wb}, λ = 2 430 kJ/kg
Specific heat of air $(T_A)_{in}$, c_g = 1 005 kJ/kg °C

To find the constant-rate exit air temperature $(T_g)_{out}$ it is necessary to calculate the fall in temperature ΔT of the air during its passage through the dryer. This is obtained from the following heat balance:
Heat taken out of air during the operation = $GA'c_g\Delta T\theta_c$.
Similarly heat absorbed in evaporating the water

$$= W(M_s - M_f) = R_c A\lambda\theta_c$$

whence

$$\Delta T = \frac{R_c\lambda}{Gc_g}\cdot\frac{A}{A'} = \frac{0{\cdot}816\times 2430\times 10{\cdot}9}{1{\cdot}36\times 1005} = 15{\cdot}8°\text{C}$$

and $(T_g)_{out}$ = 70 − 15·8 = 54·2°C
To find the heat-transfer coefficient it is necessary to calculate the log mean constant-rate temperature driving force (see Note d)

$$(T_g - T_{wb})_{lm} = \frac{\{(T_g)_{in} - T_{wb}\} - \{(T_g)_{out} - T_{wb}\}}{2{\cdot}303\log_{10}\dfrac{(T_g)_{in} - T_{wb}}{(T_g)_{out} - T_{wb}}}$$

$$= \frac{15{\cdot}8}{2{\cdot}303\log_{10} 42{\cdot}8/27{\cdot}0} = 34{\cdot}2°\text{C}$$

Then heat-transfer coefficient

$$h_c = \frac{R_c\lambda}{(T_g - T_{wb})_{lm}} \quad \text{(from equation 2.27)}$$

= 58·3 W/m² °C

To obtain the constant-rate humidity $(H_g)_{out}$ in kg/kg it is necessary to add to the inlet air humidity the quantity of water evaporated during the cycle into each kg of air.

The air quantity through the dryer during the cycle is $GA'\theta$ kg. The quantity of water evaporated during the cycle is

$$(M_s - M_c)W = R_c A\theta \text{ kg}$$

Hence water evaporated into each kg of air is

$$\frac{R_c\times 10^{-3}}{G}\cdot\frac{A}{A'}\,0{\cdot}00656 \text{ kg/kg}$$

The humidity of the heated air to the dryer is 0·0050 kg/kg. Hence constant-rate exit humidity $(H_g)_{\text{out}}$

$$= (H_g)_{\text{in}} + \frac{R_c \times 10^{-3}}{G} \cdot \frac{A}{A'} = 0{\cdot}01156 \text{ kg/kg}$$

From psychrometric tables, $(p_v)_{\text{in}} = 0{\cdot}00796$ bar
$(p_v)_{\text{out}} = 0{\cdot}01822$ bar
The saturation partial pressure at T_{wb}, $p_{wb} = 0{\cdot}0354$ atm whence the log mean constant-rate partial pressure driving force,

$$(p_{wb} - p_v)_{\text{lm}} = 0{\cdot}02175 \text{ bar}$$

The constant-rate drying coefficient on the partial pressure basis is then

$$(K_p)_c = \frac{10^3 \, h_c}{\lambda} \cdot \frac{(T_g - T_{wb})_{\text{lm}}}{(p_{wb} - p_v)_{\text{lm}}}$$

$= 37{\cdot}8$ g/s m^2 bar (from equations 2.16 and 2.27)

The constant-rate drying coefficient $(K_H)_c$ can also be obtained from humidities.
Saturation humidity at T_{wb}, $H_{sw} = 0{\cdot}02308$ kg/kg
Log mean constant-rate humidity driving force $(H_{sw} - H_g)_{\text{lm}}$

$= 0{\cdot}01458$ kg/kg (see Note e)

Then constant-rate drying coefficient $(K_H)_c$ on humidity basis

$$= \frac{h_c}{\lambda} \cdot \frac{(T_g - T_{wb})_{\text{lm}}}{(H_{sw} - H_g)_{\text{lm}}}$$

= 56 g/s m^2 unity humidity difference (see Note f).

(from equations 2.17 and 2.27)

(*b*) *Falling Rate*
A similar set of calculations is made for a series of drying rates corresponding to moisture contents during the falling-rate period. The key steps are:

Mean temperature driving force $(T_g - T_s)_m = \dfrac{\mathrm{d}W}{\mathrm{d}\theta} \cdot \dfrac{1}{A} \dfrac{\lambda}{h_c}$ (from equation 2.6)

(see Note g)

Exit air temperature $(T_g)_{\text{out}} = (T_g)_{\text{in}} - \dfrac{\mathrm{d}W}{\mathrm{d}\theta} \dfrac{1}{A} \dfrac{\lambda}{G c_g} \dfrac{A}{A'}$

(see Note h)
Take arithmetic mean air temperature as $(T_g)_m = \frac{1}{2}\{(T_g)_{\text{in}} + (T_g)_{\text{out}}\}$
Then mean surface temperature $(T_s)_m = (T_g)_m - (T_g - T_s)_m$

Now the exit air humidity $(H_g)_{\text{out}} = (H_g)_{\text{in}} + \dfrac{\mathrm{d}W}{\mathrm{d}\theta} \dfrac{1}{G} \dfrac{A}{A'}$

Take the exit air humidity as $(H_g)_m = \frac{1}{2}\{(H_g)_{\text{in}} + (H_g)_{\text{out}}\}$
From psychrometric tables obtain the mean saturation humidity $(H_{ss})_m$ corresponding to the mean surface temperature $(T_s)_m$

Then the mean humidity driving force $(H_{ss}-H_g)_m = (H_{ss})_m-(H_g)_m$ and the drying coefficient on the humidity basis

$$K_H = \frac{h_c(T_g-T_s)_m}{(H_{ss}-H_g)_m} \text{ (from equation 2.17)}$$

Obtain also from psychrometric tables the partial pressure values corresponding to the humidities, i.e. the mean partial pressure of water vapour in

Table 2.1 Values of drying coefficients for worked example 1 on convection drying

Moisture content	*Instantaneous drying rate*	*Mean temperature driving force*	*Mean surface temperature*	*Mean partial pressure driving force*	*Mean humidity driving force*	*Drying coefficients*	
						Partial pressure basis	*Humidity basis*
M (kg/kg)	$\frac{dW}{d\theta}\cdot\frac{1}{A}$ (g/s m²)	$(T_g-T_s)_m$ (°C)	$(T_s)_m$ (°C)	$(p_s-p_v)_m$ (mb)	$(H_{ss}-H_g)_m$ (kg/kg)	K_p (g/s m² bar)	K_H (g/s m² *unit humidity difference*)
*0·890	0·821	34·2	27·2	20·3	0·01458	40·3	56·1
*0·285	0·821	34·2	27·2	20·3	0·01458	40·3	56·1
0·25	0·726	29·4	34·3	41·7	0·02761	17·2	26·2
0·20	0·582	23·6	41·2	65·6	0·04492	8·8	12·9
0·15	0·447	18·3	47·8	98·4	0·06953	4·6	6·4
0·10	0·313	12·6	54·1	145·2	0·08375	2·2	3·6
0·05	0·157	7·6	60·8	193·3	0·15425	1·3	1·2
0·025	0·123	5·0	63·9	225·8	0·18651	0·54	0·65
0	0	0	70·0	299·6	0·27320	0	0

*Constant-rate values

air, $(p_v)_m$, corresponding to $(H_g)_m$, and the mean saturation partial pressure of water vapour in air $(p_s)_m$ corresponding to $(H_{ss})_m$.
Then the mean partial pressure driving force $(p_s-p_v)_m = (p_s)_m-(p_v)_m$ and the drying coefficient on a partial pressure basis

$$K_p = \frac{10^3\, h_c(T_g-T_s)_m}{\lambda(p_s-p_v)_m} \text{ (from equation 2.16)}$$

Finally by plotting the drying coefficients against moisture content we obtain curves which can be taken as unique for the particular material; these are shown in Fig. 2.8. (see also Table 2.1).

Notes on worked example 1

(a) Curves for both drying coefficients are worked out in this example in order to demonstrate procedure. Generally only one coefficient is required and the humidity based coefficient K_H is to be preferred.
(b) The critical moisture M_c is obtained from a graph of the rate of drying, $dW/d\theta$, against moisture content M.

(c) Calculation in terms of bed area A' suffices to calculate the drying rate of the same material under different conditions. An estimate of true surface area A is used in this example to give the numerical values clearer meaning; this is 10·9 m^2 per m^2 of bed surface area.
(d) The calculation has been simplified for the air/water system because the wet-bulb temperature, which is the same as the adiabatic saturation temperature, is constant throughout the bed – see Section 2.3.8.
(e) The drift correction has been taken as negligible – see Section 2.3.2.

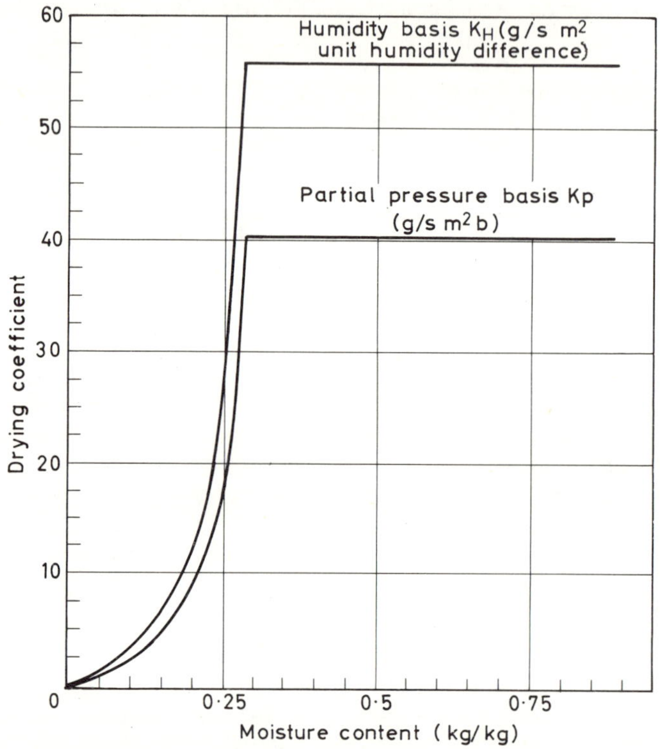

Fig. 2.8. Drying coefficients for granules in a convection dryer, example 2

(f) As a check we have for the air/water system from equation 2.33 $(K_H)_c = 0{\cdot}92\,h_c = 54$ g/s m^2.
(g) The latent heat to be used is that for the mean of the wet-bulb and inlet air temperatures, i.e. that at 48·6°C = 2 380 kJ/kg.
(h) The specific heat, c_g, to be used is that at the mean temperature at the critical moisture content and the final air temperature, i.e. 66·1°C = 1 090 kJ/kg°C.

Example 2. Calculation of drying time

Granules of the material used in example 1 are to be dried with an inlet air temperature of 90°C under otherwise identical conditions. Calculate drying time from the humidity-based drying coefficient, K_H, given in Fig. 2.8 (see Note a). Assume that the critical moisture content remains unaltered at 0·285 kg/kg.

In the application of the equations used in these calculations, it is necessary to observe whether expressions involving mass have been given in terms of grams or kilograms.

(a) Constant-rate Period

Wet-bulb temperature from psychrometric tables = 31·4°C

Saturation humidity at T_{wb} (from tables), H_{sw} = 0·0301 kg/kg

Exit air humidity, $(H_A)_{\text{out}} = (H_A)_{\text{in}} + \dfrac{R_c}{G} \cdot \dfrac{A}{A'}$

$= 0{\cdot}005 + 0{\cdot}00805\ R_c$ (i)

Constant drying rate, $R_c = K_H(H_{sw} - H_g)_m$ (from equation 2.27)

$= 56{\cdot}1\ (H_{sw} - H_g)_m$ (ii)

$= 1{\cdot}15\ \text{g/s m}^2$

Latent heat of evaporation at T_{wb}, λ = 2 430 kJ/kg

Specific heat of air at T_{wb}, c_g = 1 010 kJ/kg °C

Exit air temperature $(T_g)_{\text{out}} = (T_g)_{\text{in}} - \dfrac{R_c\lambda}{Gc_g} \cdot \dfrac{A}{A'}$

$= 67{\cdot}7°\text{C}$

Log mean constant-rate temperature driving force, (see Note b)

$$(T_g - T_{wb})_{\text{lm}} = 46{\cdot}5°\text{C}$$

Hence heat-transfer coefficient,

$$h_c = \frac{R_c\lambda}{(T_g - T_{wb})_{\text{lm}}} = 60{\cdot}3\ \text{W/m}^2\ °\text{C} \qquad \text{(iii)}$$

(b) Falling-rate period

Similar calculations are made for a series of drying coefficients K_H corresponding to moisture contents during the falling-rate period.

Mean temperature driving force, $(T_g - T_s)_{\text{lm}} = \dfrac{\lambda}{h_c A} \cdot \dfrac{\mathrm{d}W}{\mathrm{d}\theta}$ (from equation 2.6)

$$= \frac{39{\cdot}2}{A} \cdot \frac{\mathrm{d}W}{\mathrm{d}\theta} \text{(see Note c)} \qquad \text{(iv)}$$

Exit air temperature, $(T_g)_{\text{out}} = (T_g)_{\text{in}} - \dfrac{\lambda}{6c_g} \cdot \dfrac{1}{A} \cdot \dfrac{A}{A'} \cdot \dfrac{\mathrm{d}W}{\mathrm{d}\theta}$

$$= 90 - \frac{18{\cdot}8}{A} \cdot \frac{\mathrm{d}W}{\mathrm{d}\theta} \text{ (see Note d)}$$

Mean air temperature, $\dfrac{(T_g)_{\text{in}} + (T_g)_{\text{out}}}{2} = 90 - \dfrac{9{\cdot}4}{A} \cdot \dfrac{\mathrm{d}W}{\mathrm{d}\theta}$

Mean surface temperature, $(T_s)_m = (T_g)_m - (T_g - T_s)_m$

$$= 90 - (39{\cdot}2 + 9{\cdot}4) \,.\, \frac{1}{A} \cdot \frac{\mathrm{d}W}{\mathrm{d}\theta} = 90 - \frac{48{\cdot}6}{A} \cdot \frac{\mathrm{d}W}{\mathrm{d}\theta} \qquad \text{(v)}$$

$$\text{Exit air humidity, } (H_g)_{out} = (H_g)_{in} + \frac{1}{GA} \cdot \frac{A}{A'} \cdot \frac{dW}{d\theta}$$

$$= 0{\cdot}005 + 0{\cdot}00804 \,.\, \frac{1}{A} \cdot \frac{dW}{d\theta}$$

$$\text{Hence mean air humidity, } (H_g)_m = \tfrac{1}{2}\{(H_g)_{in} + (H_g)_{out}\}$$

$$= 0{\cdot}005 + 0{\cdot}00402 \,.\, \frac{1}{A} \cdot \frac{dW}{d\theta} \qquad \text{(vi)}$$

The mean saturation humidity, $(H_{ss})_m$ can be read from psychrometric tables for each value of the mean surface temperature $(T_s)_m$.
The mean humidity driving force is $(H_{ss} - H_g)_m = (H_{ss})_m - (H_g)_m$
The drying rate and humidity-based drying coefficient are related by the expression (equation 2.6)

$$\frac{1}{A} \cdot \frac{dW}{d\theta} = K_H(H_{ss} - H_g)_m \text{ (from equation 2.6)} \qquad \text{(vii)}$$

Expressions (iv) to (vii) can now be solved by trial and error to give the drying rate $1/A \,.\, dW/d\theta$ for each value of the drying coefficient, K_H, at each of a series of values of moisture content, M. The calculated values are shown in Table 2.2, including those for the constant-rate period.

The drying rate is then plotted against moisture content to give a curve which can be integrated graphically to give a graph of moisture content against time. Table 2.3 gives times to dry to each moisture content read from this last graph. It will be seen that raising the air inlet temperature from 70°C to 90°C reduces the drying time from 21 to 13 min.

Notes on Example 2

(a) A similar calculation can be made from the drying coefficient, K_p, based on partial pressures.
(b) The calculation is simplified for the air/water system because the wet-bulb temperature (= adiabatic saturation temperature) is constant throughout the bed – see Section 2.3.8.
(c) The latent heat to be used is that for the mean of the wet-bulb and final or inlet air temperature, i.e. the latent heat at 60·7°C. This is $\lambda = 2\,361$ kJ/kg.
(d) The specific heat to be used is that for the mean temperature of the critical and final air temperatures, i.e. the specific heat at 84·5°C. This is $c_A = 1\,005$ kJ/kg °C.

2.4 HEAT TRANSFER BY CONDUCTION

2.4.1 Comparison with convection drying

In convection drying the wet solid is heated via a wall so that the liquid is evaporated and leaves the apparatus as vapour. The dryers are frequently operated under reduced pressure which serves to reduce the boiling point of the liquid and thereby increase the temperature difference between the heating medium and the solid.

The process is essentially one of evaporation and the mechanism is different from that of direct drying by means of a stream of heated gas (cf. Section 2.3). The solid tends to assume the boiling point of the liquid and remain at that temperature. It will rise towards the wall temperature at the end of drying and the temperature may not be uniform, for example, with relatively

Table 2.2 Convection drying experiment (Example 2). Relation between moisture content and drying coefficient on a humidity basis

Moisture content	*Drying coefficient* K_H (g/s m^2 *unit humidity difference*)	*Drying rate* $\frac{dW}{d\theta} \cdot \frac{1}{A}$ (g/s m^2)
*0·890	56·0	1·15
*0·285	56·0	1·15
0·25	26·2	1·02
0·20	12·9	0·88
0·15	6·4	0·70
0·10	3·7	0·51
0·05	1·2	0·31
0·025	0·65	0·24
0	0	0

Table 2.3 Convection drying experiment (Example 2). Relation between moisture content and elapsed drying time

Moisture content M (kg/kg)	*Elapsed drying time* θ_c (s)	*Drying times for stages* (min)
*0·890	0	Constant-rate drying time θ_c 5·6
*0·285	335	
0·25	355	Falling-rate drying time, θ_f 7·4
0·20	395	
0·15	4355	
0·10	4900	
0·05	5690	Total drying time, θ_t 13·0
0·025	6275	
0	7775	

*Constant-rate values

thick static beds of material of low conductivity. It is thus possible to distinguish an initial fast stage of drying and a slower rate of drying which may be likened to the constant- and falling-rate periods in convection drying. The resemblance is only fortuitous, however, and there is no clear-cut distinction in mechanism as in the convection case, the stages depending more upon the operating conditions and design of the machine than on the material being dried. The drying rate will, in general, decrease progressively as drying proceeds. The rate of drying is therefore most conveniently considered from the viewpoint of variation of the heat-transfer coefficient and the temperature differential.

2.4.2 Heat transfer

Consider a typical case in which a wet solid is placed in an enclosed container which is heated externally and provided with means for egress of the vapour. Then at any instant the rate of drying will be dependent upon the heat input to the material, i.e.

$$\frac{\mathrm{d}W}{\mathrm{d}\theta} = \frac{10^3}{\lambda} \cdot \frac{\mathrm{d}Q}{\mathrm{d}\theta} \tag{2.37}$$

The rate of heat supply at any instant is in turn dependent upon the temperature difference between the heating source and material, the area available for heat transfer and the overall heat transfer by conduction. Hence

$$\frac{\mathrm{d}Q}{\mathrm{d}\theta} = 10^{-3}\, UA_h(T_h - T_s) \tag{2.38}$$

This equation can be re-written in the more useful integrated form over the whole drying cycle as

$$R_m = 10^3 \frac{(M_s - M_f)W}{A_h\theta} = \frac{U_m(T_h - T_s)}{\lambda} \tag{2.39}$$

2.4.3 Surface temperature

Consider now the value of the temperature at the surface of evaporation, (T_s)—the overall heat-transfer coefficient can be analysed in terms of the resistance of the various films to conduction. Thus

$$\frac{1}{U} = \frac{1}{h_h} + \frac{1}{h_w} + \frac{1}{h_f} + \frac{1}{h_m} + \frac{1}{h_e} \tag{2.40}$$

Analysing the conductivity coefficients in more detail we have

$$h_w = k_w/t_w \tag{2.41}$$

$$h_m = k_m/t_m \tag{2.42}$$

Typical values of the film coefficients that might be expected are given in Table 2.4.

It can be seen that in general the effect of factors other than the transfer of heat from the wall to the material and through the material is small. If these are neglected equation 2.40 can be simplified to

$$\frac{1}{U} = \frac{1}{h_f} + \frac{1}{h_m} \tag{2.43}$$

This means that we can take the temperature of the wall of the dryer as equal to that of the heating medium and the temperature of the surface at which evaporation occurs as that of the boiling point of the liquid. It is seldom necessary to take into consideration the elevation of the boiling point of a saturated solution. Equation 2.39 can then be simplified to

$$R_m = \frac{U_m(T_h - T_b)}{\lambda} \tag{2.44}$$

2.4.4 Conduction-drying coefficient

This simple equation cannot hold over the whole drying cycle however, since at the end of drying the temperature will tend to rise towards that of the dryer wall (i.e. to approach T_h). The average temperature differential over the whole cycle will thus be a little less than $T_h - T_b$ and we can write this as

$$(T_h - T_s) = a(T_h - T_b) \tag{2.45}$$

where a = constant approaching unity.

Substituting this modified temperature difference in equation 2.44 and treating (aU_m) as a valid constant we have

$$R_m = \frac{aU_m(T_h - T_b)}{\lambda} \tag{2.46}$$

The quantity (aU_m) refers to the complete drying cycle and will be substantially constant in given apparatus for a given material; it can be taken

Table 2.4 Typical values of film coefficients in conduction dryers

Coefficient		Upper value	(W/m² °C)	Lower value	(W/m² °C)
h_h	heating medium to wall	steam heating	11 000	liquid heating	550
h_w	conduction through wall	3 mm thick mild steel	14 000	resistant lining	1 100
h_f	wall to material	fluid paste 50% solids	280	dry solid	2·5
h_m	conduction through material	3 mm thick walls of medium conductivity	1 100	25 mm thick wall of low conductivity	5
h_e	evaporation from surface		1 100		1 100
U	overall coefficient		120		2
	overall coefficient based on h_f and h_m only		140		2

as the conduction-drying coefficient for the system. The overall conduction-drying equation can then be rewritten as

$$R_m = \frac{10^3\,(M_s - M_f)W}{A_h\theta} = \frac{K_c(T_h - T_b)}{\lambda} \tag{2.47}$$

2.4.5 Use of general equation for conduction drying

This general equation 2.47 for conduction drying can be used to correlate drying data in given apparatus between given moisture limits. It can also be used to correlate the drying coefficient with the initial moisture content for a given final moisture content or vice-versa with final moisture content for a given initial moisture content. Its use is illustrated in the worked examples in the next section.

2.4.6 Worked examples on conduction drying

Example 3. Vacuum tray oven

An organic preparation in paste form is dried in a vacuum tray oven under a vacuum with an absolute pressure of 250 mb with steam at 2 bar gauge pressure as heating medium. The feed paste has an average composition of 30% solids and the dry product contains 2% moisture. The dryer contains 120 trays, each 75 × 40 × 3 cm. The paste charge is 900 kg and the drying time 14 h.

Calculate the overall conduction-drying coefficient and the drying time using sub-atmospheric exhaust steam heating at 90°C.

Initial moisture content M_s = 70/30 = 2·34 kg/kg
Final moisture content M_f = 2/100 = 0·02 kg/kg
Dry weight product W = 900 × 0·3 = 270 kg
Area of dryer A_h = 36 m^2

From equation 2.47, the overall drying rate

$$R_m = \frac{10^3(M_s - M_f)W}{A_h\theta} = \frac{(2{\cdot}34 - 0{\cdot}02) \times 270 \times 1000}{36 \times 14 \times 3600}$$

$$= 0{\cdot}345 \text{ g/s m}^2$$

The temperature of the heating medium T_h is that of saturated steam at 2 bar gauge pressure = 133°C.

The boiling point of water in the dryer T_b is that at 0·25 bar absolute and is 67°C.

The latent heat of water at 67°C = 2 341 kJ/kg.

The overall conduction-drying coefficient is given by equation 2.47

$$K_c = \frac{R_m\lambda}{(T_h - T_b)}$$

$$= \frac{0{\cdot}345 \times 2341}{(133-67)} = 12{\cdot}2 \text{ W/m}^2\ {}^\circ\text{C}$$

If exhaust steam at 90°C is used for heating, T_H = 90°C

Then $$R_m = \frac{K_c(T_m - T_b)}{\lambda}$$

$$= \frac{12{\cdot}2 \times 23}{2341} = 0{\cdot}124 \text{ g/s m}^2$$

And new drying time

$$\theta' = \frac{10^3(M_s - M_f)W}{R_m A_h}$$

$$= \frac{2{\cdot}32 \times 270 \times 1000}{0{\cdot}124 \times 36} = 1{\cdot}4 \times 10^5 \text{ s} = 38{\cdot}9 \text{ h}$$

The above calculation method shows the influence of the several variables. When, however, no changes are made except to the heating steam, the new drying time can be calculated by simple proportion of the temperature differences and $\theta' = 14 \times 66/23 = 40$ h.

Example 4. Integral jacketed tray oven

A heat-sensitive paste is dried in a vacuum tray dryer with integral heating jacket construction under a vacuum of 50 mb absolute using sub-atmospheric steam at 48°C as heating medium. The dryer has a tray area of 0·3 m^2. The

Table 2.5 Overall conduction drying coefficients K_c in Example 4

Moisture content M (kg/kg)	*Total drying time* (h)	(s)	*Drying coefficient* K_c (g/s m^2)
0·473	4·06	14 600	42·6
0·253	2·51	9 040	36·2
0·057	1·01	3 620	20·8
0·014	0·29	1 043	5·2
0·010 (M_f)	0	0	0

initial charge is 3·7 kg of paste of 47·3% moisture dry basis. The weight of dry solids is 3·7/1·473 = 2·5 kg. The moisture content after various drying times is

1 h 33 min	25·3% d.b.
3 h 3 min	5·7% d.b.
3 h 46 min	1·4% d.b.
4 h 31 min	1·0% d.b.

Determine the relation between initial moisture content and overall conduction-drying coefficient and calculate the drying time from an initial moisture of 40% to 1% d.b.

From Steam Tables, saturated steam at 50 mb absolute has a temperature of 33°C and a latent heat of 2 425 kJ/kg.

The overall conduction-drying coefficients K_c for drying to 1% moisture (dry basis) is calculated by use of equation 2.47 and the data above. These values of K_c are given in Table 2.5. Then from a graph, the value of K_c for an initial moisture content of 0·4 kg/kg is found to be 41·5 g/s m^2.

The drying time from 40% to 1% dry basis is then found from equation 2.47:

$$\theta = \frac{10^3\,W\lambda(M_s - M_f)}{A_h K_c (T_h - T_b)}$$

$$= \frac{2{\cdot}5 \times 2425 \times 0{\cdot}39 \times 1000}{0{\cdot}3 \times 15 \times 41{\cdot}5}$$

$$= 12\,650\ \text{s} = 3\ \text{h}\ \ 31\ \text{min}$$

Example 5. Agitated batch dryer

An organic paste which does not aggregate to hard balls during the process is dried in a pilot-size agitated batch vacuum dryer under a reduced pressure of 0·27 bar absolute. The heating medium is steam at 2 bar gauge. The dryer has a diameter of 25 cm and diameter of 30 cm. It is completely jacketed except for the ends. The paste charge is 13·5 kg, initial solids content = 50%. After 6 h drying time, the final moisture is 0·5% (dry basis).

Calculate the drying time in a full-scale dryer of 1·2 m diameter and 2·6 m long with the same initial filling ratio and under the same conditions except that the heating steam is supplied at 3 bar gauge pressure. Assume that the agitation conditions are the same on both scales.

Laboratory dryer

For the small-scale data we have:

Initial moisture	$M_s = 1{\cdot}0$ kg/kg
Final moisture	$M_f = 0{\cdot}5\% = 0{\cdot}005$ kg/kg
Dry weight of product	$W = 6{\cdot}75$ kg
Effective area of dryer	$A_h = 0{\cdot}236\ \text{m}^2$
Drying time	$= 6\ \text{h} = 2{\cdot}16 \times 10^4$ s

Hence the overall drying rate by equation 2.47 is

$$R_m = \frac{10^3\,(M_s - M_f)W}{A_h\theta} = \frac{0{\cdot}995 \times 6{\cdot}75 \times 10^3}{0{\cdot}236 \times 2{\cdot}16 \times 10^4}$$

$$= 1{\cdot}32\ \text{g/s m}^2$$

Temperature of heating medium T_h = temperature of saturated steam at 2 bar gauge = 134°C

Boiling point of water at an absolute pressure of 0·27 bar = 67°C

Latent heat of evaporation of water at 67°C = 2 340 kJ/kg

Whence average overall conduction-drying coefficient from equation 2.47

$$K_c = \frac{Rm\lambda}{T_h - T_b} = \frac{1{\cdot}32 \times 2340}{67} = 46\ \text{W/m}^2\ {}^\circ\text{C}$$

Full-scale dryer

The ratio of the volumes of the large to small-scale dryer is 202. Hence charge to full-scale dryer for the same filling ratio is 2 750 kg wet basis, 1 375 kg dry basis. Area of the full-scale dryer,

$A_h = 9{\cdot}8\ \text{m}^2$

Temperature of heating steam $= 144^\circ\text{C}$

Mean drying rate R_m $= K_c(T_h - T_b)/\lambda$

$= 46 \times 77/2\,340$

$= 1{\cdot}51\ \text{g/s m}^2$

Drying time on full scale $\theta = 10^3(M_s - M_f)W/R_m A_h$

$$= \frac{0{\cdot}995 \times 1375 \times 10^3}{1{\cdot}51 \times 9{\cdot}8} = 9\,240\ \text{s} = 25{\cdot}5\ \text{h}$$

2.5 HEAT TRANSFER BY RADIATION

Radiation is occasionally the dominant mode of heat transfer in dryers (e.g. infra-red dryers) but more usually it represents a relatively minor enhancement of the transfer by convection or conduction. Radiation is therefore considered as a correction to the mechanisms for pure convection and pure conduction discussed in Sections 2.3 and 2.4.

The effect of radiation onto the surface of the material being dried is to increase the drying rate above that determined by the predominant mechanism of convection or conduction.

2.5.1 General equation for heat transmission by radiation

The general expression for the net transfer of heat by radiation between two grey surfaces is:

$$\frac{dQ}{d\theta} = 10^{-3}\,F_r A_r(\sigma)\,(T_r^{*4} - T_s^{*4}) = 10^{-3}\,F_s A_s(\sigma)\,(T_r^{*4} - T_s^{*4}) \qquad (2.48)$$

where (σ) is the Stefan–Boltzmann constant $= 5{\cdot}75 \times 10^{-8}\ \text{Wm}^2\ \text{K}$.

Note that T^* is absolute temperature.

The factor F_r represents the fraction of the total radiation leaving the surface A_r which is absorbed by the surface A_s; and similarly the factor F_s represents the fraction of the total radiation leaving surface A_s which is absorbed by surface A_r. It can be shown that $F_r A_r = F_s A_s$ so that the net transfer of heat from A_r to A_s is given by equation 2.48. Charts and equations are given in References 2 and 5 for evaluating F factors for a number of elementary configurations of pairs of surfaces and the values for complex systems can be synthesised from this basis.

In many types of dryer the drying surface which is receiving the radiation forms one side of an enclosure and the source of radiation is either the remaining walls of the enclosure, or heated surfaces (e.g. steam coils) in or near these walls. In either case the radiating and receiving surfaces can be treated as infinite parallel planes without serious error. Under these conditions the geometrical shape factor is given by

$$F_g = F_r = F_s = \frac{1}{1/\varepsilon_r + 1/\varepsilon_s - 1} \tag{2.49}$$

Where no special arrangements for radiation are made and the heat received by the material is a small proportion of the total heat, we can simplify further by taking

$$\varepsilon = \varepsilon_r = \varepsilon_s \longrightarrow 1$$

and from equation 2.49

$$F = \frac{\varepsilon}{2-\varepsilon} \longrightarrow \varepsilon \tag{2.50}$$

Since the effective area can also be taken as that of the drying material, $A = A_r = A_s$.

We can now simplify equation 2.48 to

$$\frac{dQ}{d\theta} = 10^{-3}\, A\varepsilon(\sigma)\,(T_r^{*4} - T_s^{*4}) \tag{2.51}$$

Equation 2.51 is a gross simplification of equation 2.48 but is adequate for the case where radiation is a minor effect in a dryer designed for heat transfer by conduction or convection. If attempts to provide radiation have been made deliberately, evaluation of the F factor by equation 2.49 should be adequate where the radiation is still a minor part of the total heat transmission.

2.5.2 Radiation heat-transfer coefficient

It is inconvenient to combine an expression of the type of equation 2.51 involving temperature to the fourth power with the equations for convection and conduction drying which involve temperature to the first power (cf. e.g. equations 2.6, 2.27, 2.28). This difficulty can be resolved by treating the radiant heat transmission in terms of a fictitious radiation heat-transfer coefficient defined as

$$h_r = \sigma \frac{\{(T_r + 273)^4 - (T_s + 273)^4\}}{T_r - T_s} \tag{2.52}$$

Equation 2.51 then becomes

$$\frac{dQ}{d\theta} = 10^{-3}\, A\varepsilon h_r (T_r - T_s) \tag{2.53}$$

It should be noted that the value of the radiation heat-transfer coefficient, h_r, is strongly dependent on the temperatures of the surfaces (cf. equation 2.52). Thus if it is to be used for a dryer with substantial temperature gradient it is desirable to divide the dryer into sections and evaluate h_r for each section.

2.5.3 Effect of radiation in convection drying

In the case of drying by pure convection we have at any instant

$$\frac{dW}{d\theta} = \frac{dQ}{d\theta} \cdot \frac{10^3}{\lambda} \text{ (from equation 2.37)}$$

and

$$\frac{dW}{d\theta} = \frac{h_c A(T_g - T_s)}{\lambda} = K_H A(H_{ss} - H_g) \text{ (from equation 2.6)}$$

Hence in the case of combined convection and radiation we have

$$\frac{dW}{d\theta} = \frac{A}{\lambda}\{\varepsilon h_r(T_r - T_s)\} = \frac{A}{\lambda}\{h_c(T_g - T_s)\} = K_H A(H_{ss} - H_g)$$

or

$$\frac{\varepsilon h_r(T_r - T_s) + h_c(T_g - T_s)}{\lambda} = K_H(H_{ss} - H_g) \qquad (2.54)$$

The surface temperature T_s at an instant can be obtained from equation 2.54 by trial and error knowing h_r, h_c and K_H. The drying data can then be analysed as described in Section 2.3 using the corrected value of surface temperature.

Now the radiation heat-transfer coefficient h_r is a function of the surface temperature T_s as is apparent from its definition above (equation 2.52). However in dryers where the proportion of heat received by radiation is small and the variation in surface temperature from the constant-rate period to the end of drying is relatively small, it can be assumed that h_r is substantially constant throughout the drying cycle (for constant ambient conditions). It has already been shown that both the convection heat-transfer coefficient h_c and the drying coefficient K can be taken as constant (Section 2.3.5); and significant change in the emissivity ε is unlikely with a thermally stable material. The latent heat λ will decrease as the surface temperature increases but again the error introduced by taking a constant value is relatively small. Laborious calculation for the falling-rate period can therefore be avoided without serious error by calculating the surface temperature by equation 2.54 for the constant-rate period and then taking the difference between this temperature and the wet-bulb temperature (which would appertain for pure convection) as a constant-surface temperature elevation which holds throughout the drying cycle. Thus

$$\Delta T_{sr} = T_s - T_{wb} \qquad (2.55)$$

The drying coefficients (Section 2.3) are independent of the drying conditions and values calculated for pure convection drying (as in Example 1 in Section 2.3.10) are equally valid where radiation is taking place. The effect of radiation in convection drying is thus to raise the surface temperature and the degree of rise can be calculated from equations 2.54 and 2.55. Furthermore it can be assumed that this surface temperature elevation will be constant throughout the drying cycle and it need therefore only be calculated for the simpler constant-rate case. Having established the surface temperature correction, calculation of drying time then follows the procedure for pure convection as described in Section 2.3.

Use of this procedure is illustrated in Example 6.

2.5.4 Worked example, on convection drying assisted by radiation

Example 6

The through-drying experiment described in the worked example 1 on convection drying (Section 2.3.10) is repeated with additional heat provided by radiation from a surface at 150°C. Conditions are such that the entire effective surface of the material receives the radiation and the emissivity of the material can be taken as 0·5. Calculate the drying time under these conditions. Assuming that the critical moisture content is unchanged at 28·5% dry basis.

From Section 2.3.10 and psychrometric tables we have for convection drying without radiation:

Wet-bulb temperature	T_{wb}	= 27·2°C
Wet-bulb humidity	H_{sw}	= 0·02308 kg/kg
Constant-drying rate	R_c	= 0·816 g/s m^2
Inlet air temperature	$(T_g)_{in}$	= 70°C
Exit air temperature	$(T_g)_{out}$	= 54·2°C
Heat-transfer coefficient	h_c	= 58·3 W/m^2 °C
Inlet air humidity	$(H_g)_{in}$	= 0·005 kg/kg
Exit air humidity	$(H_g)_{out}$	= 0·01156 kg/kg
Constant-rate drying coefficient on humidity basis	$(K_H)_c$	= 56 g/s m^2

Now temperature of radiating surface	T_r	= 150°C
Assume constant-rate temperature of drying surface	T_s	= 30°C
Radiation heat-transfer coefficient from equation 2.52 or Fig. 10.10 of Reference 2	h_r	= 8·52 W/m^2 °C
Latent heat of evaporation λ at 30°C		= 2 450 kJ/kg

From equation 2.54

$$\frac{\varepsilon h_r(T_r - T_s) + h_c\{(T_g)_m - T_s\}}{\lambda} = K_H\{H_{ss} - (H_g)_m\}$$

so

$$\frac{4{\cdot}76\,(150 - T_s) + 58{\cdot}3\,(62{\cdot}1 - T_s)}{2450} = 56\,(H_{ss} - 0{\cdot}0083) \text{ (see Note a)}$$

Solving by trial and error for T_s with psychrometric tables we obtain Constant-rate surface temperature $T_s = 29{\cdot}5°C$.

Hence temperature elevation due to radiation

$$\Delta T_{sr} = T_s - T_{wb} = 2{\cdot}3°C \tag{2.55}$$

The mean surface temperatures for pure convection have already been calculated in Example 1 and we can now increase these by 2·3°C for the effect of radiation.

We can thus calculate the values of mean surface temperature corresponding to each of a series of moisture contents and the corresponding drying coefficient as in Section 2.3.10. From this we can directly obtain the mean surface humidity $(H_s)_m$ by psychrometric tables.

Now exit air humidity $(H_g)_{out} = (H_g)_{in} + \dfrac{dW}{d\theta} \cdot \dfrac{1}{A} \cdot \dfrac{A}{A'} \cdot \dfrac{1}{G}$

$$= 0{\cdot}005 + \frac{8{\cdot}05}{A} \cdot \frac{dW}{d\theta}$$

and mean air humidity $(H_g)_m = 0{\cdot}005 + 0{\cdot}00545 \dfrac{dW}{d\theta} \cdot \dfrac{1}{A}$

From this we obtain the mean humidity driving force

$$(H_{ss} - H_g)_m = (H_{ss})_m - (H_g)_m$$

And finally from equation 2.6, the drying rate is given by

$$\frac{dW}{d\theta} = K_H (H_{ss} - H_g)_m$$

Solving these relations by trial and error, plotting the drying rate against moisture content and graphically integrating to give the time to dry to each moisture content (as in Example 2) we obtain the data shown in Table 2.6.

Table 2.6 Data derived on radiation drying in Example 6

Moisture content M (kg/kg)	*Humidity based drying coefficient*† K_H (g/s m² unit humidity difference)	*Mean surface temperature* $(T_s)_m$ (°C)	*Drying rate* $\frac{dW}{d\theta} \cdot \frac{1}{A}$ (g/s m²)	*Drying time* $\theta/60$ (min)	
*0·890	56·1	29·5	0·788	0	Constant-rate drying time θ_c = 6·1 min
*0·285	56·1	29·5	0·788	6·1	
0·25	26·2	34·6	0·732	6·5	Falling-rate drying time θ_f = 10·6 min
0·20	12·9	43·5	0·624	7·3	
0·15	6·4	50·1	0·505	8·3	
0·10	3·6	57·0	0·339	9·5	
0·05	1·2	63·1	0·206	11·6	Total drying time θ_t = 16·7 min
0·025	0·65	66·4	0·137	12·7	
0	0	72·3	0	16·7	

*Constant-rate values
†From Table 2.1

Supplying radiation from a surface at 150°C thus decreases the drying time from 21 to 17 min in this particular case.

Note a. To simplify the calculation, $(T_g)_m$ and $(H_g)_m$ are taken as the arithmetic mean of the inlet and exit conditions for the pure convection case. This is not strictly true as the outlet air temperature will be lower and the humidity higher in the combined radiation and convection case, but the error is small and is partly corrected by using an arithmetic instead of a logarithmic mean.

2.5.5 Effect of radiation in conduction drying

In the case of drying by pure conduction we have at any instant from equations 2.37 and 2.38

$$\frac{dW}{d\theta} = \frac{UA_h(T_h - T_s)}{\lambda} \tag{2.56}$$

The overall conduction drying rate is given in equation 2.47.

The instantaneous equation for drying by radiation derived from equation 2.53 is

$$\frac{dW}{d\theta} = \frac{\varepsilon h_r A(T_r - T_s)}{\lambda} \tag{2.57}$$

In this case A = area receiving radiation = area of drying material.

As explained in the case of radiation and convection, we can take the surface area receiving radiation as that of the material, and in the case of conduction dryers this can be taken as the conducting surface A_h. Since the process is essentially one of evaporation, the surface temperature of the drying material can be taken as the boiling point of the liquid, T_b, for most of the drying cycle. Furthermore, since the radiation is a relatively small part of the total heat supply, the end effect when the material surface temperature rises above the boiling point can be neglected. The instantaneous equation for radiation can therefore be written as

$$\frac{dW}{d\theta} = \frac{h_r A(T_r - T_b)}{\lambda} \tag{2.58}$$

Now if the temperature is constant, the latent heat (λ) will also be constant and it is reasonable to assume that there will be no significant change in the emissivity (ε) as drying proceeds. The overall drying rate with conduction and radiation is therefore

$$R_m = 10^3 \frac{(M_s - M_f)W}{\theta A_h} = \frac{\varepsilon h_r(T_r - T_b) + K_c(T_h - T_b)}{\lambda} \tag{2.59}$$

The effect of radiation in conduction drying is thus to increase the amount of heat transferred to the material without affecting its temperature. Hence the rate of drying can be calculated by adding a constant radiation factor as shown in equation 2.59.

2.5.6 Worked example, on radiation in a conduction dryer

Example 7

In the vacuum tray dryer using steam heating at 2 bar gauge pressure given in Example 3 in Section 2.4.6 additional heating is provided by radiation from the body of the dryer at 100°C. The emissivity of the material can be taken as 0·6. Calculate the drying time to 2% moisture under these conditions.

Radiation heat-transfer coefficient $h_r = 7{\cdot}96\ \mathrm{W/m^2\ °C}$ (from equation 2.5.2 on Fig. 10.10 of Reference 2). From equation 2.59.

Drying time

$$\theta = \frac{10^3\lambda(M_s - M_f)W}{A_h\{\varepsilon h_r(T_r - T_b) + K_c(T_h - T_b)\}}$$

$$= \frac{2340 \times (2{\cdot}34 - 0{\cdot}02) \times 270 \times 1000}{36\{0{\cdot}6 \times 7{\cdot}96(100 - 67) + 12{\cdot}2 \times 66\}} = 42\,400\ \mathrm{s} = 11{\cdot}7\ \mathrm{h}$$

The effect of radiation in this particular case is thus to decrease the drying time from 14 to 11·7 h.

CHAPTER THREE

Classification of dryer types

3.1 INTRODUCTION

There are many types of dryer on the market for chemical manufacture, each with its own speciality and a few which are sufficiently versatile to dry more than one type of wet material. They can be classified according to method of operation or duty required as indicated in Tables 3.1–3.4, which will be discussed in detail in succeeding sections. In the end, however, the choice before the process design engineer for a particular duty generally narrows down to not more than three types in terms of performance and economics.

The first big distinction is that between the requirements of the heavy inorganic and to a lesser extent the heavy organic chemical industry, where plant for continuous drying at a rate of many tonnes per hour is required; and small scale manufacture of high priced products at the rate of perhaps less than 100 kg per day, and where batch plant is sufficient or even essential on grounds of close control of each batch to a rigid specification.

3.2 METHOD OF OPERATION

Table 3.1 classifies dryers by method of operation. The first subdivision is by method of heat transfer.

(a) Conduction heating } Distinction sometimes blurred
(b) Convection heating }
(c) Infra-red heating, i.e. all forms of radiation heating
(d) Dielectric heating

Freeze drying may be classified as a special case of conduction heating.

The next main sub-division is by type of drying vessel—tray, rotating drum, fluidised bed, pneumatic, spray. The table also shows which types of vessel are more readily adapted to operation under vacuum or with inert atmospheres. Fig. 3.1 shows diagrams of most of these types of dryer.

3.3 PHYSICAL FORM OF FEED

It must first be emphasised that purely mechanical means should be used to reduce the moisture content of the wet feed to as low a figure as possible, because with few exceptions, processes such as evaporation, filtration and

Table 3.1 Classification of dryers based on method of operation

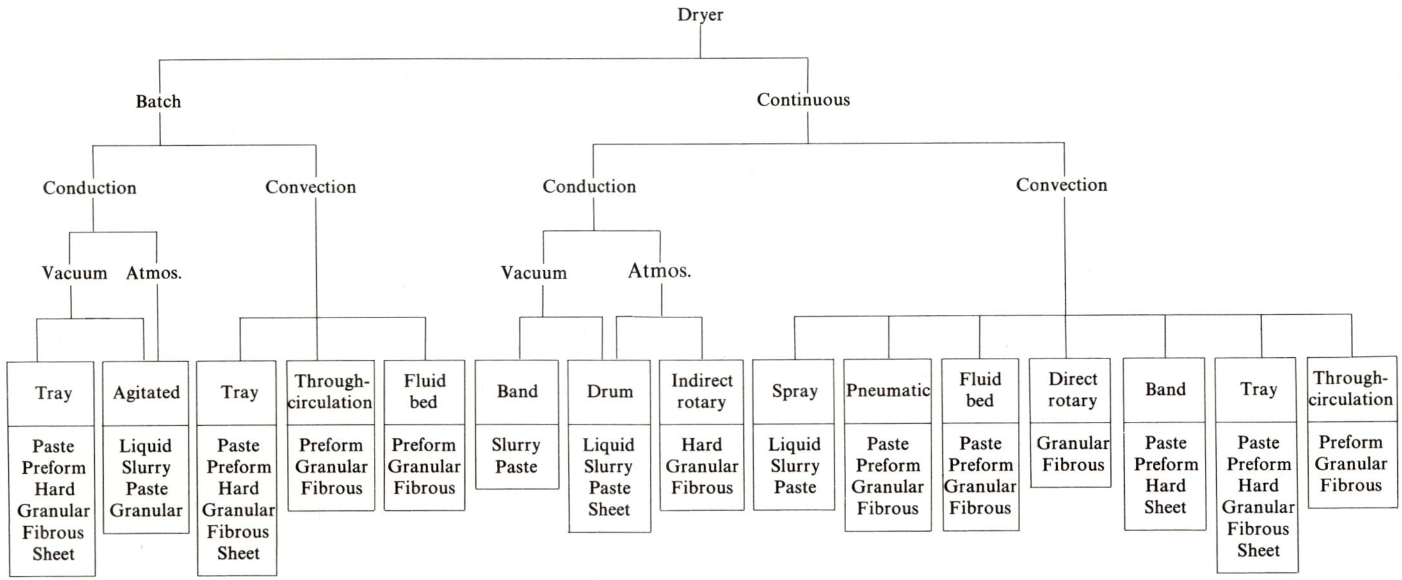

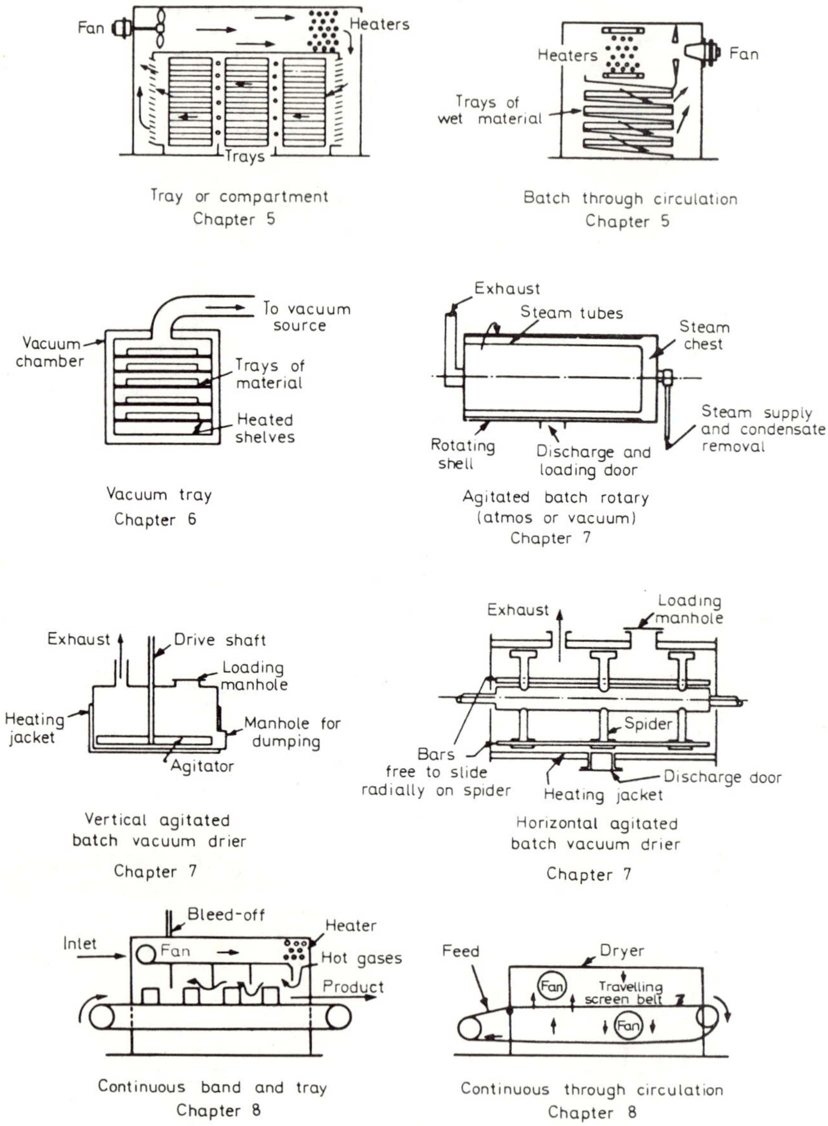

Fig. 3.1. Classification of dryers by method of operation

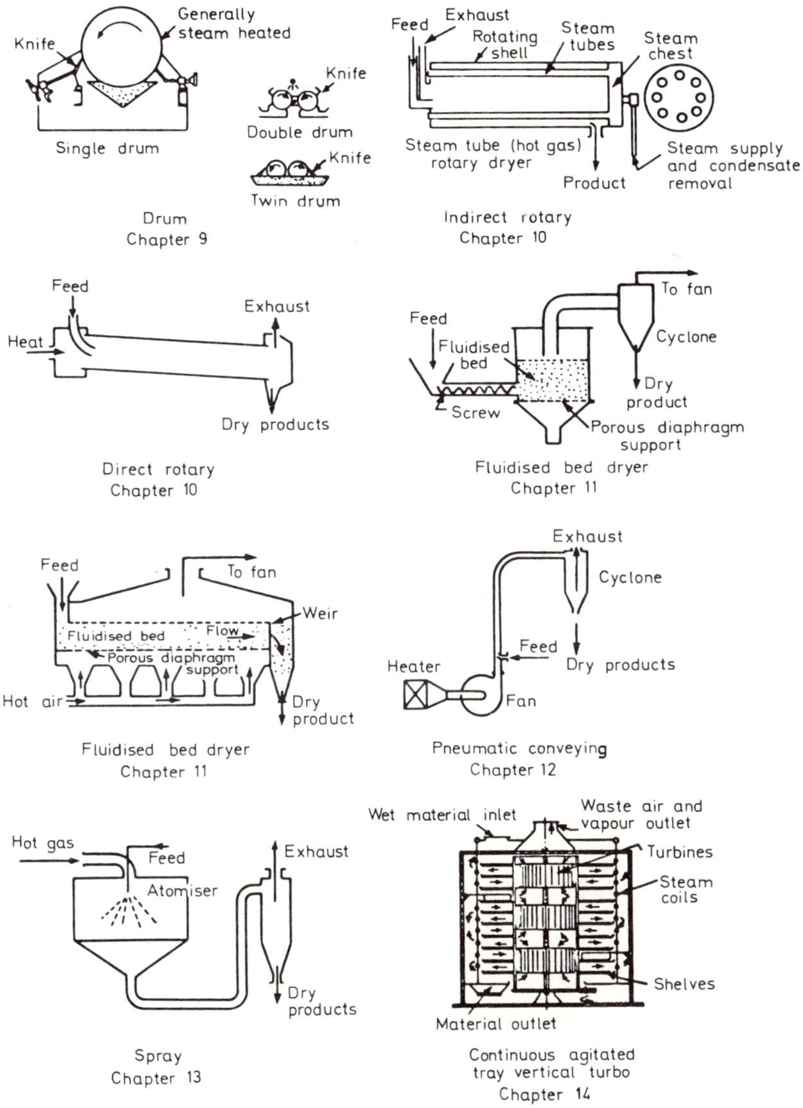

Fig. 3.1. Classification of dryers by method of operation—continued

Table 3.2 Classification of dryers based on physical form of feed

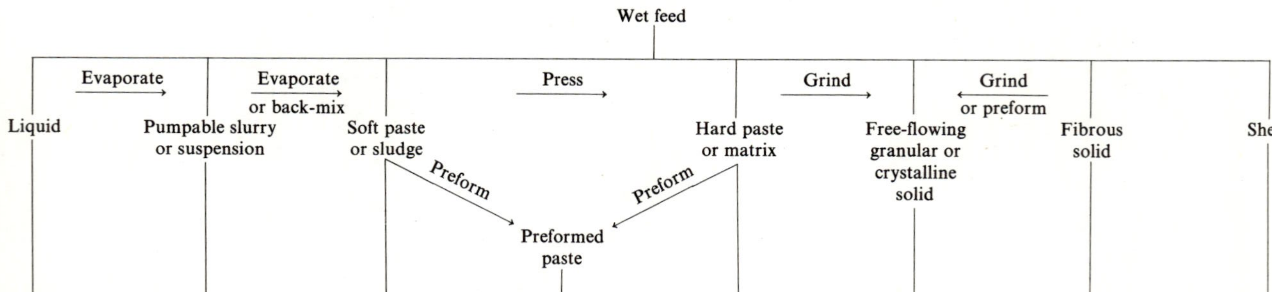

Liquid	Slurry	Paste	Preform	Hard	Granular	Fibrous	Sheet
Agitated batch	Agitated batch	Vacuum tray	Vacuum tray	Vacuum tray	Vacuum tray	Vacuum tray	Vacuum tray
Drum	Vacuum band	Agitated batch	Convection tray	Convection tray	Agitated batch	Convection tray	Convection tray
Spray	Drum	Convection tray	Batch through-circulation	Indirect rotary	Convection tray	Batch through-circulation	Drum
	Spray	Fluid bed	Fluid bed	Cont. tray	Batch through-circulation	Fluid bed	Cont. tray
		Vacuum band	Pneumatic		Fluid bed	Indirect rotary	
		Drum	Convection band		Indirect rotary	Pneumatic	
		Spray	Cont. tray		Pneumatic	Direct rotary	
		Pneumatic	Cont. through-circulation		Direct rotary	Convection band	
		Convection band			Cont. tray	Cont. tray	
		Cont. tray			Cont. through-circulation	Cont. through-circulation	

centrifuging are cheaper and faster than equivalent processes in drying plant.

Table 3.2 classifies types of dryer in terms of the physical form of the material to be fed to the plant.

Raw pastes and sludges are difficult to handle into dryers and the drying rate quickly slows down through the formation of superficial skins having a low permeability to vapour. This form of feed therefore requires pre-treatment by 'preforming' into pellets about 5 mm cube or by forming granules with mixed-back fines.

3.4 SCALE OF OPERATION

Table 3.3 shows the types of available dryer classified according to scale of

Table 3.3 Classification of dryers by scale of production

Process

Small scale to 20/50 kg/h	Medium scale 50 to 1000 kg/h		Large scale tonnes/h
Batch	Batch	Continuous	Continuous
Vacuum tray	Agitated	Fluid bed	Indirect rotary
Agitated	Through-circulation	Vacuum band	Spray
Convection tray	Fluid bed	Indirect rotary	Pneumatic
Through-circulation		Spray	Direct rotary
Fluid bed		Pneumatic	Fluid bed
		Band	
		Tray	
		Through-circulation	

production. It will be seen that the number of types for continuous large scale drying is much more limited than for medium scale outputs.

3.5 SPECIAL FEATURES

We then come to classification in terms of overriding features such as toxic or heat sensitive materials, special form of dry product etc.; these are shown in Table 3.4.

3.6 CAPITAL AND OPERATING COSTS

Dryers may also be classified in terms of the labour, heat, flexibility and capital cost of their operation. Mechanical maintenance cost can only be estimated by study of the designs provided by individual manufacturers. The former items are more conveniently discussed in the next chapter—on selection of dryers.

Table 3.4 Classification of dryers by suitability for special features

Process							
Hazards			Sensitive product			Special form of product	Low capital cost per unit of output
Dust	Toxic	Flame	Temperature	Mechanical	Oxidation		
Agitated batch Vacuum band Indirect rotary	Agitated batch Vacuum band	Agitated batch Vacuum band	Agitated batch Through-circulation Fluid bed Vacuum band Pneumatic	Through-circulation Vacuum band Cont. band Cont. tray	Vacuum tray Vacuum band Spray	Through-circulation Vacuum band Spray Drum	Fluid bed Direct rotary

CHAPTER FOUR

Selection of drying plant

4.1 DIFFICULTIES IN SELECTION

Design and operation of a complete drying installation includes consideration of many ancillaries to the basic chemical engineering unit for heat and matter transfer. These may include storage and feeding of the wet material, also of the dried product, equipment for simple pre-forming of wet sludge to 5 mm pieces or for preparing stronger granules after mixing back some dried fine material, heat supply including effective air vents and steam traps vacuum plant or refrigeration plant.

In the preliminary project stage the plant designer for a new process sometimes starts with the handicap that he is asked by his research and development colleagues to give an estimate for the size and cost of a drying plant on the basis of a sample of a few hundred grams of sticky material. He may be told that it had been centrifuged and that it could be dried by leaving in a low-temperature vacuum oven for 24–30 h and that it must not be heated above 60°C. The maximum permitted moisture in the dried product may not be known, and hours may be spent in explaining that the cost of removing the last few hundredths of a percent of moisture rises exponentially, but that moisture can diffuse from the centre of apparently dry crystals during storage with embarrassing results in storage or during bulk transport, and in competitive marketability. The specification for dryness is likely to be more strict when the final product will be subjected to wide cyclic changes of temperature during transport and in storage at its final point of use. Many preliminary project estimates have gone awry through the inability of the development chemists to supply sufficiently large samples for thorough evaluation of drying characteristics and effects of residual moisture in the dried product.

At a later stage in the development of a project for a new product the market development people will tell the design staff that samples from a pilot plant have been given more thorough trial by potential customers and that, as a result of their comments, it is necessary to reduce the amount of fines in the product – fines which have been produced by size-degradation in the type of dryer which had been proved to give a product of the required consistent dryness.

The above are only initial difficulties. Further difficulties in selection become more intense when the final plant design is being worked out and orders for equipment must be placed. These may be summarised as follows:

(a) There is no standard set of systematised laboratory tests, using standardised apparatus, to provide key data on drying characteristics of materials. These tests would form the initial basis for discussion between purchaser and supplier of a dryer.
(b) The real mechanics of the removal of a liquid from a solid during drying are by no means thoroughly understood.
(c) It is not known to the satisfaction of chemical engineers how many dryers really work.
(d) A systematic comprehensive classification of existing types of dryer has yet to be agreed throughout the chemical engineering world.
(e) There is lack of reliable procedure for scaling-up laboratory data and even pilot plant data for some types of dryer.

There is, of course, the other side of the picture. There are thousands of successful dryers in operation which are giving a product to a required competitive specification. Success with new products and with new designs of dryer can usually be attributed to close collaboration between the purchaser and manufacturer of dryer and associated equipment.

4.2 PAST EXPERIENCE OF CHEMICAL MANUFACTURERS

Designers of complete chemical plants can always learn much from the past experience of others. Criticisms that have been levelled against past choices of dryers are:

(a) Product spoilage by overheating.
(b) Inability to obtain design flowsheet output.
(c) Difficulties with feeding material into dryers and of extracting dried material.
(d) Build-up of material inside the dryer; this can reduce output and degrade heat sensitive products by the prolonged heating of the deposit.
(e) Poor access for cleaning, especially of batch plant which has to dry several products.
(f) General cleanliness of plant.
(g) Bad working conditions for operators.
(h) Poor thermal efficiency.
(i) Excessive maintenance of mechanical parts.
(j) Under-design of some ancillaries.

Some of these difficulties could have been avoided if equipment manufacturers had had reasonably sized pilot plants and the purchaser had been able to make sufficient representative wet material for trial, and had been willing to contribute towards the cost of sufficiently thorough trials. Often there is not time, or time is wasted, because equipment manufacturers have not known sufficient of the limitations of their own equipment. It is necessary, however, to be realistic on these criticisms. Dryers are only a small part, though an essential part, of a complete plant. Too much money and time can be spent on designing a first drying plant which is perfect, and it may be wiser to leave space in the plant for replacement dryers of larger or different design when laboratory or pilot plant experiments indicate that difficulties may arise on the full scale.

4.3 STATEMENT OF A DRYING PROBLEM

Before attempting to select a dryer and any associated cooler, the process plant designer should have in front of him the following information where relevant; some may be difficult to obtain.

4.3.1 Flowsheet quantities

(a) Dry solid, kg/day and kg/plant running hour (max. and min.). Dry solid refers to solid containing, if required, water of crystallisation, e.g. Na_2HPO_4, $12H_2O$.
(b) Total liquid to be removed, max. kg/day and max. kg/plant running hour.
(c) Source of wet material, batch or continuous. If batch, range of weights of batches, kg dry solid. Frequency of supply of batches.

4.3.2 Raw feed—physical

(a) Source of raw feed, e.g. *XY* process. Batch plant in particular may have to dry several materials.
(b) Previous dewatering stage, e.g. filtration, pressing, centrifuging. Technical possibility of reducing further the liquid content of wet material and the total capital and running cost of removing more liquid, cost per kg extra liquid removed.
(c) Storage capacity for raw feed, if any.
(d) Method of supplying material to dryer.
(e) Approximate particle size distribution in wet feed.
(f) Physical characteristics, handleability. In the case of pastes, the ease of making preformed pellets.
(g) Abrasive properties of wet and dry material.

4.3.3 Raw feed—chemical

(a) What are the toxic hazards – wet and dry material, solvent?
(b) Is there an odour problem – dry material and solvent?
(c) Can the material be dried in contact with hot combustion gases, containing H_2O, CO_2, SO_2, traces of oxides of nitrogen, traces of partly burnt hydrocarbons?
What is the maximum safe concentration of SO_2?
(d) Are there fire or explosion hazards? (Consider both solvent and fine dust)?

(e) What are the temperature limitations in the dryer? What are the maximum temperatures to which the material may be heated for periods of seconds, and for periods of days as for deposits on heated surfaces?
(f) What are the temperatures at which any relevant changes of phase take place—fusion, loss of water of crystallisation at given partial pressures of water vapour, change of crystal form?
(g) What are the corrosive properties of the wet material including changes through prolonged heating?
(h) What are known suitable materials of construction for the dryer in order of corrosion resistance? What materials must be avoided?

4.3.4 Dry product—specification

(a) Moisture content on dry weight basis—mean percent and range in individual batches or hourly from continuous plant?
(b) Methods of determination of solvent in the research laboratory, in the process control laboratory (including methods of sampling and use of automatic control—analysis instruments) and methods of determination stated in the sales specification?
(c) Has all solvent odour to be removed in the dryer?
(d) Particle size distribution including tolerances for maximum and minimum sizes—research and process control methods to be stated, also method stated in the sales specification.
(e) Bulk density—methods of determination to be stated as in d.
(f) Maximum percent of impurities permitted in dry product—including named decomposition products and contaminants from materials of construction of dryer and conveyors and from drying gas, cf. Section 4.3.3c.

4.3.5 Dry product—properties

(a) Liability to breakage of crystals or desired granular product.
(b) Flow properties of freshly dried material when hot and when cold, particularly in respect of flow out of transit hoppers.
(c) Temperature to which dried material must be cooled before storage in bunkers or marketing containers in order to avoid caking.

4.3.6 Available drying data

(a) Details of laboratory drying data, including description of apparatus used.
(b) Details of any pilot plant drying data.
(c) Details of drying performance of similar materials on full-scale plant.
(d) Ease of pre-forming wet material or mixing-back dried fines to form granules.

4.3.7 Solvent recovery

When water is not the liquid to be dried off, the cost of the solvent; also flash point and explosive limits in air.

4.3.8 Product losses

(a) Plant cost of product.
(b) Would a dust loss to atmosphere constitute a hazard or only a nuisance?

4.3.9 Site conditions

(a) Available sources of heat and their cost—steam, gas, fuel oil, hot waste gases.
(b) Details of electricity supply and its cost.
(c) Need to purify the air to dryer and cooler.
(d) Need to remove odour from exhaust gases discharged to atmosphere.
(e) Need to minimise noise, vibration etc.
(f) Space restrictions.
(g) Quality of process supervision, labour and maintenance facilities.
(h) Availability and quality of analytical control and instrument maintenance.

4.4 LABORATORY TESTING FOR DRYING PROPERTIES

This section gives preliminary guidance on the experimental procedure to be adopted to ascertain both design and operating data, and also the suitability of a dryer for a particular application. Detailed test procedures are given in the appropriate chapters.

On certain types of dryer, full-scale tests are the only way of establishing reliable design and operating data but in other cases, e.g. through-circulation dryers, there are techniques which have been developed for establishing reliable data from laboratory tests. Mechanical handling is a major consideration with respect to large-scale dryers and some caution must therefore be exercised in deducing throughput from small-scale tests.

For simplicity and as far as test techniques are concerned the dryers which are considered in this book (excluding specialised techniques) are categorised into groups.

4.4.1 Group 1—tray or compartment (atmospheric or vacuum, continuous or batch, see Chapters 5 and 6)

A laboratory test in this instance is relatively simple to perform and it should be carried out under similar conditions to those which are likely to be met in practice. The scale of the experiment is not of critical importance providing that the depth of material in the tray or compartment is kept the same as that proposed for the full-size dryer. In the case of batch atmospheric tray dryers it may be difficult to simulate the wide variations in conditions as drying proceeds and more convenient to carry out trials on a full-scale machine.

4.4.2 Group 2—through circulation (batch or continuous, see Chapter 8)

In this group laboratory tests are also simple to perform and are easy to scale up to full-size dryers. The scale of the experiment and the loading of the tray are not critical, but it is important that the material under test has the same physical characteristics, particularly with regard to granule size distribution, as that to be fed to the full-scale dryer.

4.4.3 Group 3—agitated (batch or rotary, atmospheric or vacuum, see Chapter 7)

Simple apparatus resembling the full-size dryer can be constructed and the data obtained from such tests scaled up directly to a full-size dryer.

4.4.4 Group 4—spray (see Chapter 13)

Tests can be carried out in a small laboratory unit but they cannot be accurately scaled up, neither will the product from a small-scale test be necessarily physically the same as from a large unit. For a spray dryer the only reliable test is on a full-scale unit of, say, 200 to 500 kg/h evaporative capacity.

4.4.5 Group 5—pneumatic conveying (see Chapter 12)

Tests can be carried out on a laboratory apparatus which provides heat- and matter-transfer data under conditions applicable to the full-scale design. The difficulty with these tests is in accurately measuring temperatures, in particular that of the dried solid.

As a preliminary to even carrying out any laboratory tests it is advisable if practicable to carry out drying tests on a simple through circulation dryer in order to establish the magnitude of the drying coefficient. It is essential to have a thorough knowledge of the particle size analysis so that the free falling speeds can be calculated as this data alone may result in the exclusion of this dryer from the list of candidates.

4.4.6 Group 6—rotary (direct and indirect, see Chapter 10)

The technique in this case is to ascertain the drying coefficient in the normal manner by a simple through-circulation drying test. This will establish whether the material is likely to be dried in this type of equipment. A test should then be carried out in a rotary dryer of at least 30 cm diameter, either batch or continuous.

4.4.7 Group 7—drum (see Chapter 9)

Whether a material can be drum dried is best established by carrying out tests on a drum preferably not less than 30 cm diameter × 30 cm long. Tests on a smaller scale are valid but dubious as regards output rate.

4.4.8 Group 8—fluidised bed (see Chapter 11)

Here again it is important to carry out tests on as large a size of dryer as is possible as a scale up from laboratory to full size can be extremely difficult.

Whether the material will fluidise or not can be established satisfactorily on a small scale. Mechanical handling problems, particularly those associated with feeding the wet material to the fluidised bed, may well be critical and cannot be assessed on a laboratory scale.

4.5 PRELIMINARY SELECTION OF DRYER—BASED ON NATURE OF THE FEED

4.5.1 Special cases not considered

The following drying operations will not be considered in this chapter for the reasons given:

(a) Small-scale batch operation – under these conditions most materials can be dried in atmospheric or vacuum tray dryers.

(b) Vacuum-freeze drying—this is expensive and should only be considered for materials which are damaged in normal drying processes and which can bear the cost. The technique is discussed in Chapter 15.

(c) Dielectric or infra-red heating, which may be useful in specialised cases. The technique is discussed in Chapter 15.

(d) Drying of large solids, special forms and shapes. Such materials, e.g. ceramics, dry very slowly in drying ovens, heated rooms or in drying tunnels. Thus blocks of common salt evaporated by the obsolete open-pan process are dried economically in chambers the floors of which are heated by the flue gases from the fires under the pans.

4.5.2 Pilot plants

Given the information set out in Section 4.3, it is usually possible to select candidate dryers from the Tables in Chapter 3, and arrange for some laboratory tests for drying properties by one of the relevant methods given in Section 4.4. After careful consideration of the following paragraphs on dryers for different types of feed it may be necessary to arrange for pilot plant trials. They are essential with some forms of dryer as indicated in later sections. With the present state of knowledge of drying characteristics they are also advisable for large inputs of awkward materials.

Since pilot plant trials are expensive and time consuming, it is important to obtain the maximum possible information from laboratory trials in order to work out a well-ordered plan for the pilot trials. This will also reduce the considerable amount of material required for trials; re-wetted dry material is never suitable for final trials. In drawing up the plan of campaign use should be made of a pictorial flowsheet prepared with the aid of Section 4.3 and of the list of possible weaknesses of the type of dryer disclosed in the chapters on individual plants. Criticism on the lines of a 'devil's advocate' after inspection of a full-scale plant drying different material serves to eliminate many potential troubles by minor alterations to the pilot plant before it is run.

4.5.3 Liquids (including colloidal solutions and emulsions)

There is little difficulty in handling this class of feed and the choice of equipment will usually be limited to one of the following:

(a) Spray dryer.
(b) Drum dryer (atmospheric or vacuum).
(c) Agitated batch vacuum dryer.

A spray-dried product will usually have different physical properties compared with the product from other dryers, and these properties may in themselves be desirable; a trial on a laboratory scale spray dryer will provide material for evaluation purposes, and data on both the critical inlet and outlet air temperatures. The laboratory product may not, however, have the same physical properties as that from a full-scale dryer.

Other considerations which may influence the final choice are:

(a) The need for small product losses and consequently a cleaner plant.
(b) Solvent recovery or the need to use an inert atmosphere, in which case an agitated batch vacuum dryer may be preferred.
(c) Temperature Sensitivity of the material. Then the choice may lie between the agitated batch vacuum dryer with a long residence time, the through-circulation dryer with moderate temperature and moderate residence time, the drum with a high mean temperature but short contact time, and the spray dryer with a very short contact time. The effective temperature would be adjusted according to each case from comparatively low to comparatively high.

4.5.4 Pumpable suspensions of fine solids (excluding pastes)

The remarks for liquids apply equally well to this class of feed and the list of candidate dryers is the same. There is less chance, however, of producing a unique kind of product from a spray dryer.

4.5.5 Pastes and sludges

In this class of feed the solids will usually be in a finely-divided state and for continuous drying the dust nuisance will often be the major consideration. Handling problems will, in general, vary and their solution will often involve pilot plant trials. The choice between a batch or continuous dryer is not so straightforward because difficulties in continuous operation are often balanced by disadvantages of batch operation, whilst an overriding factor will be the scale of the operation.

Batch operation

Candidate dryers for batch operation are:
- (a) tray – atmospheric or vacuum
- (b) agitated batch – atmospheric or vacuum;
- (c) rotary – atmospheric or vacuum.

With vacuum operation, any of the above would be suitable for cases necessitating solvent recovery, or when toxic or fire hazards exist, or when temperature limitations demand it.

Continuous operation

Candidate dryers for continuous operation are:
- (a) spray — where the paste would have to be atomised; this in itself may involve considerable experimental work.
- (b) Fluidised bed — where the difficulty is in dispersing the feed into a deep fluidised bed.
- (c) Continuous band through circulation — this dryer necessitates some preforming technique for the feed, and is particularly suitable if a dust-free product is required.
- (d) Pneumatic — provided part of the dry product is first mixed with the raw feed to facilitate satisfactory dispersion of the wet solid in gas immediately on entering the dryer.
- (e) Continuous rotary, direct or indirect — blending of the raw feed with some of the dry product will be essential to successful handling in a rotary dryer. As this class of feed normally contains fine particles, the indirect rotary dryer is to be preferred.

4.5.6 Reasonably free-flowing wet powders (particle size <300 microns (μm))

The feed may be in this condition initially, or it may be the result of blending dry solid with a raw feed as discussed in the previous section, or it may even be in the form of a friable filter cake. In any of these cases the remarks and the candidate dryers quoted in Section 4.5.5 apply but with the inclusion of the vertical rotating shelf dryer (Chapter 14), which is particularly useful when crystal breakage is an important factor.

The complete list of candidate dryers is therefore:

Batch:
(a) Tray – atmospheric or vacuum
(b) Agitated batch – atmospheric or vacuum
(c) Rotary – atmospheric or vacuum

Continuous:
(d) Fluidised bed
(e) Continuous band through circulation
(f) Pneumatic
(g) Rotary – direct or indirect
(h) Vertical rotating shelf
(i) Spray

4.5.7 Granular and crystalline solids or small pre-forms

For particles up to 300 μm the remarks in Section 4.5.6 apply, though for particles >300 μm a direct-heat rotary dryer is widely used for particles larger than 300 μm. With this type of dryer crystal breakage is sometimes a problem which can be minimised by suitable flight design.

A further candidate is a through-circulation dryer using a moving band or vibrating screen; this can be used for particles larger than, say, 25 mesh, including preforms.

Candidate dryers are therefore those quoted in Section 4.5.6 and in addition a through-circulation dryer.

4.5.8 Fibrous solids

Such materials often hold considerable amounts of water but dry quite easily. Even when broken down to small flakes their flow properties are inferior to those of powders, leading to bridging in bunkers and similar equipment and rendering fluidisation very difficult or impossible.

These materials are often highly temperature-sensitive due to their high specific surface and consequently more than usual care must be taken to ensure that the air temperature is kept well down. Through-circulation drying tests at various temperatures should establish the need, if any, to avoid overheating.

Apart from the above points, fibrous materials should be treated in a similar way to other types of solid previously discussed, in particular those discussed in Section 4.5.7.

4.6 FINAL SELECTION

The final selection will usually represent a compromise between the total cost of the operation, the quality of the product, safety considerations, and convenience of installation. Points for comparison between candidate dryers may be summarised as follows:

(a) Installed cost.
(b) Is installation convenient or are there any hidden charges?
(c) Running cost: fuel, power, labour and maintenance.
(d) Material losses: solid or solvent.
(e) Safety considerations.
(f) General amenity considerations.
(g) Is the product in the form required? If not, what is the cost of further processing? (Grinding or pelleting for example).
(h) If dryers requiring different forms of feed are being compared, what is the comparative cost of preparing the alternative feeds? (e.g. various mechanical dewatering units, mixing or preforming equipment).
(i) Is the dryer sufficiently flexible to deal with changes in load?
(j) Is control of the drying operation important? If so, does the equipment present any control problems?
(k) In the case of small dryers forming part of a multi-purpose chemical manufacturing plant, is the dryer sufficiently versatile?
(l) An analytical report on the product for comparison with the specification should be given. If the final choice must be approved by a Design Committee then a full statement along the above lines for each candidate dryer, preferably submitted with a sample of the product from each type of dryer, will make the decision easier.

CHAPTER FIVE

Batch atmospheric tray dryers

5.1 GENERAL DESCRIPTION

The batch atmospheric tray dryer is frequently referred to as a tray or truck or compartment oven or stove and not infrequently as a Grinnel stove, this latter name deriving from that of an early German proprietary design. It consists essentially of one or more chambers or compartments in which are placed trays containing the material to be dried. The trays are frequently placed on racks which can be moved as a unit into or out of the drying chamber. Drying is effected by hot air (or other gas) blown over the trays by means of a fan or fans. In some cases the trays are perforated and the drying air is blown through the material on the trays. Natural convection is occasionally used to circulate the air. Various arrangements are used for air distribution and recirculation.[1–5]

Heat is usually supplied by steam or hot water coils which may be either inside or outside the drying compartment itself. Electric heating is used occasionally, as also is direct heating with gas, oil or flue gas.

The drying chamber may be constructed of any convenient material, e.g. steel, brick, asbestos sheet, aluminium, etc. Similarly the trays may be of steel, enamelled steel, austenitic steel, aluminium, glass, plastics, etc.

Ancillaries include spare sets of racks and trays to make change from one product to another easy, trucks for moving racks of trays to and from the dryer, special arrangements for loading paste uniformly onto the trays and special draughted areas (or dedusting booths) for discharging the dried product from the trays. Equipment for washing the trays between batches or on transfer to different products is necessary.

5.2 FIELD OF APPLICATION

This type of dryer can be used to dry almost any material. Their operating costs are high due to the relatively high labour requirement and in general their use is on this account restricted to output rates of less than 50 kg/h dry material. They are particularly useful in those cases where a large number of products is to be handled, the production of each being relatively small, e.g. dyestuffs and pharmaceuticals.[5] Being batch dryers they are very suitable to take paste from the batch operated filter presses or nutsches common in such processes. Cleaning is straightforward, being usually merely a matter of washing the trays away from the dryer, and there is no difficulty in construction in corrosion resistant materials where necessary.

5.3 THEORETICAL CONSIDERATIONS

This section is limited to consideration of the application of Chapter 2 to the design and operation of atmospheric tray dryers.

5.3.1 Limitations of theoretical treatment

It is possible to make a reasonable estimate of the performance during the constant-rate period by using the well-known expressions for mass and heat transfer and a simple mass balance for the particular type of air circulation. It is however virtually impossible to make any predictions for the falling-rate period without experimental drying measurements on the material itself. Even with such data calculation of the full-scale performance on a particular material requires a number of simplifying assumptions and is arithmetically complicated because of the changing operating conditions over the drying cycle (cf. Section 5.3.6). On the other hand it is a fairly simple matter to carry out full-scale drying trials in a simple flexible dryer of this type and it is recommended that this should be done wherever possible. If it is not possible to carry out actual trials, performance can be deduced by analogy from similar materials on which large-scale performance is known; the data in Section 5.5 may be used as a guide in this way. In general any attempt to calculate the performance from theoretical considerations or from laboratory data only must be regarded as unreliable and tedious in the present state of knowledge and is not to be recommended.

The theoretical considerations summarised below are therefore mainly useful in providing an understanding of the workings of the dryer so as to enable intelligent choice to be made of the best operating conditions in given circumstances.

5.3.2 Effect of temperature and humidity

In general heat is conveyed to the material by hot air (or other gas) and the water (or other liquid) is removed as vapour in the air stream. When (as is frequently the case) the heat transfer is entirely by convection, the drying rate during the constant period can be quantitatively expressed in terms of the temperature and humidity of the drying gas as follows (see equation 2.27):

Constant-rate convection

$$R_c = \frac{10^3(M_s - M_c)W}{A\theta_c} = \frac{h_c(T_g - T_{wb})}{\lambda} = K_H(H_{sw} - H_g) \qquad (5.1)$$

In the case of air/water systems transposition of the heat-transfer and drying coefficients may be simplified by the use of the empirical relationship:

Air/water

$$h_c = SK_H \tag{5.2}$$

If part of the heat is supplied by radiation, e.g. from steam coils located inside the dryer body, the surface temperature in the constant-rate drying period will exceed the wet-bulb temperature T_{wb}. The true surface temperature can be calculated by trial and error solution of the heat and matter balance. A simplified version for the air/water system is:

Constant-rate, air/water, convection and radiation

$$T_g - T_s + \frac{\varepsilon h_r(T_r^* - T_s^*)}{h_c} = \frac{\lambda(H_{ss} - H_g)}{S} \tag{5.3}$$

A similar type of correction can be made for heat transfer by conduction, e.g. in the case where a tray containing the wet material is located in the gas stream so that the underside is exposed to the hot drying gas as well as the upper surface from which evaporation occurs. Nevertheless, with normal cake thicknesses of 2·5 cm and materials of normal thermal conductivity, the effect on the evaporating surface temperature can be neglected.

It is thus possible to calculate the drying rate in the constant-rate period but unfortunately this is not true of the subsequent falling-rate period which in practice covers the greater part of the drying of many products. Clearly, even in the falling-rate period, the material will tend to dry faster at a higher temperature and lower humidity (except perhaps where case hardening with formation of a surface layer of low permeability occurs) but the drying rate is actually controlled by the rate at which moisture is transferred to the surface at which evaporation occurs. The surface temperature will increase from the wet-bulb to the dry-bulb temperature of the drying gas as drying proceeds, and the convection heat transfer to that surface will be correspondingly diminished even though the coefficient remains constant. Heat transfer by radiation and conduction may not diminish so rapidly and may be important factors in the later stages of drying.

The influence of the humidity of the drying gas will similarly diminish as drying proceeds though, as will be shown below, this is to some extent counteracted by the fact that in normal operation with constant air recirculation, the humidity of the air in the dryer will decrease as the rate of drying decreases.

5.3.3 Effect of velocity of drying gas

The velocity of the drying gas stream over the material will affect the heat- and matter-transfer coefficients so that, other factors being equal, the greater the gas velocity the greater the rate of drying. In the constant-rate period the usual heat-transfer relations may be used to compute the effect of air velocity and a useful empirical expression for air/water systems with air flow parallel to the surface is given in Reference 4. It is:

Constant rate air/water parallel flow

$$h_c = 0{\cdot}057G^{0.8} \tag{5.4}$$

It is frequently more convenient to use the linear velocity of the drying gas (V) instead of the mass velocity in terms of cross-sectional area normal to the direction of flow (G). The two are related by:

$$G = \rho v \times 10^3 \tag{5.5}$$

In the falling-rate period the effect of heat transfer and thus of air velocity will diminish as drying proceeds and the result is not readily predictable (cf. Section 2.2). It should also be noted that convection currents will play an important part at low induced air velocities and that in such cases it may not be easy to predict the true coefficient even for the constant-rate period.

5.3.4 Effect of cake thickness

The constant drying rate depends only on the conditions at the surface of the material and the cake thickness has therefore no influence on its value. In the falling-rate drying period, however, the rate of drying is controlled by the rate of migration of water to the surface at which evaporation occurs and clearly an increase in thickness will cause a decrease in the mean falling drying rate. It will also have the effect of increasing the critical moisture content and thus further decreasing the overall drying rate by reducing the range over which this highest rate occurs. Quantitative expression of the effect of cake thickness thus depends on quantitative knowledge of the changes in the slope and shape of the falling-rate curve and is not predictable in the present state of knowledge. The qualitative effect is clear however; the overall drying rate per unit area will decrease with increasing cake thickness. Now the weight of material in the dryer will increase directly with cake thickness and there is obviously an optimum cake thickness or tray loading which will correspond to maximum output rate of dry material for given drying conditions. It is important to recognise the existence of this optimum loading in specifying the operating conditions of an atmospheric tray dryer.

5.3.5 Through-circulation drying

The discussion above has been concerned with the usual type of atmospheric tray dryer in which the drying gas is blown over the surface of the material being dried. In the batch through-circulation dryer, the drying gas is blown through a batch of granular or permeable material supported on a perforated tray. The general effect of the external conditions is unchanged and equation 5.1, for example, can be applied directly to the constant-rate drying period where A is the effective surface area of the material. The effect of conditions in the falling-rate period is still difficult to predict but as drying is generally more rapid their importance may not be so great. Furthermore the effects of heat transfer by radiation and conduction may be regarded as negligible. The bed thickness as such has now no influence on the distance which water has to migrate to the surface (the relevant parameter being the particle diameter) and there is therefore no critical tray loading in that sense. The maximum tray loading is in fact limited merely by the pressure drop and/or the available heat in the drying gas.

A useful empirical equation for constant-rate drying by through-circulation for air/water systems is given in Reference 3 and is essentially similar in form to equations 5.1 and 5.4. It is:

Through-circulation, constant rate, air/water

$$R_c = \frac{10^3(M_s - M_c)W}{A\theta_c} = \frac{146G^{0.59}(H_{ss} - H_g)}{d_m} \qquad (5.6)$$

5.3.6 Variation in conditions over the drying cycle

The atmospheric tray dryer is batch operated and the drying conditions can therefore vary over any cycle of operations. As shown above the important parameters are the velocity, temperature and humidity of the drying gas and weight of material per unit area. Of these only the velocity and condition of the drying gas are likely to vary during a given operation in given apparatus.

The air velocity is usually controlled by a fan and is substantially independent of the temperature and composition of the air. It is also largely independent of the degree of recirculation which merely serves to vary the composition of the air and, in general, it may be assumed that no substantial variation will occur during the drying cycle unless it is deliberately imposed.

Atmospheric tray dryers are normally provided with some form of thermostatic control to maintain the air temperature more or less constant and independent of the rate of drying. The temperature can thus also be considered to be constant over the cycle except where variation is deliberately imposed.

The humidity on the other hand depends on the proportion of air recirculated inside the dryer and the rate of drying at any time. It is theoretically possible to control the recirculation over the drying cycle so as to maintain a constant humidity in the exit gas approaching saturation. This means increasing the degree of recirculation as the rate of drying decreases over the cycle and would correspond to maximum thermal efficiency of operation with maintenance of a substantially constant humidity inside the dryer. In practice it is not usual to control the recirculation in this way as it involves the use of expensive control instruments operated by humidity or manual control with close supervision, either alternative mitigating against the essential simplicity and flexibility of this method of drying. Normal practice is to set the recirculation by trial and error to some convenient compromise value which is kept unchanged throughout the drying cycle. The relationship between the air humidity, the instantaneous drying rate and the degree of air recirculation can be worked out by a simple mass balance for any arrangement of dryer and two examples, covering typical designs, are as follows:

Side-to-side circulation as Section 5.4.6 and Fig. 5.2.

$$H_g = \frac{R_d A x}{G_f} + H_0 \tag{5.7}$$

Back-to-front circulation as Section 5.4.6 and Fig. 5.5.

$$H_g = \frac{R_d A(x+\frac{1}{2})}{G_f} + H_0 \tag{5.8}$$

These equations enable the humidity in the dryer during the constant-rate drying period to be calculated and they can also be used to determine the variation in humidity likely to occur over the falling-rate period.

5.3.7 Steam consumption

The heat required for drying a given product can be calculated directly from the heat required for evaporation, the heat loss by radiation and convection from the dryer body and that lost in the exit gas. Thus for heat supplied by steam coil we have:

$$S_d = \frac{10^3(Q_\lambda + Q_l + Q_s)}{\lambda_{st}} \tag{5.9}$$

where the heat loss follows directly from data on the overall heat-transfer coefficient through the dryer body to atmosphere, i.e.

$$Q_l = 10^{-3} U A_b (T_g - T_o) \tag{5.10}$$

and the heat for evaporation follows directly from the amount of water evaporated, i.e.

$$Q_\lambda = \frac{\lambda(M_s - M_f)W}{\theta_t} \tag{5.11}$$

The heat loss in the exit gas can be obtained from the thermal balance against ambient temperature conditions as datum, i.e.

$$Q_s = 10^{-3}(G_e c_g + W_e c_v)(T_1 - T_0) \tag{5.12}$$

5.4 DETAILED DESCRIPTION

5.4.1 Classification

The most convenient way of systematising the batch atmospheric tray dryer is to classify the variations in the method of air flow inside the dryer thus:

(a) Circulation by natural convection.

(b) Forced circulation over the trays from side to side.
(c) Forced circulation over the trays from back to front.
(d) Forced circulation through perforated trays.

5.4.2 Air circulation by natural convection

This is the most elementary form of the dryer. Efficiencies are low and the use of this procedure is limited to old primitive installations. One arrangement

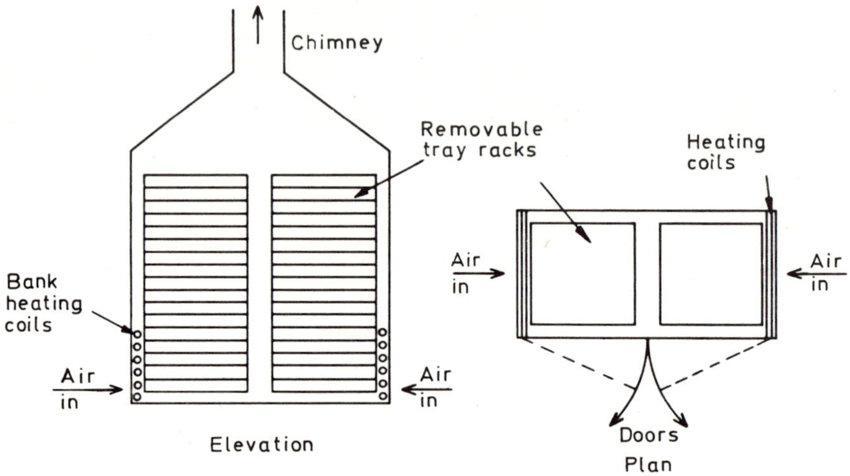

Fig. 5.1. Natural circulation atmospheric tray dryer

using two sets of removable tray racks with steam heating coils at floor level along the side of the dryer, is illustrated diagrammatically in Fig. 5.1.

5.4.3 Forced circulation over the trays from side to side

This method is commonly used in proprietary dryers.[5] A typical arrangement is shown in Fig. 5.2. Adjustable louvres are provided to assist in obtaining reasonably uniform air flow over the trays and the proportion of air recirculated through the stoves is controlled by an adjustable damper in the exit. A two-rack stove can be made to a similar design.

An arrangement for a three-rack stove is shown in Fig. 5.3. In this case the fan is made reversible so that the direction of air flow may be reversed during the cycle.

Yet another arrangement for a two-rack stove with the fan mounted vertically between the racks is shown in Fig. 5.4.

5.4.4 Forced circulation over the trays from back to front

This method is particularly convenient when it is desired to place the heating coils out of sight of the material being dried. It also lends itself to the method of construction of brick wall dryers. A typical arrangement of a four-rack stove is shown in Fig. 5.5.

5.4.5 Forced circulation through perforated trays

In this arrangement the drying air passes through a bed of the material placed on a perforated tray instead of across it. In a typical example the arrangement

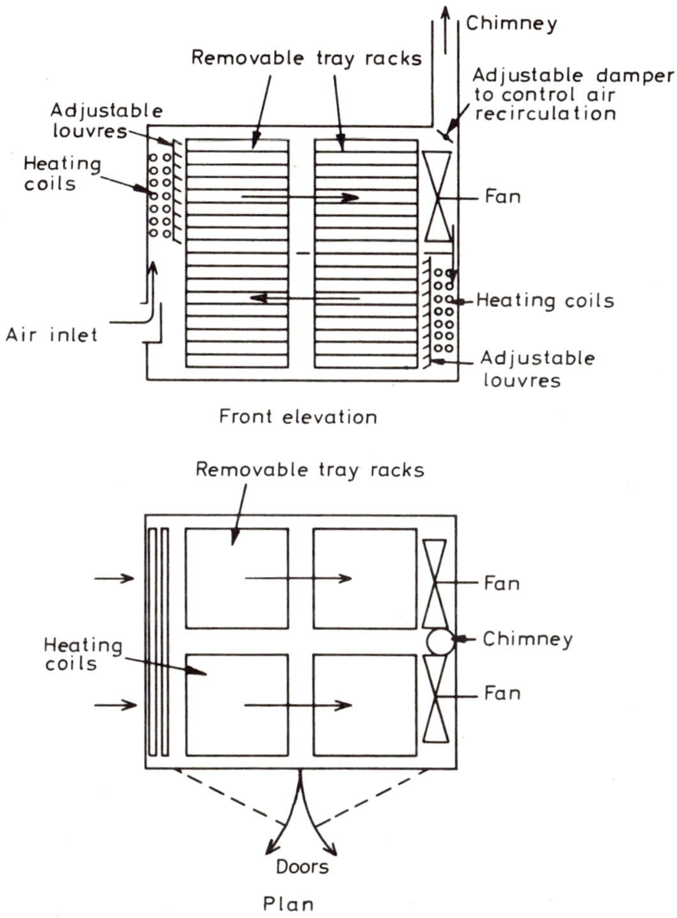

Fig. 5.2. Typical atmospheric four-rack tray dryer, with side-to-side air circulation

is similar to the normal tray dryer with provision of suitable baffling. Fixed racks are usually used to avoid the difficulty in locating movable racks to provide a good air seal. The trays may be either fixed or movable. A typical arrangement is shown diagrammatically in Fig. 5.6.

The use of batch through-circulation dryers has been surveyed and it appears that the field of application is small. Pre-forming as applied to continuous dryers is essentially a continuous operation and in general involves undue complication for batch dryers. Their use is therefore limited to granular or crystalline products. It is essential that the design avoids short-circuiting by the air and this means that cleaning and charging are more difficult and expensive. The advantage lies in a shorter drying time but the gain is small unless the drying time is long. Many granular or crystalline

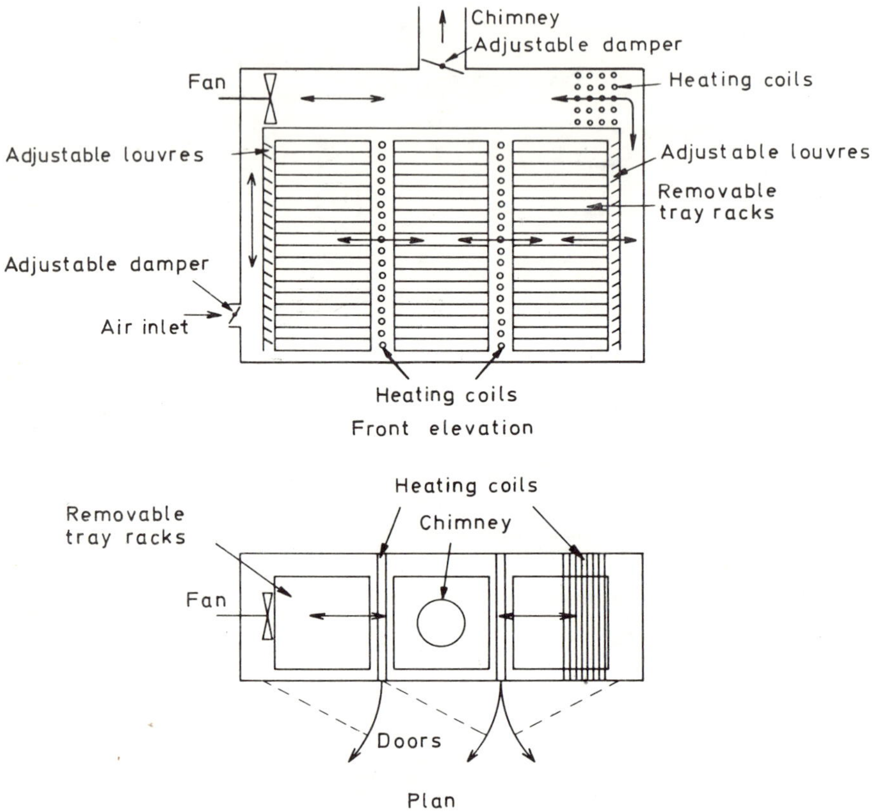

Fig. 5.3. Arrangement of atmospheric three-rack tray dryer with side-to-side air circulation

products dry rapidly in any case and the field of application of the batch through-circulation dryer is thus severely limited.

5.4.6 Details of preferred designs

The number of possible arrangements is legion as can be seen from the classification and diagrammatic illustrations above. Two preferred designs are described in more detail below, viz:

(a) A two-door four-rack stove with side-to-side air circulation (Fig. 5.2).

(b) A two-door, two-compartment, four-rack stove of brick construction with back-to-front air circulation (Fig. 5.5).

Standard racks and trays

In both cases the dimensions have been standardised on the basis of a standard movable rack holding 40 standard trays each $800 \times 400 \times 30$ mm deep arranged in 20 tiers spaced 75 mm apart. The racks are constructed of 40 mm bore mild steel tubing, are of overall dimensions $925 \times 810 \times$ 1 525 mm high and weigh 115 kg. The usual finishes are aluminium paint, galvanising or

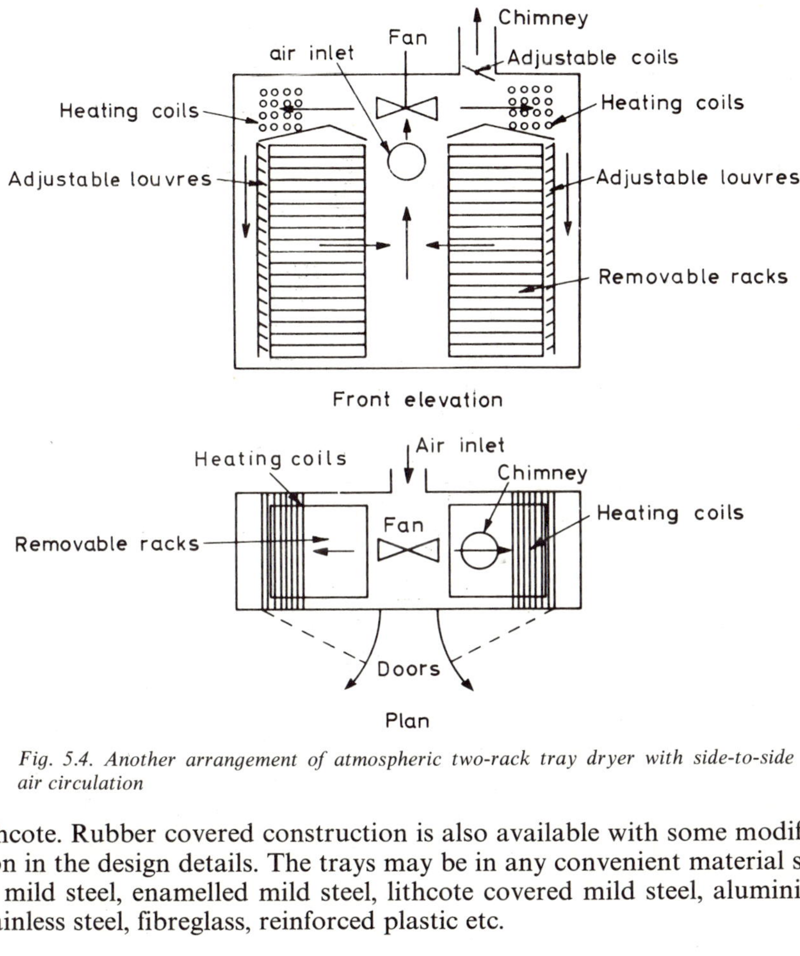

Fig. 5.4. Another arrangement of atmospheric two-rack tray dryer with side-to-side air circulation

lithcote. Rubber covered construction is also available with some modification in the design details. The trays may be in any convenient material such as mild steel, enamelled mild steel, lithcote covered mild steel, aluminium, stainless steel, fibreglass, reinforced plastic etc.

Side-to-side air circulation dryer

There are a large number of proprietary designs of this type of dryer. The arrangement is shown diagrammatically in Fig. 5.2.

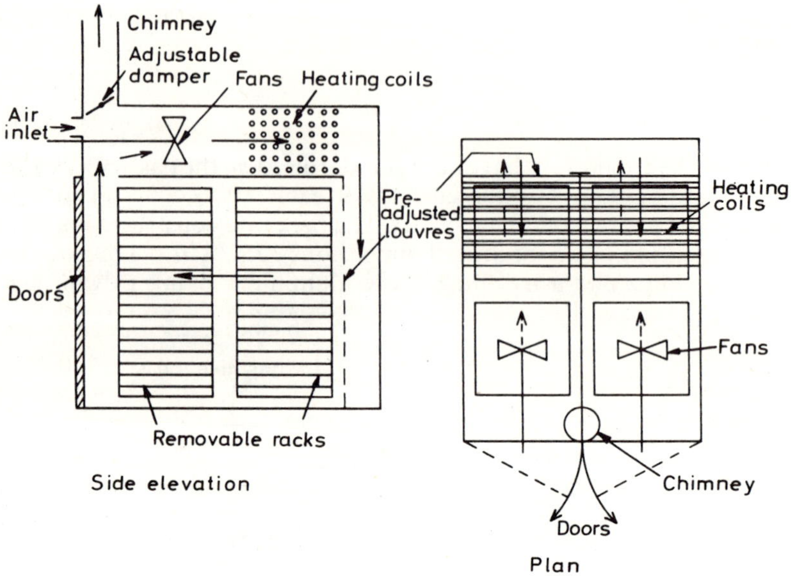

Fig. 5.5. Typical atmospheric four-rack tray dryer with back-to-front air circulation

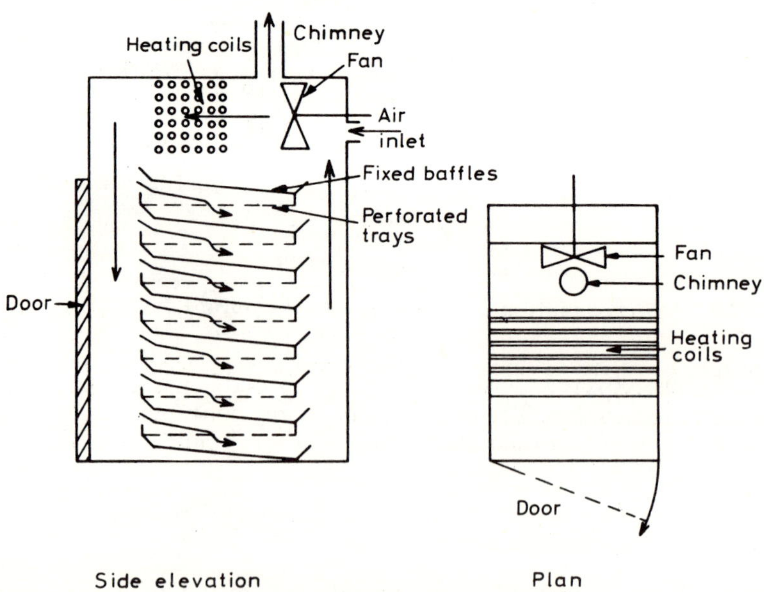

Fig. 5.6. Typical atmospheric dryer with air circulation through perforated trays

The dryers are usually constructed of mild steel and lagged on the outside. Two fans are fitted driven by a common 1.5 hp motor. The air circulation is sideways over the bottom and top halves of the racks in series as shown in Fig. 5.2 and there are two separate heating coils on each side of the dryer to heat the air as it passes on to the racks. The total area of both coils is 17.5 m^2. Similarly two sets of adjustable louvres are provided to assist distribution of air over the trays, one set on the downstream side of each heater bank. Air recirculation inside the dryer can be adjusted by control of the dampers on the air inlets at the bottom left hand side and of the butterfly valves in the exhaust ducts on the top above the fans. The total area of the 160 trays is 53 m^2 and the surface area of the dryer body (including floor) is 29 m^2.

Back-to-front air circulation stove of brick construction

Many designs of dryer of brick construction use back-to-front air circulation as shown diagrammatically in Fig. 5.5. In one design the body is constructed of double brick and divided into two compartments, each of which contains two standard racks, by a central brick partition. The heating coil of 43·5 m^2 area is located outside and on top of the drying compartments. Two fans are fitted at the front of the dryer driven by a common 2 hp motor. Air circulation is from the fans at the front of the dryer, over the heating coil, down the back of the dryer and thence over the trays in the two compartments from back to front. A fixed louvre plate, designed by experiment, is located at the air entrance at the back of the stove and serves to direct air uniformly over the trays. Air recirculation inside the dryer can be adjusted by controlling dampers on the inlet side of the fan at the front of the stove. Pre-settable dampers are also provided on the fresh air inlet to the fans and the exhaust ducts are proportioned so as to impose a minimum pressure at the front of the stove. The total area of the 160 trays is 53 m^2 and surface area of the dryer body (including floor) is 4·5 m^2.

Economy in materials and reduction in heat losses is obtained when batteries of these stoves are being used by installing them in terrace fashion side-by-side and/or back-to-back.

5.4.7 Instrumentation

Conventional practice is to govern the drying conditions in terms of the dry-bulb temperature of the air circulating in the dryer by means of an indicating temperature controller operating a control valve on the steam (or hot water) supply to the heating coils. The controller should be capable of being manually re-set by the operator so that the stove may be quickly adjusted for a wide range of drying temperatures. The location of the temperature sensitive element is a matter of some importance as there is a considerable drop in the air temperature as it passes across the product. It is usually best to measure the temperature at the highest temperature point in the dryer, thus affording a factor of safety when drying heat-sensitive products. Choice of a suitable single point presents some difficulty in the side-to-side circulation dryers in which the heating coil is split into two sections (one on each side of the dryer) and the air circulation pattern is fairly complex. A usual compromise position established by trial and error is just beyond the fans towards the front of the dryer and close to the right hand side heater (cf. Fig. 5.2). In general measurements at this point correspond reasonably well to the maximum air temperature inside the stove. There is less difficulty in the back-to-front circulation dryers and it is usual to locate the temperature-sensitive element in the air stream at the back of the stove after it has passed over the heater and before it passes on to the product (cf. Fig. 5.5). Due care should be taken to see that the temperature-sensitive element is shielded from direct radiation from the heating coils themselves, particularly in stoves in which the heating coils are inside the actual drying chamber. The usual maximum operating temperature in atmospheric tray dryers is up to 120°C and mercury in steel thermometers are convenient.

It is theoretically possible to operate the dryer at maximum thermal efficiency by controlling the proportion of re-circulated air in terms of the humidity in the chimney or air exit and arranging matters so as to maintain the exit air humidity as high as possible. This is not done in practice because of the lack of a simple, cheap, reliable hygrometer capable of working in dusty atmospheres and the expense of the control equipment required.

5.4.8 Ancillaries

Other ancillary equipment frequently used includes:

Movement gear for racks

A standard rack fully laden may weigh up to half a tonne and mechanical aids for their movement are necessary. The original method was to mount the racks on castors so that they could be pushed about. This is not mechanically very satisfactory and adds to the expense of the rack. Another arrangement is to fit the rack with wheels and provide rails into each stove along which the rack can be pushed. This is expensive and severely limits the flexibility of the movement of the racks within the drying area unless the installation is made unduly elaborate.

Modern practice is to use a hand-operated or motor driven lifting truck, which can be pushed under the rack to move it. The truck is removed when the rack has been placed in the stove. This system is simple and flexible and a standard rack has been designed for use in this way.

Discharge of trays

The discharge of the dry products from the trays is normally carried out manually and may be a dirty and hazardous operation with toxic, dusty and coloured products. Enclosed and ventilated de-traying booths are therefore necessary to comply with modern working standards.

Charging of trays

It is essential that the trays are charged uniformly to ensure that the optimum weight of product is charged to the dryer and that the level on the trays is not such as to interfere with the air circulation in the dryer. Due attention should also be paid to the question of optimum loading (cf. Section 5.3.4). The correct loading level for the standard tray/rack assembly is 4 cm depth.

Charging is normally carried out by hand. Various mechanical aids on the extruder principle have been tried from time to time but their use has involved additional expense in capital and labour which has outbalanced any potential advantages deriving from more even tray loading.

Washing of trays

Means for washing the trays rapidly is desirable in large installations to avoid contamination between different products and to ensure a rapid turn-round of trays, thus enabling a minimum stock to be carried.

5.5 METHOD OF OPERATION

Before discussing performance data it is desirable to summarise typical methods of operation of dryers of this type.

The dryers are normally operated as individual units and batteries of large numbers of stoves are frequently located in a common drying area. Several stoves may be handling the same product or one stove may be handling a different product on each drying cycle. The drying temperature and drying time will have been determined by experience. Drying is frequently assisted by raking over the partly-dried product on the trays or by removing the trays and manually breaking up the product on the trays once or more during the drying cycle. Table 5.1 gives a typical drying schedule for dyestuff on a standard four-rack dryer of the type shown in Fig. 5.2.

5.6 PERFORMANCE DATA

The actual performance of any dryer is to a great extent conditioned by the nature of the product being dried; the major factors being the initial and final moisture contents, the permissible drying temperature, the proportion of drying which takes place in the falling-rate period and the shape of the falling-rate drying curve. This dependence of the overall performance on the product is particularly noticeable in the case of the atmospheric tray dryer which is capable of handling material in almost any physical form. It is thus difficult to present overall performance data in a simple comprehensive manner. The performance of the dryer as a machine can however be adduced by consideration of constant rate drying alone which is essentially independent of the material being dried (cf. Section 5.2). With these limitations, the performance of the two main types of atmospheric tray dryer (air circulation over the material and air circulation through the material) is considered below.

5.6.1 Air circulation over the material

Data on the constant-rate drying of wet sand on a variety of designs of dryer using a standard test technique have been reviewed in detail and a method of calculating the performance for a wide range of proportions of

Table 5.1 Typical drying schedule with a filter pressed dyestuff cake

Time (h)	*Operation*
2	Charge paste to trays by hand. Total wt. 1 500 kg
0·5	Charge loaded racks to dryer. Close doors. Start fans. Apply full steam to coils
3·5	Dryer temperature rising to 90°C. Switch on temperature controller set at 90°C
21	Drying proceeds under automatic control
2	Shut off fans. Remove each rack in turn from dryer and break up product on the trays by hand
14	Drying proceeds under automatic control
0·5	Sample material and check for dryness
0·5	Shut off dryer and remove racks
1	Discharge trays of dried material to drums. Dry wt. = 21 kg
3	Clean trays, tidy up section of plant

Four-rack side-to-side air circulation dryer	160 trays
Drying cycle	48 h
Drying temperature	90°C
Paste	18 % solids
Dry Product	1 % moisture
Actual dryer occupation	42 h

air circulation and operating temperature developed. The information can be conveniently presented in the form of characteristic curves showing the constant-drying rate and steam consumption versus the operating temperature for various proportions of air recirculation. Typical curves for a side-to-side air circulation design as shown in Fig. 5.2 are given in Fig. 5.7. Similar curves for the back-to-front air circulation design as shown in Fig. 5.5 are given in Fig. 5.8.

Important features shown in Figs 5.7 and 5.8 are that the optimum proportion of air recirculation for maximum efficiency is in the region of 80 % to 90 %; and that the size of the heating coil has a considerable influence on the maximum temperature achievable and hence the maximum drying

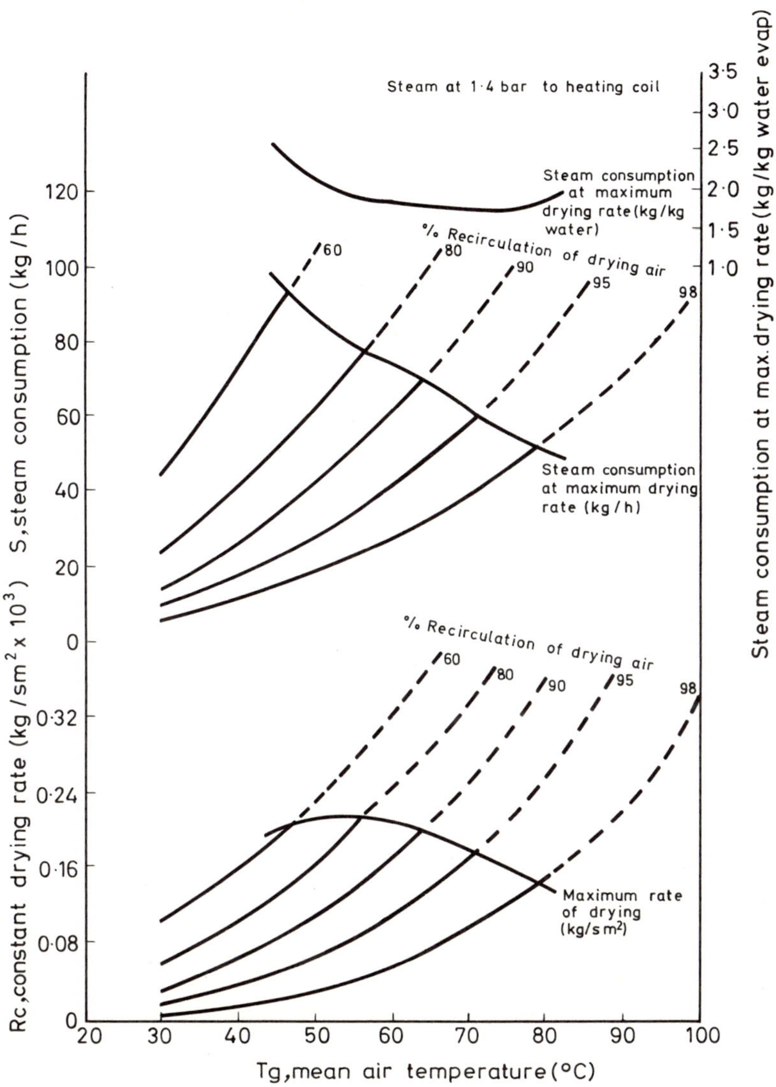

Fig. 5.7. Performance of a proprietary four-rack atmospheric tray dryer to design indicated in Fig. 5.5. Characteristic curves for constant-rate drying under steady state conditions, 160 trays, area 53 m², steam gauge pressure to heating coil 1·4 bar. Ambient conditons — 15°C and humidity 0·0107 kg/kg

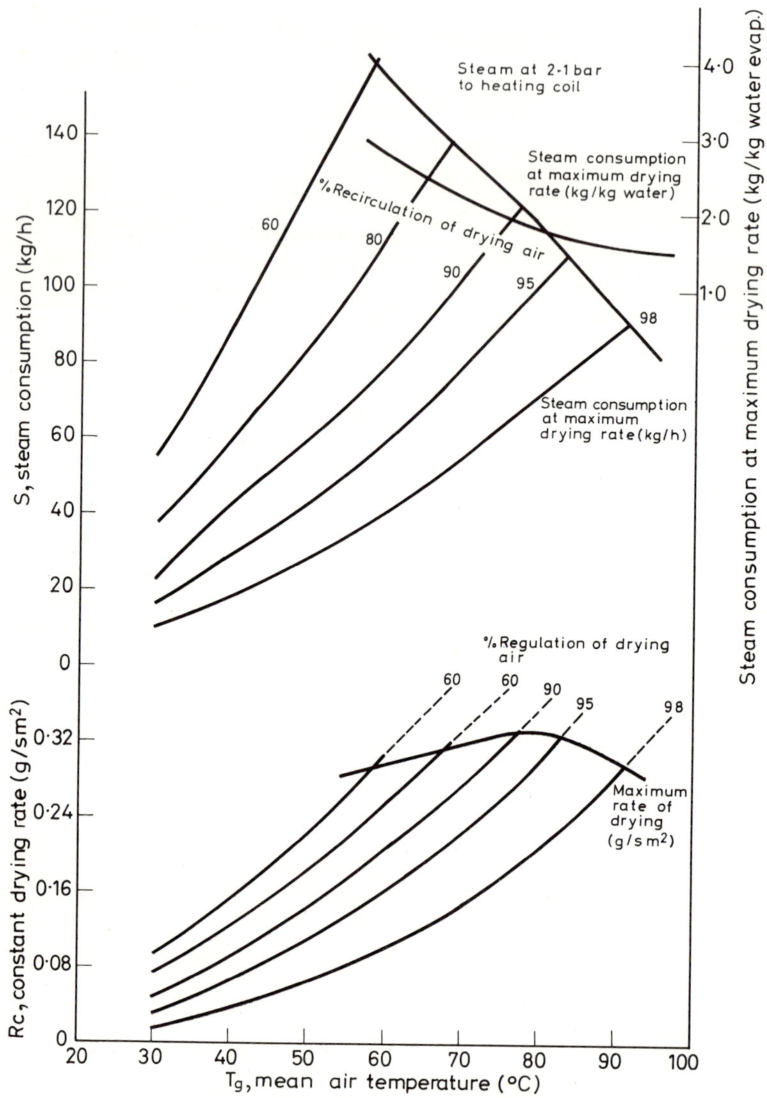

Fig. 5.8. Performance of a four-rack atmospheric tray dryer to design indicated in Fig. 5.7. Characteristic curves for constant-rate drying under steady state conditions (isolated dryer). 160 trays, area 53 m^2, steam gauge pressure to heating coil 1·4 bar

rate. These features apply to the constant-rate drying period. In the falling-rate period, where the rate of drying is limited by the material, an even higher proportion air recirculation is desirable but the influence of the heating coil is less. These considerations have been taken into account in the preferred design of dryer with back-to-front air circulation (Fig. 5.5) which uses a simple pre-adjustable baffle to cover the range 80% to 95% recirculation and incorporates a large heating coil of 43·5 m^2 surface area.

The air velocity over the trays will also influence the rate of drying (cf. equation 5.4). An upper limit is fixed by the danger of entrainment of material from the trays into the air stream. This will obviously vary with the nature of the material; a value of 1·5 m/s has been chosen as reasonably safe for most materials. The uniformity of the air stream is also important as bad air distribution will cause poor tray-to-tray drying uniformity; this will give a disproportionate increase in the drying cycle due to the practice of not discharging a dryer until all the material is dry. Poor air distribution also requires velocities to be high in some parts of the dryer to maintain a given average value, with consequent danger of entrainment of material in the air stream. Most designs of side-to-side air circulation dryer are poor in respect of air distribution and drying uniformity and much superior results

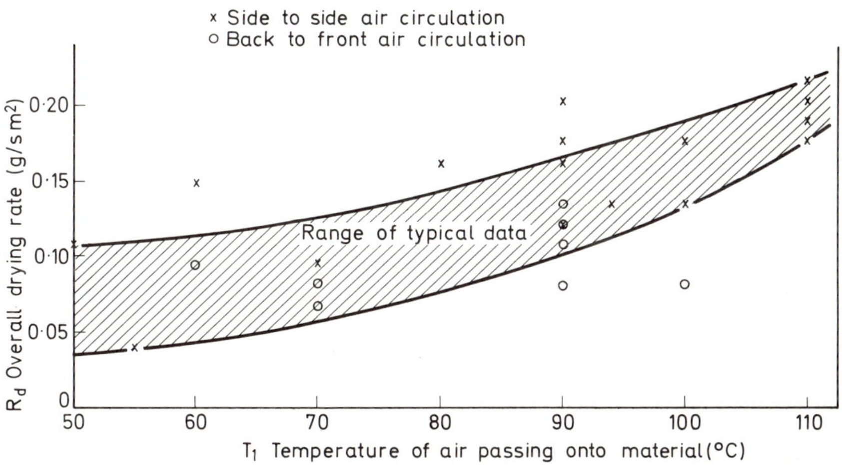

Fig. 5.9. Typical overall drying rates on atmospheric tray dryers for filter cakes of mud-like consistency. Initial moisture contents in the range 50–500% dry basis. Final moisture contents in the range 0·5–3%. Air recirculation in the range 80–95%. Depth of mud on tray about 4 cm

can be obtained in the design with back-to-front air circulation. Distribution of the air on to the trays is through a specifically designed fixed louvre plate.

Obviously overall drying performance is very much dependent on the nature of the material being dried. Typical data are given in Fig. 5.9 in the form of overall drying rate against temperature of air passing on to the material and this can be used as a guide to the type of performance that can

Table 5.2 Atmospheric tray dryers with air circulation over the material. Summary of leading data on preferred designs

		Proprietary dryer with side-to-side air circulation (Fig. 5.2)	*Brick constructed dryer with back-to-front air circulation (Fig. 5.5)*
Racks (Note 1)		4 standard	4 standard
Trays (Note 2)		160 standard Area 53 m^2	160 standard Area 53 m^2
Air circulation	Mode	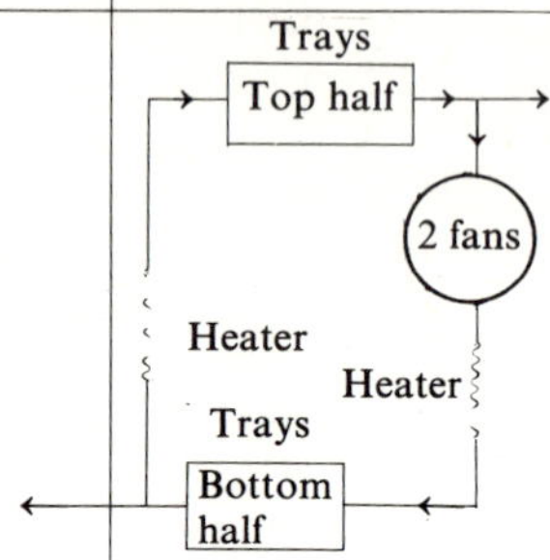	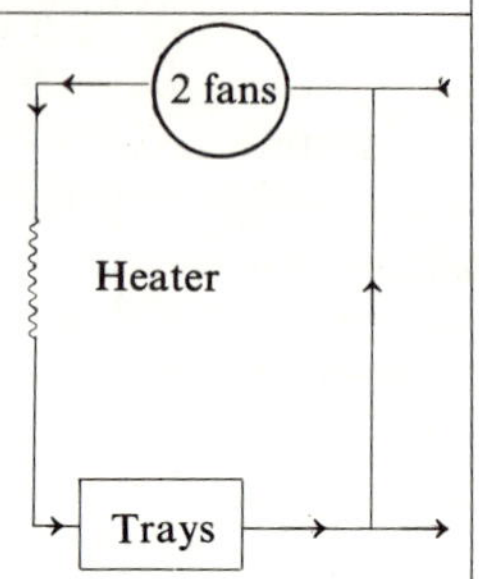
	Volume at fans	4 250 m^3/h 1·18 m^3/s	11 000 m^3/h 3·06 m^3/s
	Mean velocity over trays	1·2 m/s	1·45 m/s
	Uniformity distribution	Poor	Good
Heater	Location Medium Area Overall coefficient	Internal Steam gauge 1·4 bar 17·5 m^2 total 20·8 W/m^2 °C	External Steam gauge 2·1 bar 43·5 m^2 14·2 W/m^2 °C
Heat loss to atmosphere		177 W/°C	173 kcal/h, 206 W/°C (isolated dryer)
Installed hp (kW)		1·5	2·0
Constant rate drying at 60°C (mean) and 90% recirculation	Rate	0·635 kg/h m^2 = 0·176 g/s m^2	0·782 kg/h m^2 = 0·217 g/s m^2
	Steam consumption	1·8 kg/kg	2·0 kg/kg
Maximum constant rate drying with 90% recirculation	Temperature Rate Steam consumption	64°C 0·203 g/s m^2 1·8 kg/kg	77°C 0·339 g/s m^2 1·9 kg/kg
Typical overall drying rates at 90% recirculation	60°C	0·14–0·44 kg/h m^2 0·04–0·122 g/s m^2	
	100°C	0·50–0·43 kg/h m^2 0·14–0·119 g/s m^2	

Notes { 1. Standard rack / 2. Standard tray } as described in Section 5.4.6(a).

be expected.[1,2] The leading data on the two preferred designs is summarised in Table 5.2.

5.6.2 Air circulation through the material

Much of the information available on continuous through-circulation dryers may be directly applied to batch through-circulation plant. (cf. References 3 and 4). Typical performance data from published information is given in Table 5.3 (abstracted from Reference 4).

Table 5.3 Typical data for atmospheric tray dryers with air circulation through the material

Trays: Number	5	16	24
Dimensions, mm	915 × 915	915 × 1 040	850 × 980
Area, m^2	4·2	15·2	2·0
Loading depth, cm	5	6·5	3·8
Material	Fine crystals	Polymer crumbs	Washed vegetable seed
Drying air: Temperature °C	100	88	3·6
Superficial velocity, m/s	0·25–0·5	0·9	1·0
Overall drying rate per unit tray area, kg/h m^2	1·07	0·88	1·12
g/s m^2	0·30	0·245	0·31

5.7 CAPITAL COSTS

In this and other chapters an attempt has been made to scale up capital costs to mid-1968 prices from information collected in past years. Up-to-date costs of equipment can only be obtained from plant manufacturers, but the cost of installation varies appreciably with site conditions and the cost of connecting services such as steam. Readers can readily convert labour and service costs to local conditions from the consumptions given. The information given below should, however, indicate relative costs with reasonable accuracy.

Costs of two- and four-rack stoves of both the side-to-side and back-to-front circulation types are given in Tables 5.4 and 5.5. Similar information for the standard racks and trays in various materials of construction is given in Tables 5.6 and 5.7; these may be taken as fairly typical for any design of atmospheric tray dryer.

5.8 RUNNING COSTS

5.8.1 Services

These usually consist of steam and electricity only and the usage will depend upon the drying conditions of the particular product.

Table 5.4 Capital cost of atmospheric tray dryers with side-to-side circulation. Mild steel lagged construction (Fig. 5.2). Cost does not include racks and trays—see Tables 5.6 and 5.7

	Two-rack model	*Four-rack model*
	£	£
Purchase price including fan(s), motor and drive	1 050	1 710
Auxiliaries, including instruments, piping and lagging	250	300
Erection	50	50
Total installed cost	1 350	2 100

Table 5.5 Capital cost of atmospheric tray dryers with back-to-front circulation. Brick construction (Fig. 5.5). Cost does not include racks and trays—see Tables 5.6 and 5.7

	Two-rack model	*Four-rack model*
	£	£
(a) Economy model in Nori Brick		
Installed cost including fan(s), motor and drive	1 500	1 800
Auxiliaries, including instruments, piping and lagging	250	300
Total installed cost	1 750	2 100
(b) Acid resisting model in glazed brick		
Installed cost	1 800	2 300
Auxiliaries	250	300
Total installed cost	2 050	2 600

Table 5.6 Capital cost of Standard 40 tray racks to carry trays of Table 5.7

Material	*Each*	*Total for two-rack stove*	*Total for four-rack stove*
	£	£	£
Mild steel galvanised	45	90	180
Mild steel lithcote	100	200	400
Stainless steel	100	200	400
Timber	55	110	220

Table 5.7 Capital cost of standard 800 × 400 × 30 mm trays

Material	*Each*	*Total for two-rack stove*	*Total for four-rack stove*
	£	£	£
Mild steel enamel	1·4	112	224
Mild steel lithcote	2·4	198	396
Stainless steel	4·75	350	760
Aluminium	1·0	80	160
P.V.C.	1·2	96	192
Fibreglass/polyester	1·65	132	264

The cost of electricity is relatively small and it is sufficient to calculate it from the installed horse power on the dryer given in Table 5.2 on the basis that 1 hph requires 1 kWh at factory meter. The cost of steam is a relatively large proportion of the total cost and measurements of actual steam consumption are not always directly available. In the absence of any information a usage of between 2 and 3 kg/h per kg of water evaporated may be taken as a rough working rule. A better approximation can be made from the constant-rate data of Figs. 5.7 and 5.8 given knowledge of the actual drying time and conditions; the method is illustrated in the worked example below.

Example 1. Calculation of approximate steam consumption

A paste from a filter press is dried in a four-rack side-to-side circulation stove from 18% solids to 1% moisture in 42 h at a temperature of 90°C. The weight of dry product is 210 kg (Data of Table 5.1).

Weight of paste charged = (210–2·1)/0·18 = 1 150 kg
Water evaporated = (1 150–210) = 940 kg = 22 kg/h
Area of trays in dryer (Table 5.2) = 53 m^2

$$\text{Overall rate of drying} = \frac{940}{53 \times 42} = 0{\cdot}413 \text{ kg/h m}^2$$
$$= 0{\cdot}117 \text{ g/s m}^2$$

Assume 90% air recirculation.
If all the process continues at constant-rate drying, then from Fig. 5.7 (lower chart), the mean air temperature = 50·5°C.
And from Fig. 5.7 (upper chart), the steam consumption = 43 kg/h.
Heat loss from dryer (Table 5.2) = 152 kcal/h°C = 177 W/°C.
Hence additional heat loss for drying air temperature of 90°C = 152(90–50·5) = 6 000 kcal/h
or = 177 × 39·5W = 6 950W = 6·950 kJ/s.
For many years Callender's Steam Tables will continue to be used. The total heat of saturated steam at 20 lbf/in^2 gauge (1·4 bar) = 1 169 Btu/lb.

The steam condenses at 90°C, 194°F. Hence net useful heat in steam = 975 Btu/lb = 542 kcal/kg = 2 262 kJ/kg.
The additional steam consumption = 6 000/542 = 11 kg/h
or 6·950/2 262 = $3{\cdot}05 \times 10^{-3}$ kg/s = 11 kg/h.
The total steam consumption = 43 + 11 – 54 kg/h = 54/22 = 2·46 kg water evaporated.

Note 1
Use of the air inlet temperature (90°C) in place of the mean air temperature gives a high value to the steam consumption but this is compensated by the fact that no allowance is made for the higher temperature of the air leaving the dryer compared with the constant-rate value.

Note 2
Plant measuring meters are likely to continue to be calibrated in hourly

rates. Hence the use of kg/h and kcal/h is more convenient for this calculation than use of SI units of kg/s and kJ/s (=W). The heat loss calculation is given by both methods to assist the growing body of chemical engineers who will be more at home with SI units during the 1970s.

5.8.2 Labour

This is the other major cost of drying. Labour is required for charging and discharging the dryer (or the trays separately), supervising during drying and for operations such as breaking up the material on the trays at some point during the drying cycle (cf. Table 5.1). The labour utilisation may be estimated from the typical values in Table 5.8.

Table 5.8 Labour requirements for standard atmospheric tray dryers

Operation	*Two-rack model*	*Four-rack model*
Charge paste to trays on racks	1 man-h	2 man-h
Charge racks to dryers and start up dryer	0·25 man-h	0·25 man-h
Supervision during drying	0·1 man-h/h	0·1 man-h/h
Break up material during drying cycle	0·5 man-h/operation	1 man-h/operation
Remove racks and shut down dryer	0·1 man-h	0·1 man-h
Discharge trays	0·5 man-h	1 man-h

5.8.3 Maintenance and depreciation

Actual maintenance and depreciation depends a good deal on the conditions under which the dryer is used. The figures in Table 5.9 may be taken as typical for dryers in constant use (24 h day) and made of the appropriate material of construction for the duty.

Table 5.9 Annual maintenance and depreciation on atmospheric tray dryers

	Annual maintenance	*Approx. life*	*Annual depreciation*
Proprietary side-to-side circulation dryers of mild steel construction, Fig. 5.2	7·5 % of capital cost exclusive of racks and trays	10 years	10 % of capital cost exclusive of racks and trays
Standard back-to-front circulation dryers of brick construction, Fig. 5.5	2·5 % of capital cost exclusive of racks and trays	20 years	5 % of capital cost exclusive of racks and trays
Standard 40-tray racks	0	10 years	10 % of capital cost
Standard $32 \times 16 \times 1\frac{1}{4}$ in ($800 \times 400 \times 30$ mm) trays	0	3 years	33·1/3 % of capital cost

5.8.4 Supervision and overheads

These will depend upon local conditions and lie outside the scope of this book.

Example 2. Drying cost calculation

This example shows the use of the data in the above tables. It should not be taken as necessarily representative or typical of actual drying costs.

Problem:
A material from a filter press is dried in a proprietary four-rack side-to-side circulation stove from 18% solids to 1% moisture in 42 h at a temperature of 90°C. The weight of dry product is 210 kg. The dryer racks are of galvanised mild steel and the trays of aluminium. The dryer cycle is 48 h and available occupation is for 45 weeks/y at 168 hr/week (7 560 h/y).

(a) *Production*

Production 210/48 = 4·37 kg/h

$$= \frac{210 \times 45 \times 168}{48} = 33\,000 \text{ kg/y}$$

(b) *Capital cost*

	£	
Dryer	2 100	(Table 5.4)
Racks	180	(Table 5.6)
Trays	160	(Table 5.7)
	2 440	

(c) *Maintenance*

Dryer 0·075 × 2 100 = £157/y (Table 5.9)
= £4·8/1 000 kg product

(d) *Depreciation*

Dryer	= 0·10 × 2 100	= 210	Table 5.9
Racks	= 0·10 × 1 180	= 18	Table 5.9
Trays	= 0·333 × 140	= 53	Table 5.9
		281	

Total cost = £8·5/1 000 kg product

(e) *Steam*

Water evaporated = 22 kg/h (see Section 5.8.1)
Steam consumption = 2·46 kg/kg water evaporated
= 54 kg/h
Cost of L.P. steam at £1/tonne
Cost of steam = £12·4/1 000 kg product

(f) *Electricity*

Installed hp = 2 (Table 5.2)
Electricity usage = 4 600 kWh/1 000 kg product
Cost of electricity at £6 per MWh = £2·7/1 000 kg product

(g) *Labour*

Charging and discharging dryer	= 3·35 man h	Table 5.8
Breaking up product once during cycle	= 1·0 man h	
Supervision of dryer = 0·1 × 48	= 4·8 man h	
	9·15 man h	

Cost of labour at £0·7/hr = £30·6/1 000 kg product

The total drying cost less overheads is thus:

	£/1 000 kg *product*
Steam	= 12·4
Electricity	= 2·7
Labour	= 30·6
Maintenance	= 4·8
Depreciation	= 8·5
	59

CHAPTER SIX

Batch vacuum tray dryers

6.1 APPLICATION

The simplest indirectly-heated vacuum dryer consists of a single evacuated chamber of cast iron or steel plate construction, containing heated shelves on which the trays holding the product are supported. A more recent design is based on a number of single- or multi-shelf compartments each operating independently within the main frame of the dryer. In this case each tray is constructed with an integral heating jacket to permit higher rates of heat transfer.

These dryers are used extensively for drying temperature-sensitive or easily-oxidised materials; and also for small batches of expensive products where material loss must be avoided, e.g. in pharmaceutical and dyestuff manufacture. They are not suitable, however, for drying at very low temperatures, viz. below 40°C.

Heat transfer to the material takes place mainly by conduction through the metal surfaces and interface of the tray and heating plate, although some (relatively small) heat transfer by radiation occurs.

6.2 THEORY

As discussed in more detail in Chapter 2, Section 2.4 the process is essentially one of evaporation and the solid tends to assume the boiling-point of the liquid and remain mostly at that temperature. The temperature does, however, rise towards that of the heating medium on completion of drying.

Because the temperature of the solid varies only slightly about the boiling-point of the liquid, it is convenient to assume that the temperature differential is constant throughout the drying cycle and that the heat-transfer coefficient varies, i.e. the resistance to drying increased as drying proceeds.

The overall drying equation can then be written as follows (this is the same as equation 2.47):

$$R_m = \frac{10^3\,(M_s - M_f)W}{A_h\theta} = \frac{k_c(T_h - T_b)}{\lambda} \tag{6.1}$$

Heat transfer by radiation is neglected as it can be shown that the radiation heat-transfer coefficient, h_r, is less than 5 W/m^2°C for drying temperatures up to 100°C.

The above equation can be used to correlate drying data in given apparatus between given moisture limits, and also to correlate drying coefficient with initial moisture content for a given final moisture content, or vice versa. It should, however, be used with discretion because the material is dried in layers of 2·5 to 4 cm in depth without agitation and, with certain materials, case-hardening or skin formation can occur and give variable results.

6.3 GENERAL APPLICATION OF BATCH VACUUM TRAY DRYERS

The batch vacuum dryer is used for drying granular solids, pastes; also very infrequently slurries or solutions. It finds particular application in the pharmaceutical, fine chemical and dyestuff industries where materials sensitive to heat or air oxidation, particularly the latter, are dried at relatively low tonnage outputs. It is particularly suited to multi-product operation.

Its relatively high operating cost is mainly due to the high labour requirement and puts it in the same category as the atmospheric tray dryer. This places it at a disadvantage when drying products of very low material cost unless there are other factors of greater importance, e.g. quality. Although these dryers are normally operated as individual units, batteries of stoves are frequently located in common drying area for effective utilisation of labour.

All types of tray dryer are unsuitable for drying toxic products because of the hazard involved in charging and discharging the trays; the latter operation is inevitably a dusty operation of varying degree and must be carried out in a suitably draughted enclosure.

6.4 SINGLE-COMPARTMENT VACUUM TRAY DRYER (TYPE SVT)

6.4.1 Description

The simplest vacuum dryer is essentially an evacuated cabinet containing heated shelves on which the trays carrying wet material are placed (Fig. 6.1). In order to avoid use of proprietary names it will be designated Type SVT in this book. Models contain 14 to 20 shelves. Normal shelf dimensions are 1·5 × 2·0 m, and each shelf holds 2–8 trays depending on tray size.

The heating shelves are normally fabricated in mild steel, occasionally in stainless steel, or treated mild steel. The shell and doors are in cast iron and mild steel but may also be coated or aluminium sprayed when freedom from iron scale is essential. Two doors are fitted to the dryer to allow free access for loading and unloading the trays.

Vapour from the top of the stove passes to a tubular condenser mounted on a condensate receiver. The receiver is discharged of condensate by means of a 'tippler' trap and the incondensible gases are removed by a (300 × 300 mm) single stage dry vacuum pump. (This system obviates the possibility of flooding the drying compartment by pull-back of water from ejectors or wet vacuum pumps.) The average size of condenser per Passburg stove has 18 m^2 outside the tube area, typical dimensions being 130 tubes each 22 mm o.d. × 2 000 mm.

6.4.2 Operation

Drying is normally carried out at an absolute pressure of 0·067–0·27 bar. Hot water, sub-atmospheric steam or pressure steam may be used for heating depending on the temperature sensitivity of the material being dried. Vacuum steam is not recommended for temperatures below 60°C; the rather more

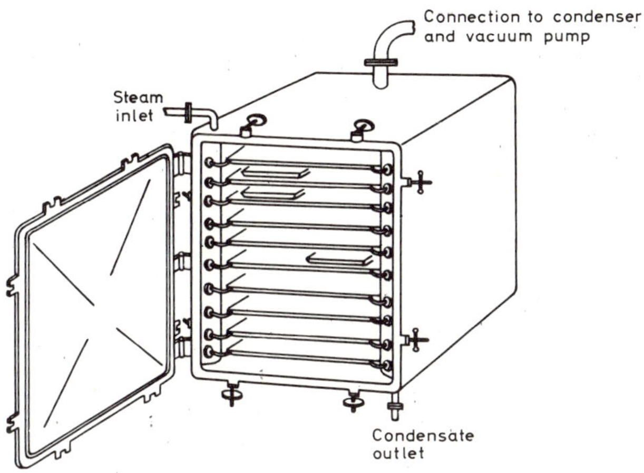

Fig. 6.1. Simple form of batch vacuum tray dryer (Type SVT)

complicated, but more controllable, hot water circulation system is preferable for the temperature range 40–70°C.

As in all other types of tray dryer, uniform loading of the trays is of primary importance. The average loading range is 10–35 kg wet paste per m^2 tray area. Breaking up the partially-dried material on the trays after a predetermined lapse of time is often practised to counteract 'case hardening', thereby effectively reducing the overall drying time. At the same time, condensate draining to the base of the stove from the (unheated) walls can be drained away. It has also been suggested that a small air bleed into the dryer is effective in reducing the amount of condensation on the walls, and also helps to keep drying times to a minimum. It is often preferable to leave the top shelf empty of trays, as material dried on this shelf is frequently not dry when the remainder of the stove-charge is ready for emptying (due to the absence of a radiating surface above the top shelf). Occasionally due to the configuration of the heating plates a 'hotspot' may exist on a particular plate (usually near the steam inlet) and appropriate precautions should be taken when drying particularly heat-sensitive materials.

Steam leakage from the union couplings of the heating plates and steam manifold is a potential weakness of this Type SVT heating system and examination of these connections should be frequent.

6.4.3 Performance

Although dryers of Type SVT are widely used, the paucity of useful (recorded) drying data is surprising. In Table 6.1 typical drying data is given for what is considered to be a reasonable range of products. However, the variability of drying rate of various products in Type SVT is probably greater than in any other form of dryer and drying-rate estimation by analogy is to be avoided.

6.5 MULTI-COMPARTMENT VACUUM TRAY DRYER (MCVT)

6.5.1 Description

A more modern design of vacuum tray dryer, designated here as Type MCVT has several separate vacuum compartments which can hold one or more trays (Fig. 6.2). It has its own low-pressure steam generator which can in turn be heated by factory steam or by gas. Trays in one design are 1 800 × 900 mm. The two- and three-tray dryers have a manually operated elevator for loading the trays; the larger dryers have a motorised elevator.

The heating arrangements are shown in Fig. 6.3. At the rear of the tray jacket is a spigot fitting into a simple rubber seat in the compartment wall.

Steam (35–95°C) is generated at sub-atmospheric pressure in the boiler at the base of the dryer and is passed via a non-return valve through the spigot and into the jacket of each tray. The condensate is exhausted via a small jet into the drying compartment at a point coincident with the vacuum exhaust point. Exhaust steam and condensate from the jacket, and vapour from the material being dried, pass through a spray condenser external to the drying chamber and discharge through the (wet) vacuum pump. Steam or hot water is also passed through coils situated around each drying compartment to prevent condensation on the inside wall.

Because the boiler and spray condenser are built on to the dryer, the only ancillary equipment required is a reciprocating or rotary vacuum pump of capacity 100 m^3/h. The water from the spray condenser passes through the pump and, to avoid flooding the dryer if the pump motor stops, the water supply to the condenser is automatically isolated by pump failure.

6.5.2 Operation

The maker's claim of a minimum operating temperature of 30°C with sub-atmospheric steam at 40 mb absolute is doubtful; in practice the lowest

Table 6.1 Typical operating data for Type SVT dryer

Type of material	*Loading* (kg *wet paste*/m² *of tray*)	*Drying conditions*		*Moisture content* (*approx.*) *of paste* (kg *water*/ kg *dry solids*)	*Moisture content of product* (kg *water*/ kg *dry solids*)	*Total drying time* (h)	*Overall drying rate* (g *water*/m² *of tray* s)	*Remarks*
		Absolute pressure (mb)	*Heating plate* (°C)					
Dyestuffs	25–30	100–200	125	2·5–4·0	0·005	20	0·13–0·27	
Organic chemicals								
(A)	12	67–130	125	1·4	0·01	16	0·12	cf. rate for same product in Table 6.2
(B)	15	67–130	125	0·4	0·01	10	0·12	
(C)	15	67–130	55	1·0	0·01	66	0·027	Heating medium – hot water
(D)	12	67–130	60	1·3	0·01	108	0·013	

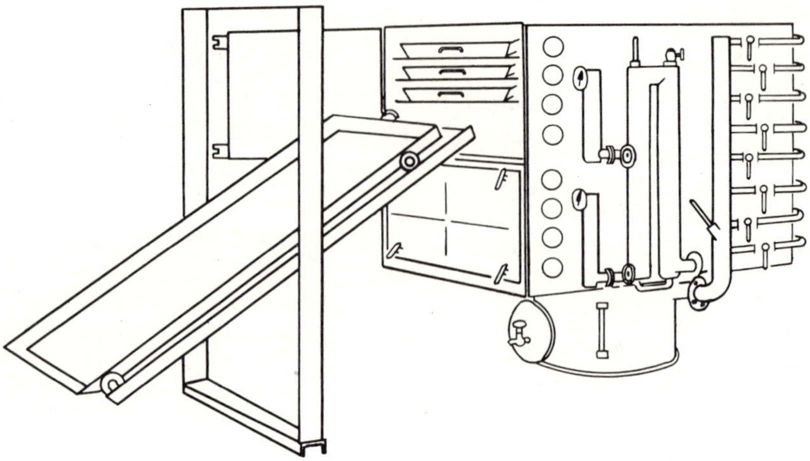

Fig. 6.2. 2 × 4 tray dryers (Type MCVT)

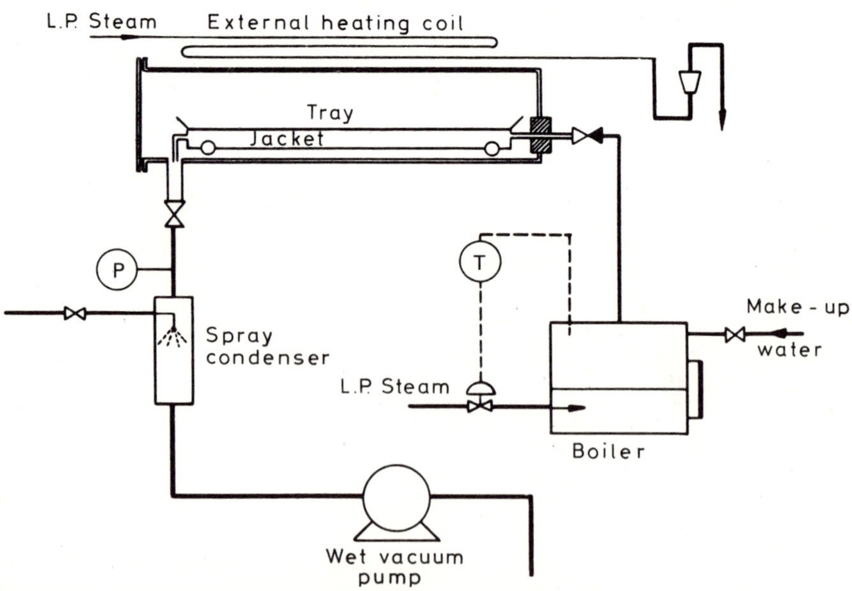

Fig. 6.3. Heating arrangement for Type MCVT dryers

operating temperature is 40°C at 130 mb absolute. The external coil heaters are kept at 10–20°C above the nominal drying temperature.

The average material loading is 25 kg wet paste (50% solids) per m^2 of tray. If a slurry is to be dried, the tray is only half filled to avoid spillage by frothing.

It is frequently necessary to utilise an air bleed into the drying compartment when a dry material of very low moisture content is required since, because of the method of steam heating, the final water content of the product is in equilibrium with the saturation water vapour pressure at the operating temperature. Hence to reduce the equilibrium moisture content still further, the partial water vapour pressure must be reduced by controlled ingress of air.

Points arising from practice are:

(a) It is essential that the dry material should not be left in the drying compartment after releasing vacuum. Residual condensate in the heating jacket may vaporise and produce a significant rise in moisture content of the dried material. It is recommended that the tray jacket should have a drainage well across the base, leading to the normal drainage point, to facilitate complete removal of condensate. Some makers now fit a baffle device in the inlet nozzle of the jacket which improves the steam distribution and reduces the uneven drying effect.

(b) The condenser outlet temperature is normally restricted to a maximum of 30°C and may be controlled automatically if considered essential.

6.5.3 Performance

Operating data for a six-tray stove drying a few materials are given in Table 6.2. Although high evaporative rates approaching 3·3 g water per second per m^2 tray, are quoted for the drying of wet sand, some intermediate chemicals have dried at the low rate of 0·04 g/s m^2. It is probable that crystalline chemical products dry at the rate of 0·13–0·27 g/s m^2 and many finely divided materials at the rate of 0·07–0·14 g/s m^2.

6.6 DESIGN REQUIREMENTS FOR BATCH VACUUM TRAY DRYERS

Primary design requirements are knowledge of optimum drying conditions related to a specific quality of product, and a mass- or heat-transfer coefficient measured at the proposed drying conditions. The maximum temperature to which the material can be heated without decomposition should already be known, or will be found during the laboratory scale tests.

The heat- or mass-transfer coefficients can be measured simply in a laboratory vacuum oven of the hot-plate type. (The jacketed type is less suitable; its use will introduce only small errors in the case of simulation of Type SVT but significant errors could be expected if the results are applied to drying on integrally-heated plates, e.g. in the MCVT type dryer.)

Table 6.2 Typical operating data for Type MCVT vacuum dryer. Note: data for (A) and (B) organic chemicals are also given in Table 6.1

Type of Material	*Loading* (kg *wet paste*/m^2 *of tray*)	*Drying conditions*		*Moisture content (approx.) of paste* (kg *water*/kg *dry solids*)	*Moisture content of product* (kg *water*/kg *dry solids*)	*Total drying time* (h)	*Overall drying rate* (g *water*/m^2 *of tray* s)	*Remarks*
		Absolute pressure (mb)	*Heating plate temp.* (°C)					
Organic chemicals (E)	15	67	50	0·8	0·001	18	0·1 (for all three)	Paste of an amorphous nature
	22	67	50	0·8	0·002	30		
	42	50	60	As above	To constant weight	50		
(A)	27	67	80	0·2	0·002	11	0·12	Fine crystals
	29	67	95	1·4	0·016	12	0·39	Somewhat thixotropic but water separates readily from paste
(B)	20	67	95	0·4	0·001	7	0·26	Similar to (A) but to a lesser extent
(F)	27	67	95	0·2	0·001	8	0·15	Very fine crystals
Inorganic materials: Iron powder (30 mesh)	Not known	(67)	35	0·1	0·004	6	0·23	Data supplied by the dryer manufacturer (for both)
Lithium hydroxide	Not known	(67)	40	0·2	0·02	$15\frac{1}{2}$	0·08	

For simulation of dryers of Type SVT the test trays should be loaded with wet paste to the equivalent of approximately 25 kg wet paste per square metre of tray, i.e. to about 2·5 cm depth. The physical form of the paste should approximate to that expected under normal drying conditions. For integrally-heated tray simulation, the paste can be tried on the hot-plate itself provided that the temperature within the plate can be effectively controlled at the desired value. Again a tray loading of about 25 kg wet paste per m^2 of tray area should be used. The drying of semi-liquid pastes or liquids in the latter case will obviously require modification of the hot-plate to retain the liquid. In both cases it is essential to have not less than 2 tiers of trays so that the effect of radiation from the upper tier is simulated.

The drying curve of moisture content against time is plotted from the results of timed weighings. (The effect of breaking-up the material at some stage in the drying cycle should also be determined to check the possibility of 'case hardening'.) Taking the specification moisture content as the lower limit, the mean drying rate is calculated for drying time by substitution in equation 6.1.

This mean drying rate makes no allowance for outage time for charging paste and discharging dry material. The required correction, calculated from Table 6.3, becomes

$$\frac{\theta}{\theta + \theta_o}$$

where θ = measured drying time and θ_o = time for charging and discharging the stove, and breaking-up the partially dried material at an intermediate stage, if necessary, all in hours. The total required drying area, A_t in m^2 is then:

$$A_t = \theta_t \frac{(M_s - M_f) \times \text{kg/y* dry material}}{28\,800\, R_m} \cdot \frac{\theta}{\theta + \theta_o} \qquad (6.2)$$

From the total drying area the required number of stoves is then calculated according to size and type chosen from Section 6.7.

6.7 CAPITAL COSTS

Tables 6.3 and 6.4 give capital costs of typical dryers.

In comparing the costs of single and multi-compartment dryers particular note should be made of the difference in performance between the two types of stove as outlined in Section 6.5.3.

6.8 OPERATING COSTS

Labour

Labour is primarily required for loading and off-loading the trays and general supervision of the drying operation is very small in comparison. (See Table 6.5.)

*For a year of 8 000 h

Table 6.3 Single compartment dryers (Type SVT) capital costs (1968)

Size 1	*Cost*	*Size 2*	*Cost*
	£		£
Oven			
C.I. body		C.I. body	
14 plates, each about 1·5 × 1·0 m	4 500	15 plates, each about 2·0 × 1·5 m	3 300
Drying area 21 m²		Drying area 45 m²	
Condenser			
Vertical C.I. condenser/receiver	950	Vertical C.I. condenser/receiver	850
Finish			
Body – aluminium sprayed		Body – aluminium sprayed	
Plates and pipes –	1 200	Plates and pipes –	850
Fescolised with 0·15 mm nickel		Fescolised with 0·15 mm nickel	
Total	6 650		5 000

Extra for dry vacuum pump, £1 300 – to serve one or two dryers

Table 6.4 Multi-compartment dryers (Type MCVT) capital costs for 1968

Type	*Approx. drying area* m²	*Capital cost £* *Type* A	*Type* B
Two-tray	31	1 400	1 700
Three-tray	47	1 800	2 300
Four-tray	63	2 200	2 350
Five-tray	79	2 650	3 500
Six-tray	94	3 050	4 000

Type A – Stainless steel trays, door frames and door plates, rest mild steel
Type B – All stainless

Additional cost	£
Water ring vacuum pump	250
or reciprocating pump	450
Water supply tank (0·25 m³).	180

Table 6.5 Average labour requirement for drying dyestuff pastes

Operation	*Labour requirement man*-h/1 000 kg (*wet*) *paste* *Type* SVT		*Type* MCVT	
	Range	*Average*	*Range*	*Average*
Traying up	1·7–3·1	2·4	1·7–3·0	2·4
Detraying	0·6–1·0	0·80	1·2–1·8	1·5
Supervision	–	1·10	–	1·1

Typical charges for stoves are:

Type SVT dryer	900–1 150 kg wet paste
Type MCVT dryer	180–220 kg wet paste

Services

Electricity requirements are 10 kW for the Type SVT dryer and 5 kW for the Type MCVT dryer.

Steam requirements for either stove may vary considerably according to nature of product and final required moisture content. In general, the Type MCVT dryer is more efficient than Type SVT and values of 1·5 kg steam/kg water evaporated for the MCVT and 2 kg steam/kg water evaporated for the Type SVT are probably of the correct order for design purposes.

Maintenance

Maintenance costs for the single compartment dryer are relatively low—approximately 2% of initial capital cost has been quoted. Reliable data are scarce however and $3\frac{1}{2}$% of total capital cost for mild steel/cast iron units and $2\frac{1}{2}$% for stainless steel units are probably more reasonable estimates of annual maintenance cost. Similar values apply for the multi-compartment stove.

Depreciation

Expected life of either type of stove is at least 20 y, hence 5% capital cost per annum should be taken for depreciation.

Approximate drying costs

As the drying data in Tables 6.1 and 6.2 indicate, the cost of drying varies considerably according to product. The average site cost (i.e. less overheads) is probably of the order of £30–60 per 1 000 kg dry product with limits of £15–200 per 1 000 kg.

CHAPTER SEVEN

Batch agitated dryers

7.1 INTRODUCTION

7.1.1 General description

The agitated batch dryer is essentially a heated vessel, either horizontal (cylinder) or vertical (pan), in which the material to be dried is agitated in contact with the heated metal wall. The moisture is removed either at atmospheric pressure or under vacuum. Agitation is provided by a paddle agitator in the vertical type and either a scroll agitator or quadrants and rollers in the horizontal type. The quadrants and rollers provide simultaneous grinding and mixing. Pressure steam is the usual heating medium for drying under atmospheric conditions; for vacuum operation the dryers are heated by pressure steam, sub-atmospheric pressure steam, or hot water. Any of the usual metals can be used as material of construction but in general, cast iron or mild steel have been used for the vertical and mild steel or stainless steel for the cylindrical type.

For operation at atmospheric pressure ancillary equipment may simply consist of a vapour duct and chutes for charging and discharging the material; for operation under vacuum the equipment will include condenser, vacuum pump or ejector, cyclone separator or heated filter, hot water circulation system or temperature controlled steam supply. Effective draughting arrangements must be provided at the charge and discharge points when toxic products are being dried.

7.1.2 General application

The agitated batch dryer can be used to dry most materials, either in paste or slurry form, at a relatively low operating cost since labour requirements are small. For a given batch size it provides, in fact, probably the cheapest form of non-continuous drying.

Generally, this type of dryer is best suited to the drying of materials which are friable and which do not pass through an extremely viscous stage during the drying process. It is particularly suitable for the drying of toxic products where total enclosure is necessary. Also, for toxic products, the horizontal type has the advantage of combined agitation and grinding thus eliminating the potential hazard of the grinding stage which often follows the vertical type. The horizontal dryer will also handle material which tends to 'ball' in the vertical type.

The cleaning of each type of machine is relatively easy. Multi-product operation in single batches is not normally employed because of the hold-up of material (approximately 2% of batch weight) on completion of drying.

7.2 GENERAL THEORY

Drying in agitated batch dryers can be considered as drying by evaporation at or below the boiling-point of the liquid phase. Calculations based on the heat flow from a heated wall to the material can thus be applied. The main resistances to heat transfer are in the material itself and across a film of solids adhering to the wall.

At any given time the heat flux can be expressed by the simple basic equation for rate of drying.

$$\frac{dQ}{d\theta} = 10^{-3} U_t A_h \Delta T \qquad (7.1)$$

where the overall heat-transfer coefficient is

$$U_t = \frac{1}{1/h_{\text{steam}} + 1/h_{\text{wall}} + 1/h_{\text{product}}}$$

During the drying period the heat-transfer coefficient U_t and the temperature difference Δt are both decreasing as the material dries. If U_t and Δt were then functions of the moisture content of the material, and if these values could be determined accurately, the total heat transferred could be determined either graphically, or mathematically assuming a relationship

$$\frac{dQ}{d\theta} = A_h f_1(M) f_2(M) \qquad (7.2)$$

where f_1 and f_2 are functions.

Unfortunately the heat-transfer coefficients and the temperature differences have not yet been determined with the required degree of accuracy. Hence a rigorous method of calculation is not yet practicable.

For normal purposes, including design, the following simplified equation is sufficiently accurate.

$$\frac{Q}{\theta} = 10^{-3} U_c A_h \Delta T_m \qquad (7.3)$$

For the types of material normally dried in an agitated batch dryer, the values of U_c fall between fairly well-defined limits, 10–85 W/m^2 °C, and are roughly proportional to the initial moisture content of the material (Fig. 7.1). They are also independent of the operating pressure. Basic design calculations employing the mean heat-transfer coefficient are given in Sections 7.7. and 7.10.

7.3 DETAILED DESCRIPTION

Agitated batch dryers can be classified into two main groups according to the form of vessel used.

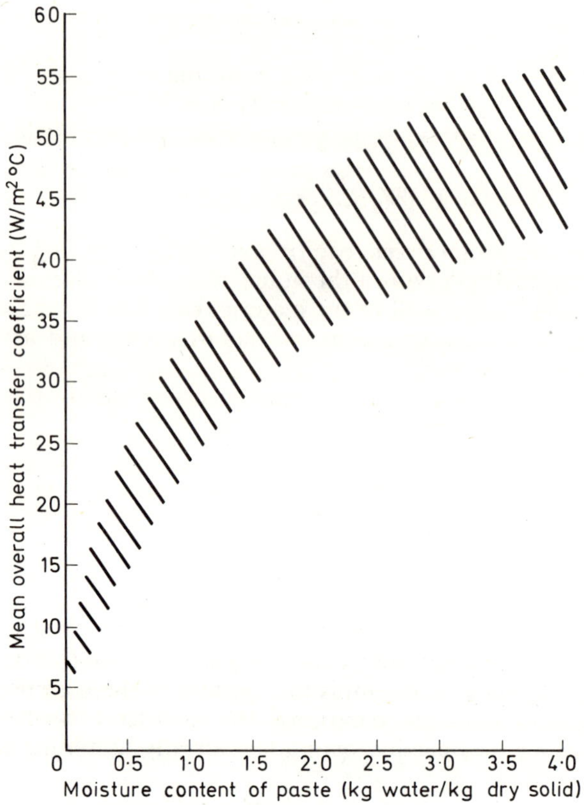

Fig. 7.1. Mean heat-transfer coefficients for the design of vertical and horizontal batch dryers

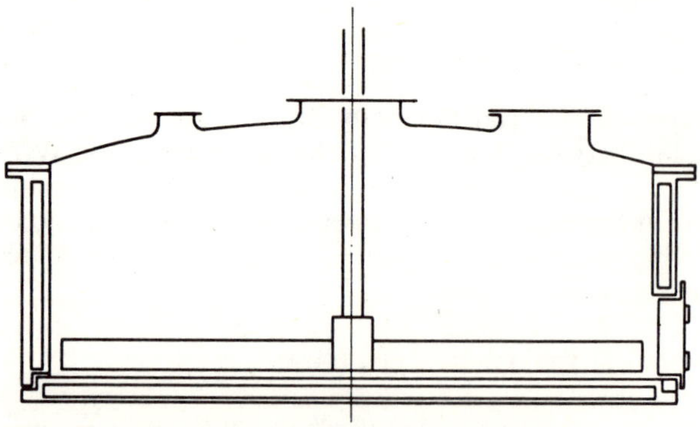

Fig. 7.2. Pan type dryer with vertical paddle—Type PVP

7.3.1 The vertical or pan dryer (Type PVP)

A pan dryer with a vertical paddle, Fig. 7.2 and designated Type PVP in this book is essentially a shallow jacketed pan of 1–3·5 m diameter fitted with a slow moving agitator. The diameter/depth ratio varies between 3 and 6. Operation may be at atmospheric pressure or under vacuum. The domed cover is fitted with a manhole for charging wet material and a hinged door is provided in the side wall for discharging the dry product. The vapour off-take pipe is of substantial dimensions (100–200 mm according to pan diameter) to give low vapour velocities and minimise entrainment loss. For steam pressures greater than 2 bar a cast-in coil replaces the standard steam jacket. A heavy agitator, usually four-bladed, is set to a fine clearance with the flat machined base of the dryer and often carries end-plates to prevent 'caking' on the vessel wall. It is driven via a large diameter shaft and reduction gear at 5–12 rev/min. The blades are set to give a slow sweep of dry product towards the side door when discharging. The base of the vessel is often a separate casting or forging, particularly when heated by a cast-in coil. Materials of construction are cast iron, mild steel or stainless steel.

7.3.2 The horizontal dryer

This dryer consists of a jacketed horizontal cylinder of length/diameter ratio $2\frac{1}{2}$–5. In one Type HCRP, Fig. 7.3, agitation is provided by radial paddle arms mounted on a central spindle which may also be hollow to provide additional heating surface. The radial arms are set in quadrants, each quadrant carrying a tube or bar acting as a grinding weight. The quadrants are

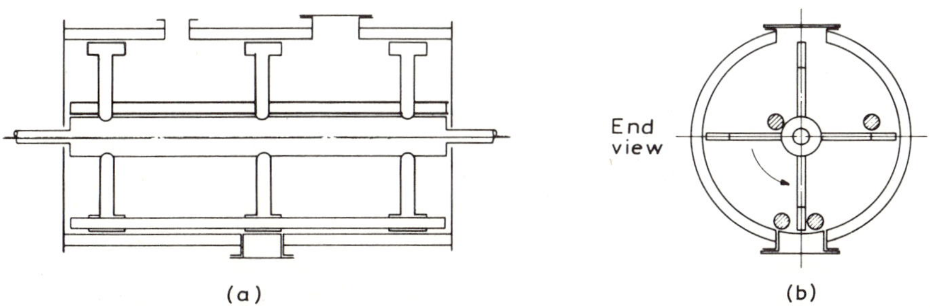

Fig. 7.3. Horizontal cylinder dryer with radial paddles—Type HCRP

arranged so that the tips of the radial arms effectively sweep the entire heat transfer surface of the dryer. This type of agitator may be driven through a reduction gear directly, or through a periodic reversal gear, so that uniform mixing is obtained.

Another horizontal cylinder design is agitated by scroll blades with opposed flights. This is shown in Fig. 7.4 and designated Type HCSB. The wet paste is charged through a manhole on the top of the horizontal cylinder and the dry product discharged through a 'top-hat' discharge door in the base. Vapours pass through a catch-vessel (often fitted with a scraper to return entrained powder to the dryer), through a water scrubber or (heated) bag filter, and to the vacuum pump. The dished ends of the dryer carry the agitator bearings and stuffing boxes, and occasionally the reduction gear and driving motor.

7.4 METHODS OF OPERATION

These dryers are normally operated as individual units. For multi-product manufacture the machine is operated campaignwise, i.e. several batches of one product are dried consecutively, to minimise loss of material by cleaning between batches. Operation at atmospheric pressure or under vacuum depends largely on the heat sensitivity of the product. With certain materials it is essential to dry initially at atmospheric pressure, and subsequently under vacuum to eliminate frothing. The filling ratio of the cylindrical dryer

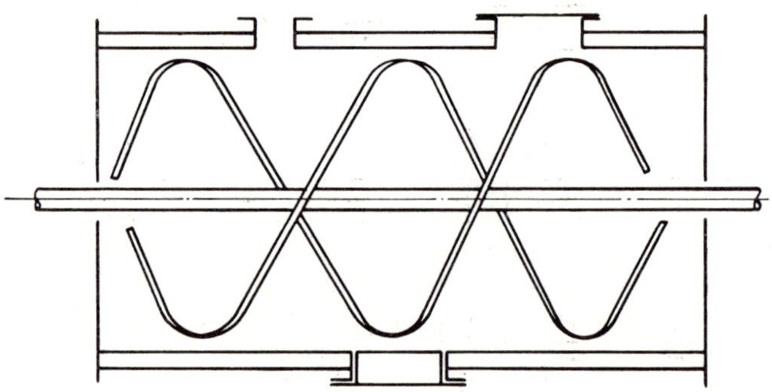

Fig. 7.4. Horizontal cylinder dryer with scroll blade agitator—Type HCSB

varies from 0·3 for heavy materials to 0·8 for thin slurries but generally the machine should be about half full. The vertical dryer is usually operated at the slightly lower filling ratio of 0·3–0·4.

For drying under reduced pressure, an absolute pressure in the range 130–300 mb, obtained by a single stage pump, is usually sufficient. A velocity limiting orifice may be fitted in the vapour off-take to minimise frothing when drying thin pastes or slurries under vacuum. Where pumps are used to maintain the vacuum it is usual practice to fit a water scrubber, discharging via a barometric leg, in the vapour line to the pump. Entrained solids are recovered from the barometric discharge by filtration. Alternatively, dust bags can be fitted in the vapour lines but unless effectively heated they tend to blind quickly from intermittent condensation and drying out.

Drying temperatures vary between 50°C and 110°C with a temperature difference of 30–60 degC between heating medium and batch. The drying time is usually based on experience; or on (a) observation of the steam flow to the jacket (this falls to minimum value equal to heat losses only), (b) observation of the temperature of the outlet water of the wet scrubber where this is fitted in the vapour line. When the final moisture content is critical the batch is sampled at intervals during the drying cycle.

The degree of grinding is not controllable in the vertical and scroll type dryers hence there is a marked tendency for 'balling' of the product; lumpy material may have to be treated in a separate grinding stage if a uniform product is required. The physical form of the dry product can however be fairly closely controlled in the horizontal paddle dryer (Type HCRP) by varying the number and weight of the grinding bars. Mild steel tubes are normally used, also aluminium bar or tube for coarse grinding, as the grinding load or shock produced by solid mild steel bars may be sufficient to cause measurable distortion of the shell. A secondary function of the bars is to prevent material sticking to the radial agitator arms and rotating

Table 7.1 Typical drying schedule for an organic chemical paste

(a) Drying conditions	
Dryer Type HCRP	5 500 mm long × 1 200 mm diameter
Drying cycle	18 h
Batch temperature	50–60°C
Jacket temperature	80–90°C
Vacuum in dryer, absolute pressure	130–250 mb
Weight of wet material	1 800 kg
Initial moisture content of material	Approx. 30% (wet basis)
Final moisture content of material	<1%
(b) Drying sequence	
Charge paste to dryer	1 h
Seal and evacuate dryer	$\frac{1}{2}$ h
Preliminary drying period	15 h
Sample product and determine moisture content	1 h
Complete drying cycle if necessary	—
Discharge dry product	1 h
Total	$18\frac{1}{2}$ h

en masse, in which case heat transfer from the dryer wall to the material becomes almost negligible.

The sequence of drying operations in each type of machine is very similar and a typical schedule for the cylindrical dryer is given in Table 7.1.

7.5 PERFORMANCE DATA

Whilst the performance of dryers must necessarily depend on the nature of the material being dried, the available data fall within relatively narrow limits, possibly because of the similarity of the materials dried in this type of dryer. Typical drying data are given in Table 7.2.

For horizontal dryers the output rates of dry solids (<1% moisture) may vary between 8–32 kg/hm^3 (based on total capacity) according to initial moisture content of paste, but generally 10–15 kg/h m^3 (2·8–4·2 g/s m^3)

Table 7.2 Typical drying data for agitated batch dyers

Material	*Type of dryer*	*Size of dryer* (mm) *Lgth.*	*Size of dryer* (mm) *Diam.*	*Driving motor* (hp)	*Wet charge* (kg)	*Filling ratio*	*Initial moisture content* % (*wet basis*)	*Absolute pressure in dryer* (mb)	*Jacket temp.* (°C)	*Drying time* (h)	*Mean overall coeff.* U_c (W/m² °C)
Organic paste	HCRP	5 500	1 200		4 000	0·36	30	200	80	15	35
Different fine aromatic organic compound crystals	HCRP	3 800	1 350	15	2 260	0·2	68	265	125	6	45
	HCRP	3 800	1 350	15	4 660	0·4	75	265	125	8	60
	HCRP	5 500	1 200		2 100	0·2	6	200	125	4	25
Anthracene (water & pyridine)	HCRP	8 900	1 800	35	37 000	0·72	76	665–1 000	170	16	75
Dyestuff paste	HCSB	2 750	1 200	10	2 000	0·3	70	265	105	14	30
Different organic pastes	PVP		1 800	15	1 080	0·4	41	1 000	125	32	35
	PVP		2 450	25	800	0·4	35	665	125	7½	25
Different dyestuff pastes	PVP		1 800	15	1 035	0·4	61	1 000	125	11	135
	PVP		2 450	20–30	2 400	0·7	64	470	125	12	115

is obtainable. For vertical dryers the output rates of dry solids are of the same order.

In some cases 'de-dusting' of the product is combined with the drying operation by adding mineral oil to the material in the dryer. This can have the effect of slightly reducing the drying rate and it is better practice to spray the oil into the machine after drying is complete, i.e. immediately before discharge. In contrast, an increase in drying rate following the addition of a surface active agent to the batch has been reported.

7.6 MECHANICAL PERFORMANCE

Because of the relatively simple design of vertical and horizontal dryers few difficulties are normally associated with their operation.

An inherent weakness of the vertical dryer Type PVR, has been metal failure by cracking in the vicinity of the charge door or where there is a sharp change in metal section. Careful attention to design detail can greatly alleviate this problem.

Mechanical weaknesses found in large horizontal dryers have usually been ascribed to insufficiently careful design of bearings and gears for 24 h/day operation, and also to inadequate allowance for expansion during the heating up period. The importance of running the agitator whilst heating or cooling in order to prevent axial scoring of the shaft has on occasion been forgotten by operators. Distortion of the grinding bars frequently occurs in Type HCRP dryers; also loss of the end-caps if these are not securely fitted to the tubes.

Frequent attention should be given to the stuffing boxes carrying the agitator shaft; excessive air leakage at these points increases dust loss during the final drying stages, throwing an extra load on the dust trap or wet scrubber when these are fitted, or lowering the yield when they are absent.

7.7 DESIGN BASIS

7.7.1 Assessment of material

The drying properties of the material may be assessed by analogy, by laboratory test, or by plant trial. Laboratory apparatus of the type shown in Fig. 7.5 gives valuable preliminary information for horizontal dryers and similar means have been used for vertical dryers. When greater accuracy is required, however, a semi-technical or plant trial is essential. Manufacturers of dryers are usually able to conduct trials in their works.

7.7.2 Choice of dryer

The type of dryer will depend to some extent on the preferred physical form of final product but there is now a definite general preference for the Type

HCRP dryer. On economic and mechanical grounds the horizontal machine shows marked advantages over the vertical type dryer and unless there is an overriding reason for choosing the vertical, the design should be based on Type HCRP.

7.7.3 Size of dryer

Horizontal

The size of dryer is a compromise between reasonable capacity and the required heat transfer area. In the absence of semi-technical plant experimental data the machine may be sized as follows:

(a) Determine the rate of supply of wet material by weight and volume and the maximum permissible drying time to fit in with other plant

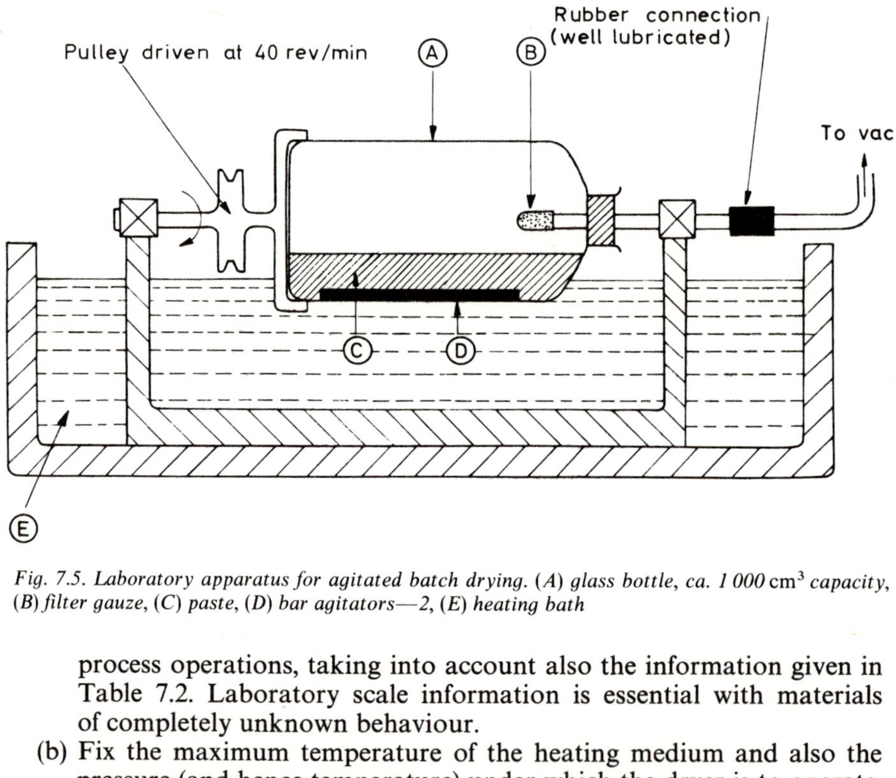

Fig. 7.5. Laboratory apparatus for agitated batch drying. (A) glass bottle, ca. 1 000 cm^3 *capacity, (B) filter gauze, (C) paste, (D) bar agitators—2, (E) heating bath*

process operations, taking into account also the information given in Table 7.2. Laboratory scale information is essential with materials of completely unknown behaviour.

(b) Fix the maximum temperature of the heating medium and also the pressure (and hence temperature) under which the dryer is to operate, giving ΔT_m.

(c) Determine the mean rate of heat flow by dividing the heat equivalent of water evaporated plus sensible heat in the dryer by the time available for heating up and drying the batch.

(d) Read the mean overall heat-transfer coefficient from Fig. 7.1. From this and the amount of water to be evaporated, calculate the average heat-transfer rate required for the permissible drying time.

(e) The volume of the dryer should be 1·3–1·7 times the volume of each batch of material to be dried, i.e. the filling ratio should normally be between 0·3 and 0·6. In addition, the length of the dryer should be about four times its diameter. Using these criteria it is possible to choose the cheapest standard size of dryer from plant manufacturers' lists. If the filling ratio is a little less than 0·3 in order to obtain the required heating area, it is desirable to use a heated hollow shaft of suitable diameter for the paddle in order to increase the heating area and to heat the falling particles directly and the water vapour in the body of the machine. Standard dimensions are given in Table 7.3.

Table 7.3 Approximate capital costs (1968) of agitated horizontal batch dryers.

Type	*Dimensions* (mm)		*Approx. total capacity* (m^3)	*Cost in mild steel* (£/m^3)
	Dia.	*Length*		
HCRP	1 070	2 440	2·0	1 480
HCRP	1 220	3 050	3·5	740
HCRP	1 370	3 720	5·5	740
HCRP	1 370	5 800	8·6	640
HCSB	1 680	5 000	11·2	600
HCRP	1 830	8 850	23·2	390

The cost in stainless steel is about 1·7 times greater per m^3.

Table 7.4 Approximate capital costs (1968) of agitated vertical batch dryers, Type PVP

Type	*Dimensions* (mm)		*Approx. total capacity* (m^3)	*Cost in mild steel* (£/m^3)
	Dia.	*Depth*		
PVP	1 220	382	4·3	2 450
PVP	2 440	457	2·1	1 200
PVP	2 440	610	2·8	1 000
PVP	3 050	457	3·3	1 300
PVP	2 440	915	4·3	800

The cost in cast iron is about 1·3 times, and in stainless steel about 1·6 times greater per m^3.

(f) Fix the agitator speed at 5 rev/min. To obtain the size of the driving motor, use the higher of the figures (i) 7 hp per 1 000 kg of wet charge and (ii) hp = 2·3 length × diameter in metres.

Vertical

The basis of design is essentially the same as that used for horizontal dryers. The required heat-transfer area is calculated from an estimated mean heat-transfer coefficient and a known temperature difference. The required volumetric capacity is calculated assuming a 40% filling ratio for wet paste. A size of dryer than chosen to satisfy heat-transfer and volumetric requirements. Standard dimensions of vertical dryers are given in Table 7.4. The limiting size is about 3·5 m diameter.

The driving motors fitted to vertical dryers have in the past varied considerably for given sizes of dryer; a very approximate estimate of power requirements is given by 11 kW/1 000 kg wet paste. (If the material has physical properties which suggest that heavy motor loads will be experienced at some stage during the drying, a plant test (by the maker) with a reduced charge is advisable before finalising the design).

7.8 CAPITAL COSTS

Approximate capital costs (1968 prices) for various sizes of agitated vacuum dryers are given in Tables 7.3 and 7.4. These are for mild steel construction and include gear and agitator. Costs vary according to manufacture and design details and hence the figures given in these tables should be for guidance only.

7.9 RUNNING COSTS

For a given batch size, drying in an agitated batch dryer is probably the cheapest form of non-continuous drying. The works cost, excluding capital charges, is around £15–30/1 000 kg dry weight (1968 costs).

7.9.1 Labour

Labour costs are low, particularly if the wet material is charged mechanically to the dryer and the dry product discharged directly to a storage hopper.

For estimation purposes, the values of Table 7.5 should be sufficiently accurate.

7.9.2 Services

Electricity requirements are best estimated on the basis of 6·5 hp (≡6·5 kW allowing for motor efficiency) per 1 000 kg wet paste. Steam requirements

Table 7.5 Labour requirements in batch drying

Charging wet material	(a) 0·55 man-h (manual) (b) 0·2–0·35 man-h (mechanical)	per 1 000 kg wet weight
Overlooking of drying	0·05–0·15 man-h per h	
Discharging dry material	(a) 0·55 man-h (manual) (b) 0·22 man-h (mechanical)	per 1 000 kg dry weight

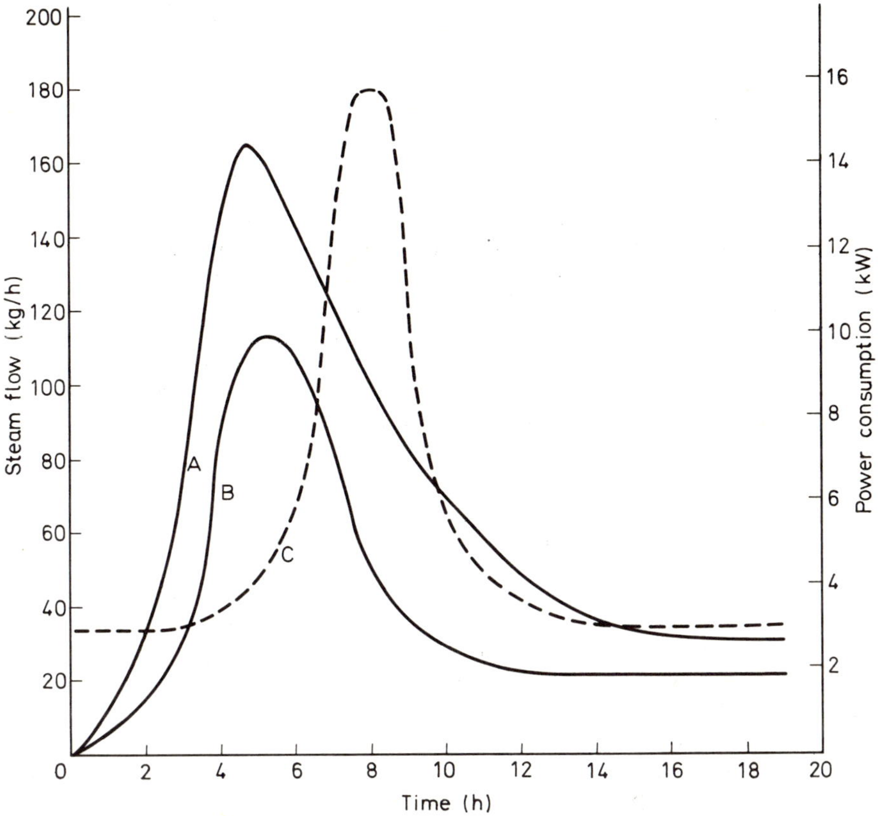

Fig. 7.6. Typical steam and power consumption curves for agitated batch dryers. (A) steam consumption for 1·2 × 5·5 m dryer, Type HCRP, (B) steam consumption for 3 m diameter dryer, Type PVP, (C) power consumption for 1·35 × 3·8 m dryer, Type HCRP

(or the equivalent of hot water) may vary between 1·5 and 3·0 kg of factory L.P. steam/kg water evaporated depending on the initial and final moisture contents of the product. 2·5 kg steam/kg water evaporated is a safe assumption for costing purposes. Additional service requirements for ancillary plant such as vacuum pumps, ejectors etc. are usually calculable with moderate accuracy.

Typical steam and power consumption curves for agitated batch dryers are shown in Fig. 7.6.

7.9.3 Maintenance

Maintenance costs vary according to type of dryer and nature of product. Assuming that ordinary maintenance is applicable, and that abnormalities are covered by a special maintenance fund, $7\frac{1}{2}\%$ of the capital cost may be taken as average annual maintenance.

7.10 WORKED EXAMPLE OF CAPITAL AND PLANT OPERATING COST

Problem

Estimate the capital cost and the plant operating costs for the drying requirement of an organic chemical paste which laboratory experiments indicate can be suitably dried in a Type HCRP steam jacketed horizontal cylindrical vacuum dryer given:

Output of dry product (1 % moisture)	250 t/y
Moisture content of wet material, % by weight	20 %
Weight of each batch wet material	wet 1 500 kg (dry basis 1 200 kg)
Maximum temperature to which material may be heated	90°C
Normal occupation time of dryer	42 weeks/y = 7 000 h/y

Time cycle per batch: With a dryer availability of 42 weeks/y, the output of dry product required per week = 6 000 kg. The batch size is 1 200 kg (dry basis). Hence the number of batches per week = 5.

The maximum permissible time cycle is therefore 168/5 h = 34 h/batch. In the calculation of dryer size, the time allowed for actual drying is taken below as 24 h/batch, even though the time taken for loading, heating-up and discharging will be far less than 10 h. Some allowance must be made for occasional poorly filtered wetter batches and for maintenance.

Heat-transfer area: From Fig. 7.1, the mean heat-transfer coefficient for a paste containing 20 % water (i.e. 0·25 kg water)/kg dry solid is 25·5 W/m^2°C. At a vacuum corresponding to an absolute pressure of 165 mb, the batch temperature will be 55°C and for a jacket temperature controlled to below 90°C, the mean temperature difference will be 35°C.

The total weight of water to be evaporated per batch in 24 h = 300 kg.

From Callendar's Steam Tables, the latent heat of evaporation of water at 55°C $= 2\,380$ kJ/kg

Heat required to evaporate the water $= 300 \times 2\,380 = 715$ MJ

Average heat-transfer rate per second required for evaporation $= \dfrac{715 \times 10^6 \text{ J}}{24 \times 3\,600} = 8.3$ kW

Hence heat-transfer area required with a temperature difference of 35°C $= \dfrac{8{\cdot}3 \times 10^3}{25{\cdot}5 \times 35} = 9{\cdot}3 \text{ m}^2$

It is to be noted that the total time per batch will be increased above 24 h by the time taken to heat up the charge and dryer after loading.

Dimensions of dryer

The maximum filling ratio of the dryer is 0·5. Hence for 1 500 kg of wet paste, the volume of the dryer must exceed 3·0 m^3. The length of the dryer should be about four times its diameter. A dryer of internal diameter = 1 m, and length 4 m will have a heating surface of 12·5 m^2 and a volume of 3·1 m^3; hence 3·5 m^3 will be a suitable standard size. The heating surface will be slightly greater if the paddle driving shaft is hollow and also supplied with steam at the pressure required to avoid heating the product above 90°C.

Capital cost: The capital cost of a Type HCRP dryer in stainless steel will be about £1 250/m^3 capacity (Table 7.3). Hence:

	£
Purchase price of dryer	4 500
Ancillary plant	2 500
Installation, including connection of services	2 000
Total capital cost	9 000

Operating cost: Labour. From Table 7.5:

	Man-h/batch
Charging 1 500 kg wet material	0·85
Discharging 1 200 kg dry material	0·65
Overlooking of heating up and drying operation during 25 h per batch	2·50
Total	5·00

Cost of shift process labour at £0·7/h = £3·5 per batch
= £2·92 per 1 000 kg (dry)

Process management: The process overhead charge for chargehand, foreman and staff is about 25% = £0·73 per 1 000 kg

Total charge = £3·65 per 1 000 kg (dry)

Steam: The water evaporated is 300 kg/batch = 12·5 kg/h

The steam consumption is 2·5 kg/kg water evaporated
= 31·2 kg/h

Adding 25% for heating-up and radiation loss and for mains losses, then total steam consumption from factory mains at about 1·5 bar (gauge)
= 39 kg/h

For a 25 h cycle, steam consumption = 975 kg/batch
= 810 kg/1 000 kg (dry)

With a works cost of £1·0/1 000 kg steam, the cost of steam
= £0·8/1 000 kg (dry)

Electricity:

Horse power of driving motor at 6·5 hp/100 kg wet paste
= 10

hp of ancillaries including vacuum pump = 10

Total 20

Assuming 80% motor efficiency and 5% losses between factory meter and plant, equivalent to 1 kWh per hph the electricity usage for 25 h per batch
= 420 kwh/1 000 kg (dry)

Hence cost at £6/MWh = £2·52/1 000 kg (dry)

Maintenance: This is about $7\frac{1}{2}\%$ of the capital cost of dryer and ancillaries
= £525/y for 250 tonnes (dry)
= £2·1/1 000 kg (dry)

Depreciation at 15% of £9 000 = £4·2/1 000 kg (dry)

Sundry direct expenses: These include analysis control, consumable stores and haulage and will vary considerably with product and location. For the purposes of this calculation of drying costs they will be taken as £2 per batch
= £1·67/1 000 kg (dry)

Total works drying cost: This is conveniently expressed as in Table 7.6.

This cost does not include general works and company overheads, which generally are allocated to the final finished packed product ready for sale.

Table 7.6

	£/1 000 kg *dried product*	%
Labour	2·92	20
Supervision	0·73	5
Steam	0·81	5
Electricity	2·52	17
Maintenance	2·10	14
Sundry direct expenses	1·67	11
Depreciation	4·20	28
Total	14·95	100

CHAPTER EIGHT

Continuous band and tray dryers

8.1 TYPES OF DRYER DESCRIBED

This Chapter deals with dryers in which the material is handled on a continuous band or tray conveyor. The conveyor is usually horizontal and may take the form of a perforated band on which particulate material (e.g. preforms) is placed and dried by blowing hot gas through the material and band (band through-circulation dryer); alternatively it may be a solid band on which the material is dried by blowing hot gas over or down on to it (convection solid-band dryer); again the material can be heated by conduction by passing the solid band over a heated surface and in this case the apparatus may be operated under vacuum (conduction-type solid-band dryer); finally the band can be replaced by a series of trays but the machine otherwise operated in accord with any of the above principles (continuous tray dryer).

The special case of the conveying screen dryer associated with filtration draining of a coarse granular product is dealt with in Chapter 14 (Miscellaneous Continuous Dryers). Dryers for handling sheet material may be very similar in construction to continuous band dryers; one type (i.e. multicylinder dryers) is described in Chapter 9 (Continuous Drum or Film Dryers). Radiation can be used to assist any continuous band dryer but the special case of radiation dryers is dealt with in Chapter 15 (Specialised Techniques).

Consequently there are many types of machines of differing construction and operating principles. There is however an essential common feature in that the material is moved through the dryer on a relatively slow moving conveyor so that it is otherwise static. These machines have thus a gentle action on the material. The detailed treatment in this Chapter is limited to those types of continuous band and tray dryers which are considered particularly applicable to the chemical industry. These comprise:

(a) *The band through-circulation dryer* in which the wet material in particulate form is fed on to a perforate band on which it is conveyed through a drying chamber where hot air is blown through the material and the band. A preformer is usually necessary to convert the wet material into the required particulate form, though in some cases the feed may be available in the right size.

(b) *The convection solid-band dryer* in which the conveyor band is solid and the hot air (or drying gas) is blown over the material or in some cases down on to it with spillage sideways.

(c) *The conduction solid-band dryer* in which the solid band is passed over a hot surface (usually a steam chest) so that it is heated by conduction through the band. Operation is normally under vacuum since at atmospheric pressure the convection solid-band dryer would be more convenient and efficient.

(d) *Continuous tray dryers* which can follow the principles of any of the band dryers ((a), (b), or (c)) with substitution of trays carried on some form of chain or slat conveyor. In practice the use of trays offers possibilities of compact (and therefore cheap) construction for through-circulation drying. They are also used for convection drying with solid trays in the form of *continuous tunnel dryers* in which stacks of trays are conveyed through a drying chamber.

8.2 THEORY

This section is limited to the direct application of Chapter 2 to the various types of continuous band and tray dryer.

8.2.1 Through-circulation drying

Drying takes place by blowing hot gas (usually air) through a thin bed of particulate material which is itself moving through the dryer. The bed is not usually more than 5 cm deep. The material is stationary with respect to the band and the system is completely analogous to air blowing through a stationary bed of material. Data on any particular material can thus be obtained quite simply in a small batch type experiment operated under the appropriate conditions. The effect of recycling the drying gas can be assessed by means of a trial-and-error heat-and-matter transfer balance.

The drying conditions are those for heat transfer by convection involving both constant- and falling-rate drying as described in Section 2.3 of Chapter 2. Continuous band through-circulation dryers are almost invariably used on the air/water system and the simplification for that system can be applied (Section 2.3.8). The special case of drying from solutions is dealt with in Section 2.3.7.

For convection drying the basic equation is the instantaneous heat and matter transfer equilibrium (equation 2.6) and in the constant rate period the drying rate is given by simple integration in terms of the wet-bulb temperature (equation 2.27).

In the case of a through-circulation dryer however the effective surface area of the granules is not clearly defined and it is more convenient to use the cross-sectional area of the bed (A'). The humidity driving force is more convenient to use than the partial pressure driving force and, since the inlet and exit air conditions are different, a logarithmic mean driving force should be used (for the air/water system calculation is simplified by the fact that the saturation humidity (H_w) does not change—cf. Chapter 2). The constant-rate drying equation 2.27 is therefore conveniently modified to

$$R'_c = \frac{10^3(M_s - M_c)W}{A^1\theta} = K'_H(H_{sw} - H_g)_{\text{lm}} = K'_H(\Delta H)_{\text{lm}} \qquad (8.1)$$

Now W/A' represents the loading of the dryer as kg dry weight/m^2 and takes into account the depth of the bed with a given material. Furthermore in a continuous dryer with a single band this loading is the same at any point in the dryer. It is thus convenient to write

$$R'_c = \frac{10^3(M_s - M_c)L}{\theta_c} \tag{8.2}$$

For design purposes it is required to calculate the dryer area A_c in terms of the required throughput of dry material.

$$\theta = 10^3 LA'/W_d$$

$$A_c = \frac{(M_s - M_c)W_d}{R'_c} \tag{8.3}$$

Whilst these equations are sufficient to describe the process, it is useful to consider the significance of the coefficient based on the cross-sectional area of the dryer (K'_H) for ease of application of the theory to the design of a plant from laboratory data as described in Appendix 8.2. From equations 2.27 and 8.1 we have

$$R_c = \frac{W_c}{A} = K_H(\Delta H)_{\text{lm}}$$

$$R'_c = \frac{W_c}{A'} = \frac{W_c}{A_c} = K'_H(\Delta H)_{\text{lm}}$$

or

$$R'_c = K_H(\Delta H)_{\text{lm}} \cdot \frac{A}{A_c} = K_H(\Delta H)_{\text{lm}} \cdot \frac{AL}{W} = K'_H(\Delta H)_{\text{lm}}$$

The coefficients based on the area of the material and the area of the dryer are thus related as

$$\frac{K'_H}{L} = \frac{K_H A}{W} \tag{8.4}$$

Now with preformed rods of material of a given shape, the surface area is proportional to the initial weight of material and similarly the weight of the dry material per unit cross-sectional area of the drying bed is proportional to the depth of the bed. Hence the loading or weight of dry material per unit area of bed should completely specify the drying area available. This is implicit in the statement of equation 8.4 and has been confirmed by the experimental observation that the constant rate per unit cross-sectional area of bed is almost directly proportional to the depth of bed (see Fig. 8.19, Appendix 8.2). Use may therefore be made of the expression K'/L as a derived constant independent of the depth of the bed and consequently equation 8.1 can be conveniently rewritten in the form

$$R'_c = \left(\frac{K'_H}{L}\right) . L(\Delta H)_{\text{lm}} \tag{8.5}$$

It is implicit in the use of equation 8.5 that the conditions are such that the drying air is not saturated before it leaves the bed. Examples of the value of K'_H/L in equation 8.5 for a few conditions of bed loading and humidity of approach air are given in Table 8.6 of Appendix 8.2. It will be seen that K'_H/L is in fact reasonably constant and independent of the humidity and bed depth, thus confirming the validity of equation 8.5 over the range of conditions covered (see also References 2, 4, 5).

These equations 8.1–8.5 define the constant drying rate in terms of the

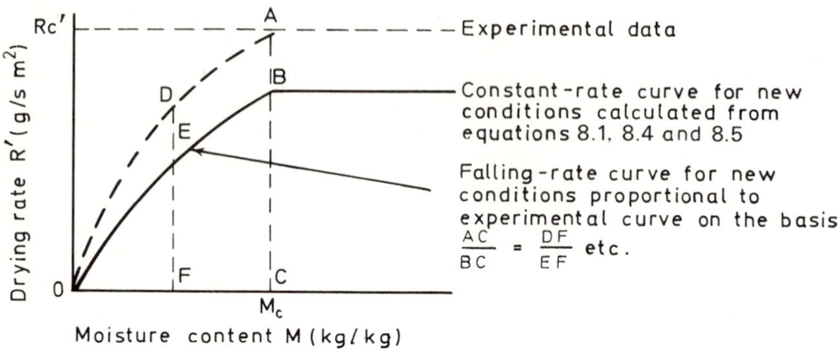

Fig. 8.1. Calculation from experimental data of constant drying rate for required design conditions

drying air conditions (humidity and temperature) for a given bed depth or loading of a given pre-formed or granular material at a given air velocity through the bed.

The effect of rod diameter on the constant drying rate has been determined experimentally for preformed rods of 1·5–4·5 mm diameter[2–5] as

$$R_c = at^{-0{\cdot}59}(H_{sw} - H_g)_{\mathrm{lm}} = at^{-0{\cdot}59}(\Delta H)_{\mathrm{lm}} \tag{8.6}$$

A few experimental figures from a laboratory dryer are shown in Appendix 8.2, Fig. 8.20.

The effect of air velocity on the constant drying rate has also been investigated over the range 900–2 000 g/s m^2 as[2–5]

$$R_c = bG^{0{\cdot}89}(H_{sw} - H_g)_{\mathrm{lm}} = bG^{0{\cdot}89}(\Delta H)_{\mathrm{lm}} \tag{8.7}$$

Experimental figures from a laboratory dryer are shown in Appendix 8.2, Fig. 8.21.

The power term 0·89 is given as 0·81 in Reference 3 but the difference is within experimental error over the range covered. Further information on air flow through granular beds is given in References 7 and 10; and on heat and mass transfer in References 6, 8, 9, 11 and 12.

Equations 8.1–8.7 can be used to transpose experimental data for constant rate drying on a given material to other conditions within the limits specified. The equations cannot, of course, be used for transposing data between different materials. The constant-rate drying time and area of dryer required can then be calculated.

Analysis of the falling-rate period can be done rigorously by calculating the change in drying coefficient with moisture content and then recalculating the drying rate for the new conditions as described in Chapter 2, Section

2.3.9. In the case of continuous through-circulation dryers with their relatively short drying time however, a simpler method of direct graphical analysis of the experimental falling-rate drying curve is quite adequate. It is described in detail in the worked example given in Appendix 8.2.

The first step is to calculate the constant drying rate for the required conditions from experimental data on the given material. The new falling-rate drying time is then proportioned to the experimental curve as shown in Fig. 8.1.

The falling rate drying curve itself can then be analysed by splitting it into a series of equivalent straight lines through the origin. Since the drying rate at any moisture content is some function of the moisture content:

$$\frac{dW}{d\theta}\cdot\frac{1}{A} = f(M)$$

Integrating over the falling rate period between the critical and final moisture contents, we have

$$\int_{\theta=0}^{\theta=f} d\theta = \theta_f = \frac{W}{A_f}\int_{M_f}^{M_c}\frac{dM}{f(M)}$$

Taking the simplest case where the falling-rate drying curve is a straight line from the origin to the critical moisture content as in Fig. 8.2, we have

$$f(M) = \frac{MR'_c}{M_c}$$

$$\theta_f = \frac{10^3 W M_c}{A_f R'_c}\int_{M_f}^{M_c}\frac{dM}{M} = \frac{10^3 W M_c}{A_f R'_c}\,.\ln\frac{M_c}{M_f} \tag{8.8}$$

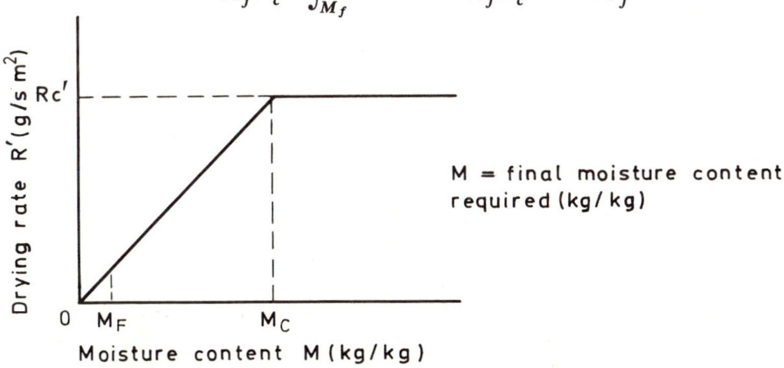

Fig. 8.2. A step in the analysis of a falling-rate drying curve

Rearranging in the form of equation 8.3

$$A_f = W_d\,.\frac{M_c}{R'_c}\ln\frac{M_c}{M_f} \tag{8.9}$$

For the mean falling rate R_f we have

$$R_f = \frac{10^3 W}{\theta_f A'} = \frac{R_c}{M_c \ln(M_c/M_f)} \tag{8.10}$$

It is thus possible in this idealised straight line case to calculate the mean falling rate from the constant drying rate and the critical and final moisture contents. An actual falling-rate drying curve can similarly be analysed as a series of equivalent straight lines as shown in Fig. 8.3.

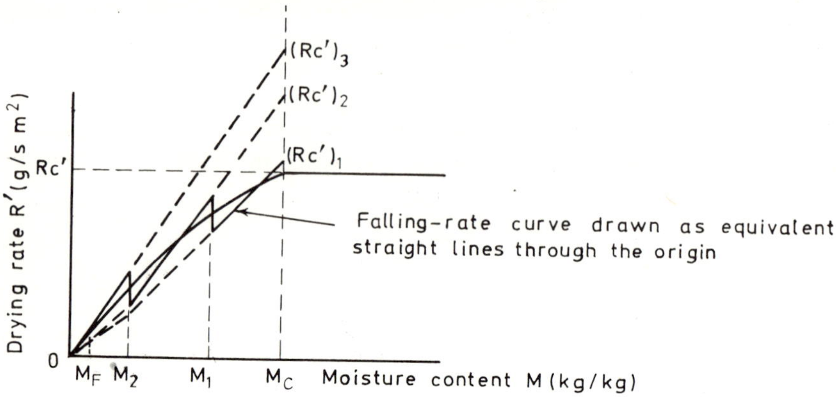

Fig. 8.3. Analysis of an experimental falling-rate drying curve

The total falling-rate drying area in this case given by

$$A_f = 2{\cdot}303\, M_c W_d \left(\frac{1}{R'_{c1}} \log \frac{M_c}{M_1} + \frac{1}{R'_{c2}} \log \frac{M_1}{M_2} + \frac{1}{R'_{c3}} \log \frac{M_2}{M_f} \right) \quad (8.11)$$

In the complex falling-rate curve shown in Fig. 8.3 and in more detail in Appendix 8.2, Fig. 8.23, the time taken to go from M_s to M_f on Fig. 8.23 under the appropriate drying conditions may be approximated by drawing a tangent to the curve. This line passes through the origin and is projected back until it crosses the vertical line through M_c at R_{c2}. Then

$$\theta_{M_3 - M_f} = \frac{10^3 W M_c}{A' R'_{c2}} \ln \frac{M_3}{M_f} \quad (8.12)$$

More construction lines are, of course, necessary than those shown on the small diagram Fig. 8.23.

For continuous dryers where a single band is moving at constant speed through the dryer, the ratio $W/\theta_{M_3 - M_f}$ in equation 8.12 refers to the throughput of dry material which is constant at all points in the dryer. Equation 8.12 may, therefore, be written in the more convenient form for calculation as

$$A'_{M_3 - M_f} = \frac{W_d M_c}{} \ln \frac{M_3}{M_f} \quad (8.13)$$

Where $A'_{M_3 - M_f}$ = area of dryer to reduce the moisture content from M_3 to M_f.

8.2.2 Convection solid-band drying

In this case drying takes place by blowing hot gas over the surface of the wet material supported on a solid band. Again the material is usually stationary with respect to the band so that the system is analogous to air blowing over a stationary layer of material (cf. Chapter 5).

Drying is by convection and the basic instantaneous heat- and matter-transfer equation holds. The area of the drying surface is in general the same as the area of the drying band so that the working equation for the constant-rate period is directly in the form of equation 2.27

$$R_c = \frac{10^3(M_s - M_c)W}{A_c\theta_c} = K_H(H_{sw} - H_g)_{\text{lm}} = K_H(\Delta H)_{\text{lm}} \qquad (8.14)$$

and the constant-rate drying area can be expressed in terms of the throughput as

$$A_c = \frac{(M_s - M_c)W_d}{R_c} \qquad (8.15)$$

The effect of air velocity on the constant-rate drying can be derived from available heat-transfer data (Reference 1). For air flow parallel to the surface it is sufficiently accurate to take

$$R_c = bG^{0\cdot 8}(H_{sw} - H_g)_{\text{lm}} \qquad (8.16)$$

Where b = constant = 0·052 for the air/water system when R_c and G are expressed in g/s m^2.

For air flow normal to the surface

$$R_c = bG^{0\cdot 37}(H_{sw} - H_g)_{\text{lm}} \qquad (8.17)$$

Where b = constant = 1·72 for the air/water system when R_c and G are expressed in g/s m^2.

Since drying is taking place at the exposed surface only, the thickness of the bed will affect the rate of drying. It should not affect the actual constant rate itself (except where additional heat transfer by conduction is involved – see below) as this is dependent only on the surface condition, but it will affect the moisture range over which constant-rate drying holds, i.e. it will affect the critical moisture content. It will also affect the shape of the falling-rate drying curve. In general, the thinner the bed the lower the critical moisture content and the greater the proportion of relatively fast constant-rate drying. It is not possible, however, to predict the effect of thickness quantitatively and experimental data should therefore be based on the thickness which it is proposed to use on the full scale.

Constant-rate experimental data can thus be calculated to different conditions of air temperature, humidity and rate. A falling-rate drying curve is then constructed by proportion as described in Section 8.2.1 (Fig. 8.1). Drying rates are generally slower than for the through-drying case and the critical moisture content is likely to change significantly with change in drying conditions for thick beds. Caution is therefore required in extrapolating much beyond the experimental drying conditions. The falling-rate drying curve can be analysed by the simplified procedure described in Section 8.2.1 and Appendix 8.2, but caution is needed and serious errors may result where there is a prolonged falling-rate period. More accurate results can be obtained by the more detailed analytical procedure described in Chapter 2, Sections 2.3.9/10. In practice it is usually more convenient to carry out experiments for the range of conditions obtaining in the various sections of the full-scale dryer and use this data directly. Small-scale batch experiments can be carried out in which the material is placed on a plate or tray with drying conditions otherwise similar to those in each section of the full-scale dryer.

Additional heat transfer by radiation is quite frequently applied in solid-band dryers. The effect can be assessed by the method described in Chapter 2 (Sections 2.5.3/4). Heat may also be transferred to the underside of the band and hence to the material by conduction. The effect will be to raise the surface temperature in a similar fashion to radiation and the new constant-rate surface temperature can be assessed by trial and error solution as suggested in Chapter 2, Section 2.5.4. It is difficult to simulate these radiation and conduction effects accurately in small-scale laboratory experiments.

8.2.3 Conduction solid-band drying

In this case drying takes place by virtue of heat conducted to the band and the dryer is usually operated under vacuum. The process is essentially one of evaporation and the constant- and falling-rate drying periods of convection drying are irrelevant. The theory detailed in Chapter 2, Section 2.4 applies and data can be correlated and scaled up in terms of the relevant overall heat-transfer equation 2.47 repeated here for convenience

$$R_m = \frac{10^3(M_s - M_f)W}{A\theta} = \frac{K_c(T_h - T_b)}{\lambda} \tag{8.18}$$

For a continuous dryer the area of the band can be expressed in terms of the output rate in the same way as in equations 8.3 and 8.15. i.e.

$$A = \frac{(M_s - M_f)W_d}{R_m} \tag{8.19}$$

The conduction drying coefficient, K_c, is in effect the mean overall heat-transfer coefficient corrected for end effects as described in Chapter 2 (Section 2.4) and is determined by experiment for a given material dried between given moisture limits. It cannot be used to calculate the drying time for moisture limits outside the range of the experimental data.

Radiation is sometimes used to assist drying in conduction type solid band dryers. Assessment of its effect is given in Chapter 2 (Section 2.5)

8.3 LABORATORY DETERMINATION OF THROUGH-DRYING PARAMETERS

The drying rate of a particular material can be determined quite simply in the laboratory (cf. Reference 2). Convenient apparatus comprises a batch through-circulation dryer based on a circular 20 cm diameter BSS sieve of aperture 250 μm. The sieve is arranged in a vertical duct so that hot air can be blown down through it and so that the sieve and contents can be weighed at intervals to determine the moisture content. One method of achieving quick weighing is to support the sieve on one arm of a balance and arrange to divert the air flow during each weighing.

The procedure is to charge the sieve with a known weight of preformed paste of the size to be used in practice. Laboratory methods of preforming are indicated in Appendix 8.1. Equation 8.6 can be used to calculate the effects of limited variations in size. The bed depth is chosen to be the same as will be used in practice (not normally more than about 5 cm). Air at the temperature that will be used in practice (usually not above 150°C) and at the velocity that will be used in practice (usually 1·25 m/s though limited variation can be calculated by equation 8.5), is then blown down through the bed of granules and the weight of the sieve and contents determined at about 2 min intervals. Drying is continued to constant weight or to the required moisture content, which usually takes 15 min to 1 h. It is usually convenient to heat up room air which can be assumed to be nearly saturated at room temperature. No attempt is made to recycle the drying air in the experimental apparatus. Humidified air can be used but is generally only necessary when investigating the drying rate of materials which tend to case harden. Fig. 8.4 shows the arrangement of an experimental drying apparatus using humidified air.

The experiment thus gives a direct plot of moisture content against time

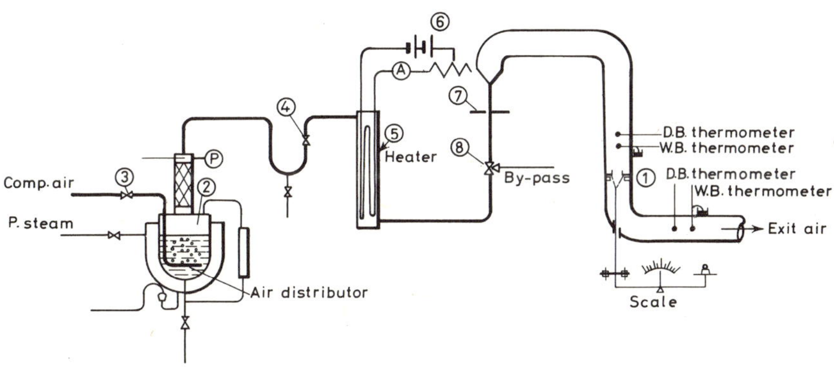

Fig. 8.4. Experimental apparatus for study of continuous through-circulation drying
(*1*) *Basket about 20 cm diameter holding material under study*
(*2*) *Humidification unit.*
(*3*) *Spray arrester.*
(*4*) *Expansion valve controlling flow of unsaturated air of known humidity.*
(*5*) *Electric heater controlled by rheostat 6.*
(*7*) *Orifice plate to measure air flow rate.*
(*8*) *By-pass to divert air stream when basket being weighed.*

for a given bed depth of granules of given size under known conditions of inlet air rate, temperature and humidity. This can be recalculated to give a graph of drying rate against moisture content for the same conditions. The drying-rate curve is then recalculated by the procedure described above for the higher inlet humidities which appertain in practice with air recirculation.

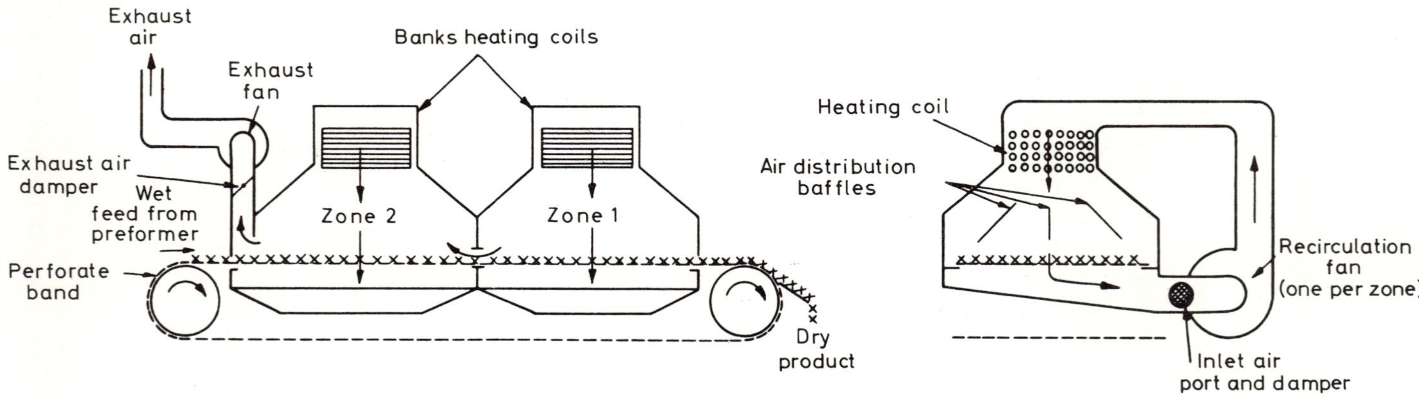

Fig. 8.5. Continuous 2-zone band through-circulation dryer

A worked example of the experimental and design procedure is given in Appendix 8.2.

8.4 DETAILED DESCRIPTIONS

8.4.1 Continuous band through-circulation dryer

The number of possible arrangements is legion and very similar proprietary machines are available from several manufacturers.

In a typical arrangement the material is fed continuously from a preforming device, which breaks it up into pieces, onto a horizontal perforate band which travels through the dryer at a speed of about 5 mm/s. Air is recirculated in two or more separate zones through the material and the band at 1–1·5 m/s. The dry material is discharged from the end of the band. Fresh air is admitted at the discharge end of the dryer and humid air is exhausted from the feed end of the dryer, thus providing a degree of countercurrency of the air circulation with respect to the drying product. The arrangement of a typical two zone dryer is shown in Fig. 8.5. The air velocity through the bed is controlled by the recirculation fans and it is usually convenient to have one drying zone per recirculation fan. The counter-current passage of air through the dryer as a whole is controlled by the exhaust fan and the damper on the air inlet to the dry zone (zone 1) recirculation fan.

In some cases it is undesirable to place the heaters directly over the drying band because of high temperature heat radiation onto the product or (usually more important) decomposed dust trapped in the heater falling onto the product and contaminating it. A typical arrangement with such an outrigged heater is shown in Fig. 8.6. A disadvantage is that the heat losses are usually higher. Axial flow recirculation fans mounted on the main

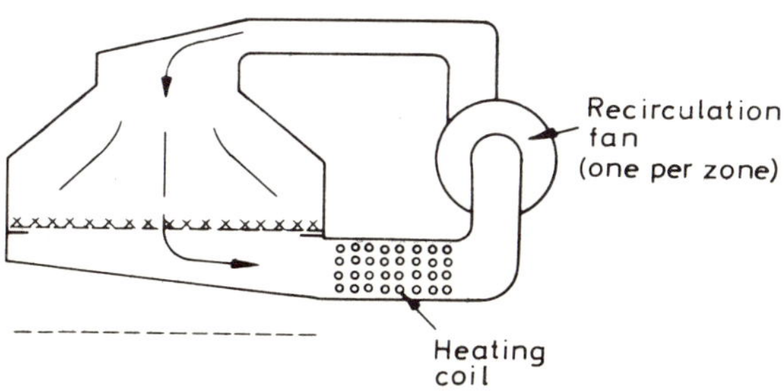

Fig. 8.6. Continuous band through-circulation dryer with outrigged heater

body of the dryer are also used as alternatives to the outrigged paddle-type fans shown in Figs. 8.5 and 8.6. It is also possible to use arrangements in which the air is alternatively blown up and down through the band. This is claimed to give better contact between the drying air and underside of the granular bed and to help to loosen the bed and prevent it settling into the perforations in the band. It is doubtful if these advantages have any practical significance but alternate up and down flow does lend itself to a neat construction, particularly as regards location of the heaters. The arrangement of a dryer on these lines is shown in Fig. 8.7.

Dryers with more than two zones merely amount to extension of those illustrated in Figs 8.5, 8.6 and 8.7. In general two zones are essential to provide a reasonable degree of counter-currency of overall air flow but additional zones show little advantage for a given total area, except when it is desirable to dry at different temperatures according to the moisture content of the material passing through the dryer. On the other hand it is constructionally convenient to make one zone correspond to one recirculation fan so that the number of zones is largely determined by the largest convenient size of suitable fan. For the usual air velocity through the bed of 1·25 m/s a maximum zone area of 5 m^2 appears to be convenient, corresponding to a fan size of 6 m^3/s. Thus a 25 m^2 dryer, for example, might well be of four or five zone construction.

Materials of construction

The shell of the dryer is generally of sheet construction on some form of framework. Any of the usual metals are suitable and aluminium sheet is frequently used. Flimsy non-metallic construction such as asbestos board is not desirable because of its poor mechanical strength and the difficulty of making the structure leak-proof. Brickwork shells may be possible in some circumstances.

A fairly large heater area is required and usually necessitates use of gilled heaters. The choice of material depends on the conditions and can be mild steel, nickel, Monel metal, austenitic steel, etc. Galvanised mild steel affords a cheap construction of heater which will withstand many corrosive conditions.

The shell area of these dryers is relatively large and heat loss by conduction and radiation amounts to some 20–30% of the total steam consumption. Thermal insulation of the shell and associated ducting is thus important; this can be achieved by double wall construction and use of adequate lagging.

Perforate band

The construction of the perforate band is important since it must retain the preformed material, offer a low pressure drop to allow passage of the drying air, bend round driving or idling rollers of reasonable dimensions, not cause breakdown or grinding of the product and in many cases be resistant to corrosion.

A heavy chain-mail of wire mesh has been used successfully in dryers up to 25 m^2 in area. The material is quite open, has a negligible pressure drop at 1·5 m/s air velocity and yet will satisfactorily retain preformed material down to 3 mm diameter. It can conveniently be driven from the

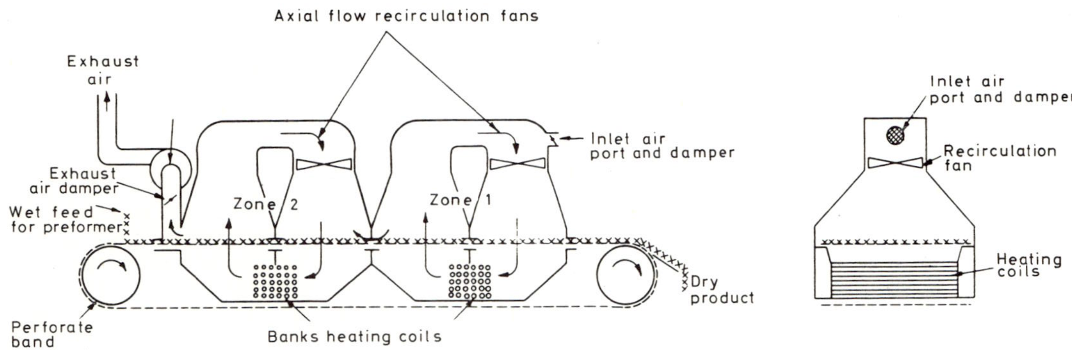

Fig. 8.7. Continuous 2-zone band through-circulation dryer with alternate up and down flow

Table 8.1 Specification of preformers

Class of paste	*Description and preliminary examination*	*Possible types of preformers (see Appendix 8.2 for description)*	*Laboratory testing of paste*	*Proving trials*
1. Fluid	Flows readily. If cut into pieces these flow together Determine whether it can be converted to a soft or hard paste (class 2 or 3) by partial drying or by mixing in dry solid	Pre-evaporate or mix in dry solid and treat as class 2 or 3		
		Grooved drum	Press into grooved plate and partly dry. Determine if a stable preform can be doctored out of the grooves. Convenient apparatus is a domestic electric flat iron fitted with a grooved base	Pilot or full scale machine
		Band	Pour on to flat plate, score with knife, and partially dry. Determine if partially dry material breaks easily to a stable preform	Pilot or full scale machine. (Probably necessary to make or improvise a machine for the specific requirements)

2. Soft	Does not flow readily and can be cut into stable pieces with a spatula	Mincing machine	Trial in domestic mincing machine	Industrial size motor-driven mincing machine
		Rotary extruder		Pilot or full scale machine
		Reciprocating extruder	Extrude through coarse mesh by reciprocating action. This can be done with a wooden mandril similar to a roller blotting pad	Pilot or full scale machine
		Direct pressure extrusion	Extrude in a grease gun or similar arrangement	Pilot or full scale trials with a suitable pump and extrusion. A major problem is getting the paste into the pump
3. Hard	Stiff cake which requires a fair amount of pressure from a knife to cut up	Pelleting machine	Trial in a proprietary laboratory machine	Full scale trials. A number of machines are only available in full size
		Size reduction	Determine whether paste can be fractured without undue formation of fines by judicious trial with a pestle and mortar	Pilot or full scale trials with a suitable mill set for coarse grinding

end rollers provided satisfactory arrangements for tracking and tensioning are made. It can be made in austenitic steel, Monel metal or nickel in addition to mild steel and is thus suitable for all the usual corrosive cases. There is very little attrition of the granules which sit on top of the mesh but on some materials which tend to disintegrate as they become dry, the powder settles into the mesh and eventually blinds the band. In such circumstances this type of band is extremely difficult to clean.

A slat construction of rigid perforate sheets or rigidly supported wire mesh panels is also used. This enables a relatively light band of rigid construction to be designed in any required aperture size; it also lends itself to provision of fitments such as side guards to prevent the granular bed falling off the side of the band. Chain drive on the side of the bands can also be used with elimination of tracking problems. Slat bands require careful detailed mechanical design to avoid undue wear and to prevent granules being trapped between the slats and broken up as they pass round the end rollers.

Another method of construction is a sandwich band comprising a coarse mesh carrier band, conveniently driven by side chains and sprockets, on the top side of which is fastened a light wire mesh or woven cloth or perforate plastic band. As an example coarse chain-link type fencing material of 2·5 cm aperture can be used as the carrier and wire gauze of 0·25 mm or a P.V.C. bolting cloth as the upper retaining band. This type of construction gives a cheap, light band which is relatively easy to clean and allows non-metallic materials to be used in contact with the product. One difficulty is the attachment of the light top band to the carrier. It is possible to run this band unattached but there is then a danger of slip with respect to the carrier.

Preforming

The preformer converts the feed material to a definite particulate form, usually in the size range 3–15 mm. It is necessary in all cases except those where the feed is already in a suitable granular form.

Details of typical preformers are given in Appendix 8.1. The general working principles used include partial drying for very fluid pastes, extrusion for moderately fluid pastes and stiff pastes, and size reduction (coarse grinding) for hard stiff pastes. The type of preformer required depends very much upon the properties of the feed paste and final choice and specification can in general only be determined by actual trial. An indication of types of machine used and possible laboratory test procedure is given in Table 8.1.

Instrumentation

Control of continuous-band through-circulation dryers is usually done by varying the temperature and the band speed (which controls the drying time and depth of the bed) with other factors kept constant. The essential instruments are therefore thermometers. It is convenient to measure the approach air temperature in each zone (i.e. above the band in the usual arrangement) and to use this to control the steam input to the appropriate heaters. Care should be taken in siting the temperature sensing elements so that they give a true reading.

The band speed should be variable and controllable by the plant operative to between half to double the nominal design value.

Field of application

The continuous band through-circulation dryer can be used on any material which can be made available in a discrete particulate form which is usually of size 3–15 mm; finer material can be handled and special dryers for large shapes can follow the same principles. It is essential that the wet particulate material should retain its shape sufficiently to allow passage of air through the bed of material and that it does not break down to dust when dry. The particular merit of this dryer is that it offers efficient rapid drying with a minimum disturbance of the product. It is thus particularly suitable for heat sensitive and delicate materials but its high efficiency gives it a quite general application. A further advantage of the through-circulation dryer is that it affords a convenient method of producing the dried product in granular form.

8.4.2 Convection solid-band dryer

There are again a large number of possible arrangements. The special case of using a solid band dryer as a partial dryer to preform fluid paste for through-circulation drying is described in Appendix 8.1.

A typical direct-fired solid band dryer is illustrated in Fig. 8.8. This comprises a solid metal band (using steel or stainless steel) which moves continuously through the drying chamber. The paste is fed at one end by some convenient means (see Appendix 8.1) and dry material is continuously discharged from the other end of the dryer. A brush is provided for cleaning the underside of the returning band. The range of metals possible for the band can be increased by using a fine mesh gauze capable of retaining the wet material: this also has the advantage of reducing the radius of bending and hence the diameter of the end rollers. Gauze bands however, have a limited life.

Heat is provided by direct gas-firing by means of burners below the band carrying the material. The hot gases are conveyed from the underside of the band to the upper side in direct contact with the material by means of recirculation fans. The dryer is divided into a number of zones (four in the example of Fig. 8.8) which can be controlled at different temperatures if required. The wet gas is exhausted from the top of the dryer via control dampers and an extract fan.

Drying is thus by a combination of conduction through the drying band and convection by direct contact of the material and drying gas—analogous to an indirect-direct rotary dryer.

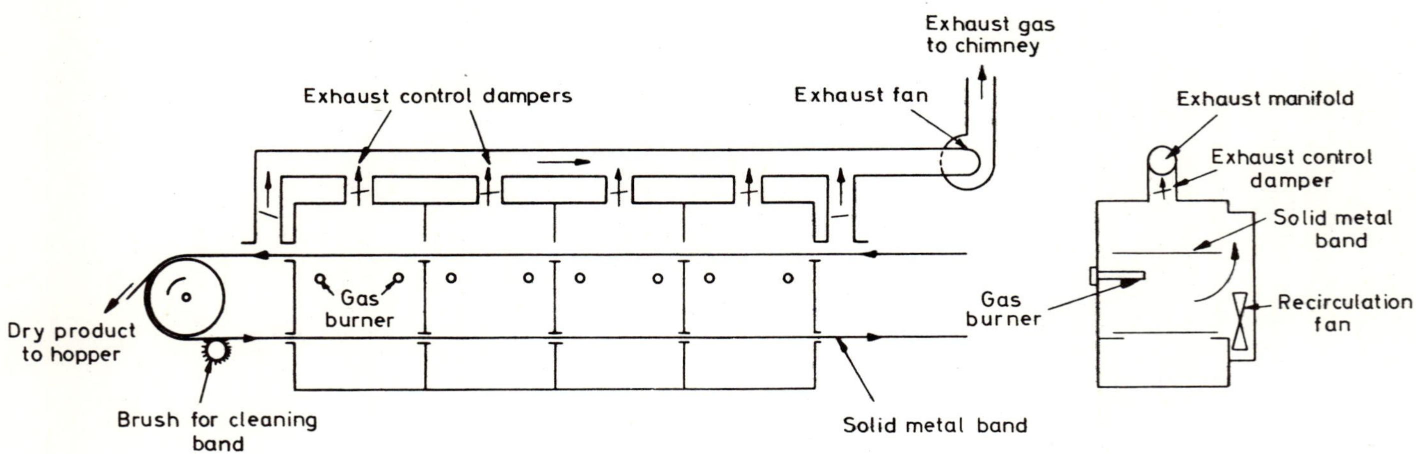

Fig. 8.8. Direct gas-fired convection solid band continuous dryer.

Dryers of this type can be operated at temperatures of several hundred degrees Centigrade and can give reasonable efficiencies under these conditions, bearing in mind the relative costs of direct heat from gas compared with heat from steam. Thermal losses in the exhaust gases may be recovered on large plants by some form of heat exchanger with the combustion air. They can be used on almost any material that can be fed continuously. The efficiency is generally lower than that of the through-circulation dryer because of the relatively poor contact between the drying gas and material. The tendency is thus to use the solid band dryer for stable sticky materials which are too soft to preform and too sticky to handle on a continuous rotary dryer but which can stand high temperatures so that a reasonable output rate can be obtained. The through-circulation dryer is to be preferred where it can be used. A possible exception is the use of solid band dryers for rapid drying of very thin layers of 3 mm or less for which high efficiency is claimed; this is however more analogous to the drum dryer described in Chapter 9.

8.4.3 Conduction solid-band dryer

This type of dryer is usually operated under vacuum. If it were not, the convection solid-band arrangement (Section 8.4.2) would be cheaper and more convenient. The usual form of machine therefore amounts to a continuous vacuum dryer and one form of construction is illustrated in Fig. 8.9. The machine comprises a number of narrow bands in a cylindrical pressure chamber operated under vacuum, the bands being fed in parallel or in series. This arrangement is used in order to reduce the dimensions of the expensive pressure chamber to the minimum for a given band area.

Feed is by pumping the material on to the drying bands via a manifold and extended orifices. The bands themselves are usually made of a fine wire mesh (aperture about 0·15 mm) rather than a solid construction in order to allow them to be bent around a sharp radius and thus keep the dimensions of the drying chamber to a minimum. The mesh should, of course, be such that the feed does not flow through it. Arrangements for centering and tracking the bands, and varying their speed, are essential and should be operable from outside the dryer.

The upper part of each band is supported on a number of steam chests which provide the source of heat by conduction through the band. Several independently connected steam chests are provided along the length of each band so that different sections of the dryer can be operated at different temperatures. If desired the final section at the discharge end can be operated as a cooling section by applying water to it instead of steam. Infra-red radiation is sometimes used to assist the drying.

Material is discharged from the end of the bands, via a coarse grinding mill or granulator (if necessary), into a bin or bins which are held under vacuum and discharged periodically. Alternatively, continuous discharge via a sealing system such as a scroll or rotary valve can be used, although it is difficult to avoid leakage of air into the dryer with such arrangements.

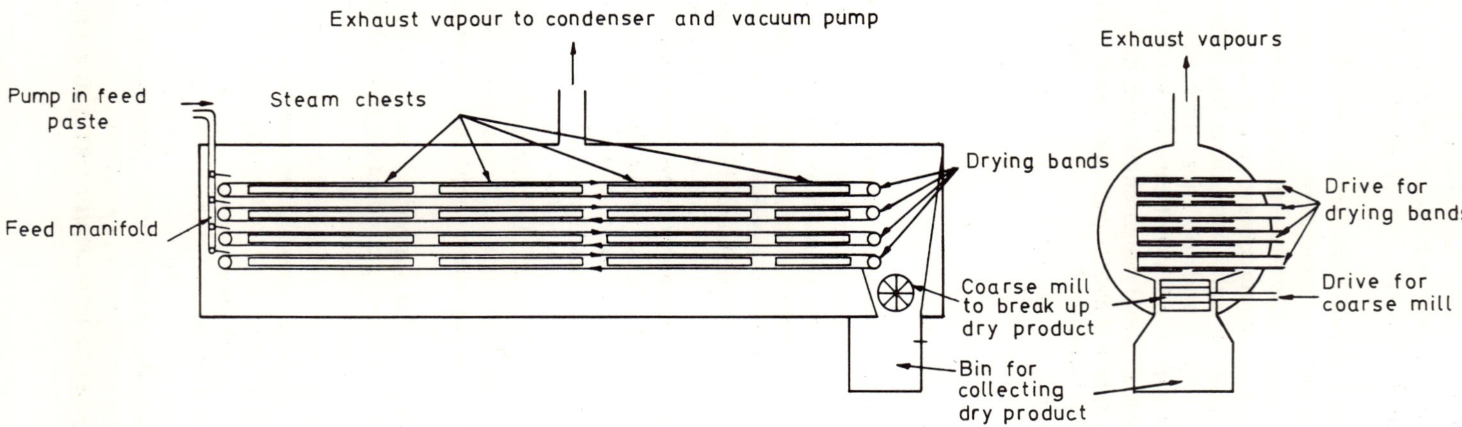

Fig. 8.9. Conduction solid band continuous dryer operating under vacuum

Vapour evaporated in the dryer is condensed and exhausted by vacuum pump in conventional fashion. Pre-evaporation of the feed in a conventional evaporator is frequently used with these dryers in order to improve efficiency and provide a pumpable feed which is not so fluid that it will run off or through the bands.

The shell can be made of any convenient metal construction which will stand the pressure conditions and can be lined with corrosion-resistant material where required. The bands themselves and the steam chests have obviously to have good thermal conductivity and metal is the only practicable material of construction.

Dryers of this type will tend to be tailor-made to a particular duty. A typical medium size dryer comprises a shell 2 m diameter, 10 m long, containing 8 drying bands, each 750 mm wide and driven on 120 mm diameter rollers.

These dryers are intrinsically expensive and have inefficient heat transfer, although this is overcome to some extent by operating under vacuum with a high temperature differential. A further disadvantage is that they are difficult to clean and are thus essentially one product machines. Again their labour usage is relatively high for a continuous dryer largely because use of so many bands requires constant supervision to preserve correct band tracking; the intermittent batch discharge which is normally used also requires more labour than a mechanised continuous discharge.

The major application is to moderately heat sensitive materials which dry badly by convection because of case-hardening. Materials which oxidise easily can sometimes be dried with advantage. It is not easy to feed coherent high-solids-content material to a vacuum dryer but pumpable solutions or pastes can readily be sucked into the vacuum chamber and distributed on to the bands. The machine can be useful for materials which are dried as a froth to give relatively delicate granules (e.g. certain dye-stuffs and chocolate crumb) as the vacuum assists frothing and the gentle handling inside the dryer avoids undue damage to the granules.

8.4.4 Continuous-tray through-circulation dryer

The continuous-band through-circulation dryer (Section 8.4.1) can be converted to continuous-tray through-circulation by substituting a series of perforate trays carried on some form of conveyor for the continuous band. This system has little merit in terms of a horizontally conveying dryer since it involves the addition of mechanism for handling and discharging the trays with little compensating advantage.

If, however, a traverse of trays is arranged in the vertical plane so that the drying air is blown through a stack of trays in series, there is then offered the possibility of a compact arrangement with reduced capital cost and reduced heat losses (cf. Section 8.5 and 8.6). Fig. 8.10 illustrates a possible arrangement. Alternatively a train of trays can be arranged to follow a spiral path in helter-skelter fashion.

Apart from reduced capital cost and heat losses, one other advantage of the tray arrangement is that it permits easy cleaning of the trays as a separate operation; also segregation of products (by using different sets of trays) in a multi-product dryer. A disadvantage is that the drying gas is blown through several trays in series. Since reheating between trays is impracticable, the overall drying efficiency is reduced and the output per unit area of perforate surface is lower than that of the continuous-band through-circulation dryer.

8.4.5 Continuous solid-tray dryer

Trays may be substituted for the continuous band in the horizontal band dryer described in Section 8.4.2, Fig. 8.8. The arrangement has nothing particular to commend it, except possibly easier cleaning and segregation of products in multi-product drying. Moving stacked tray arrangements as suggested for the continuous-tray through-circulation dryer (Section

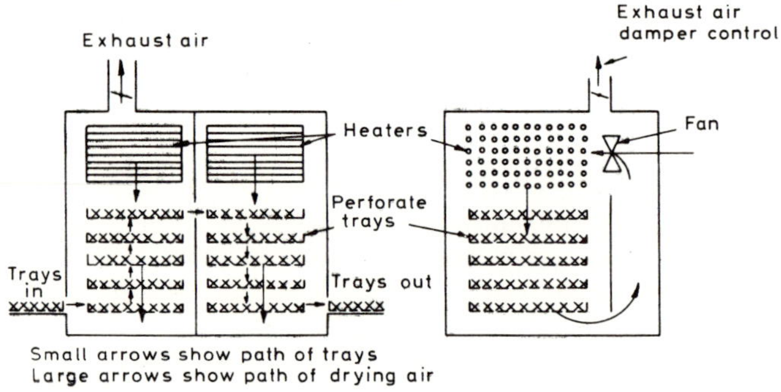

Fig. 8.10. Continuous tray through-circulation dryer

8.4.4, Fig. 8.10) are possible but it is not easy to arrange a satisfactory method of air circulation since it cannot be blown through the trays themselves. The helter-skelter arrangement effects space saving.

The most useful form of continuous solid tray dryer is the continuous tunnel in which the stacks of trays themselves are moved through the dryer with counter-current circulation of air. A typical arrangement is illustrated in Fig. 8.11. Other arrangements are described in References 1 and 22. The material is loaded on to trays which are in turn loaded into racks. The racks are moved continuously through a chamber or tunnel which has arrangements for circulation of hot air, either in an overall sense or in a series of zones similar to the band continuous through-circulation dryer (Section 8.4.1). Counter-current circulation of the drying air is usually used. Semi-continuous tunnel dryers in which the racks are intermittently moved through the dryer are also used but these can more properly be treated as a number of batch tray dryers in series (cf. Chapter 5).

Dryers of this type are normally used for materials which require slow drying under controlled conditions (timber and pottery are examples, cf.

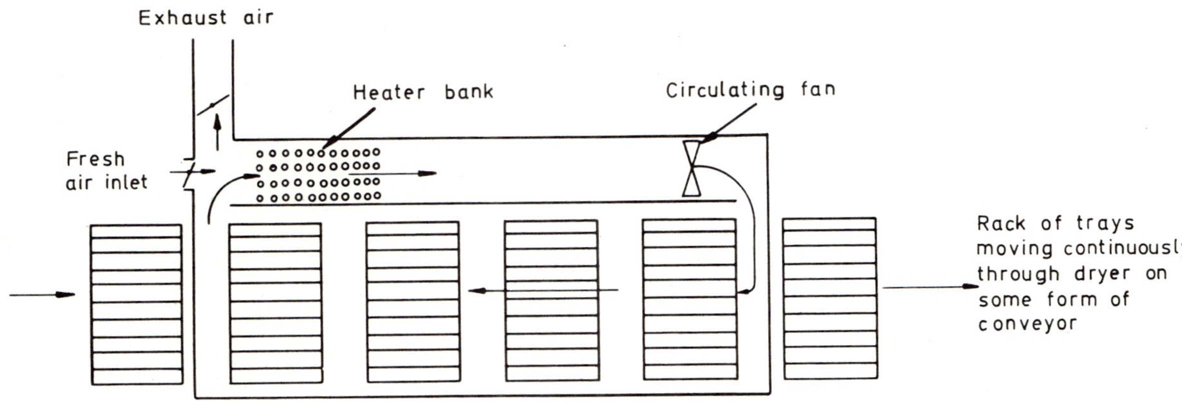

Fig. 8.11. Continuous tunnel dryer

References 1, 17). They are not in wide use in the chemical industry. Information on the design of tunnel dryers is given in References 16, 19, 20 and 21.

8.5 PERFORMANCE

8.5.1 Continuous-band through-circulation dryers

Operating conditions

The operating conditions of these dryers have to a large extent been standardised for use of a circulating superficial air velocity of 1·25 m/s (1 350 g/s m^2) through a bed 12–50 mm deep of preforms of 3–10 mm cross-section (cf. References 2, 3, 13, 14 and 15). Approach air temperatures are usually in the range up to 150°C and the air humidity is adjusted to give the maximum drying rate consistent with an acceptable sensible heat loss in the exit air (cf. Reference 2). This corresponds to typical approach air humidities in the range 0·04–0·2 kg/kg.

These more or less standardised operating conditions are largely dictated by the natural limitations of the machine. Thus the higher the air velocity the higher the drying rate, other factors being equal—see equation 8.7. However, the drying rate only increases as the 0·8–0·9 power of the air rate whereas the pressure drop increases as the square. Thus if we vary the air rate alone the pressure drop will increase as the 1·3–1·4 power of the drying rate. Pressure drops much in excess of 1·5–2·5 mb across the band are undesirable because of the tendency for air leakage round the edge of the band (which is difficult to seal) which both reduces the air flow through the product itself and may cause distortion of the air flow patterns leading to further local starvation of air and a tendency to disturb the granule bed with increase in dust or material loss. Practical experience has confirmed that 1·25 m/s seems to be about the optimum air flow for these reasons. The same argument limits the maximum depth of the bed to about 50 mm and on many soft materials which tend to coalesce it may be no advantage to operate at bed depths greater than about 12 mm.

Again the smaller the preforms, the greater the surface area and the higher the drying rate, the drying rate being proportional to the 0·6 power of the diameter for rod-like preforms (equation 8.6). Smaller preforms will, however, tend to pack tighter into the carrier, leading to reduction of surface area, increase of pressure drop and possible mechanical loss. It is thus usually not practicable to use preforms of much less than about 3 mm size and on many products sizes as great as 10 mm may be more convenient to use.

The maximum temperature is often limited by danger of thermal degradation of the material being dried but otherwise should be as high as possible. However, at very high temperatures the heat losses from the bulky shell of the dryer tend to become large as radiation effects become significant (see below). Since drying rates are relatively high in any case, it is not usual to operate these dryers at much above 150°C.

Performance data

Typical performance data in these machines is given in Table 8.2. Other data on a variety of products is available in the literature, e.g. References 1, 3, 13, 14 and 15.

In general the performance to be expected can be assessed with some precision from laboratory information on the drying curve of the particular preformed material as outlined in Section 8.2. In order to assess the steam consumption it is necessary to have knowledge of the evaporation rate, the sensible heat put into the product and the heat lost in the exhaust air; all these follow from the optimum operating condition which can be determined from the drying curve. In addition it is necessary to know the heat loss by radiation etc. from the dryer itself. Typical data for dryers ranging

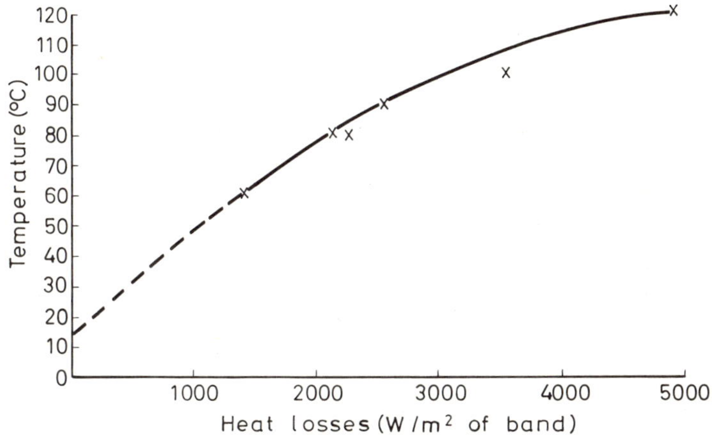

Fig. 8.12. Heat losses from continuous-band through-circulation dryers as a function of temperature — valid for dryers of band area 5–25 m²

between 5 and 25 m^2 suggest that this amounts to 25–35% of the total heat consumption for reasonable operating condition, and for most purposes it should suffice to allow about 30% extra steam for these losses. If more precise calculation is required, use can be made of the fact that heat loss on machines of this size appears to depend on the band area, probably because these heat losses include leakage and the larger machines have a greater complexity of ducting. The experimental temperature/heat loss curve in Fig. 8.12 is sufficiently accurate for most purposes.

8.5.2 Convection solid-band dryers

It is not really possible to generalise on typical operating conditions for machines of this type which are usually tailor-made to suit particular

Table 8.2 Typical performance of some continuous-band through-circulation dryers

Physical form feed	*Solids content, percent*	*Approx. shape and size of preforms* (mm)	*Drying temperature* (°C)	*Moisture content product* (%DB)	*Output* (g dry/s m²)	*Evaporation* (g water/s m²)	*Steam usage* (kg/kg water)
Wet fibrous mass	40	—	80–120	7·5	1·42	1·65	4–5
Stiff paste	25	Rectangular chips 6 × 4 × 12	150–170	3–7·5	2·74	8·07	1·9–2·1
Water insoluble biscuit-like paste	45·5 55·5	Rectangular chips 6 × 12 × 12–20	100 100	0·2 0·2	2·18 1·59	2·82 1·26	2·23 2·86
Water insoluble clay-like paste	30	Rods 4 dia. × 12–20	120	trace	2·40	5·82	
Water insoluble clay-like paste	25	Rods 4 dia. × 12–20	90	trace	2·12	4·48	
Water soluble grease-like paste	33	Rods 5 dia. × 6–12	120	3·0	1·24	2·78	

Water soluble grease-like paste	55	Rods 4 dia. × 20–25	90	0·6	1·38	1·05	2·44
Crystalline paste	50	Rectangular tablets 20 × 20 × 6	70	1·0	1·36	1·36	
Wet fibrous mass	44	Not preformed	—	0·5–1·0	3·39	4·20	
Water insoluble gritty paste	55	Rods 4 × 12–20	100	0·5	6·02	4·81	
Water insoluble doughy paste	55	Rods 4 × 12–20	70 85	0·5 trace	2·71 3·77	2·21 3·09	
Water insoluble fine paste	50	Rods 4 × 6–12	90	0·5	3·55	3·52	
Water insoluble doughy paste	55	Rods 4 × 12–20	80 90	trace trace	2·14 2·45	1·56 2·12	1·99 2·23

This performance data corresponds to actual times in the dryer between ½ and 1 h and to loadings on the drying band between 5 and 10 kg/m^2.

circumstances. The two examples quoted in Table 8.3 are probably fairly typical for drying with indirectly-heated air at moderate temperature and for use of direct combustion gases at high temperature.

8.5.3 Conduction solid-band dryers

It is again difficult to generalise on typical operating condition. Available data obtained from an 8-band conduction dryer is summarised in Table 8.4 as an example.

In general, performance of these dryers will tend to be similar to that of

Table 8.3 Performance of some solid-band convection dryers

Total drying area	7 m^2		8 m^2
Feed from	Pump to comb		Gravity to extrusion preformer
Dried by	Hot air blown down into band		Town's gas combustion products circulating across band (Significant radiation heating)
Physical form feed	Pumpable paste		Soft paste
Solids content feed	22%	24%	35%
Bed on band	Parallel strips 6 mm thick by 6–10 mm wide		Irregularly coalesced strips 12–40 mm thick
Gas velocity on to band	0·6 m/s		Approx. 0·1 m/s
Drying temperature	150°C		350°C
Moisture content product	120% DB	80% DB	43% DB
Output (dry material, g/s m^2)	2·0	1·5	4·9
Evaporation water	4·7	3·5	6·1
Thermal efficiency	Steam* 2·2 kg/kg water	Steam* 2·9 kg/kg water	18·3% gross efficiency on town gas

*Steam consumption includes through-circulation drying section (see Table 8.2).

the batch vacuum tray dryer when allowance is made for the fact that the dead time for batch charging and discharging is eliminated by the continuous operation. Reference may be made to the typical performance data in Chapter 6.

8.5.4 Continuous-tray through-circulation dryers

Performance on continuous-tray through-circulation dryers in which the trays are carried as a single layer on a horizontal conveyor will be similar to that of the continuous-band through-circulation dryer (Section 8.4.1).

Efficiency of the system in which the trays are stacked vertically and air is blown through a number of trays in series (as described in Section 8.4.4 and Fig. 8.10) will in general be inferior because the system operates in effect as a deep bed dryer and it is necessary to compromise on the drying conditions presented to any tray. Experimental data illustrating the type of difference that might be expected is summarised in Table 8.5.

8.5.5 Tunnel dryers

These dryers comprise racks of trays moving (usually slowly) through a drying chamber through which the drying gas is circulated—cf. Section 8.4.5. There is little reliable performance data for these machines for the chemical

Table 8.4 Performance of a solid-band conduction dryer

Total drying area	56 m^2 (8 bands, 9·1 m long)
Feed from	Pump from an evaporator
Physical form feed	Pumpable paste (hot)
Vacuum, absolute pressure	67 mb
Solids content feed	78 % solids
Moisture content product	2 % dry basis
Output based on 100 % dry product	2·0 g/s m^2
Evaporation of water	0·79 g/s m^2

Table 8.5 Comparison of stacked-tray continuous through-circulation dryer with continuous band through-circulation dryer

	Stacked-tray continuous dryer		*Continuous-band dryer*
Solids content feed, %	30	50	30
Number of trays in stack	4	2	Equivalent to 1
Drying rate based on 100 % dry product (g/s m^2)	1·4	4·8	2·0
Evaporation rate (g/s m^2)	3·2	4·8	4·8

Moisture content of product – Not greater than 1 %
Maximum drying air temperature – 100°C

industry although some figures for duties such as earthenware, timber and foodstuffs drying are given in the literature (e.g. References 16–22).

For practical purposes the general performance of these dryers can be taken to be very similar to that of the atmospheric tray dryers described in Chapter 5.

8.6 CAPITAL AND OPERATING COSTS

It is difficult to assemble usable cost data on continuous-band dryers because of the wide range of designs possible and the fact that these machines are very frequently tailor-made to a particular duty. The data in this section is thus only given for rough guidance. All costs have been corrected to mid-1968 prices.

8.6.1 Continuous-band through-circulation dryer

Typical basic cost data is summarised below.

Capital: £1 600/m^2 for the smaller dryer up to 10 m^2
£1 100/m^2 for the larger dryers above 20 m^2
Steam: Between 2 and 3 kg/kg water evaporated
Electricity: About 2·5 kW/m^2
Labour: $\frac{1}{2}$ man/shift to operate a dryer of any size.

8.6.2 Convection solid-band dryer

Typical basic cost data is summarised below.

Capital: £1 100–1 600/m^2
Fuel: Direct firing. Take gross thermal efficiency at 20–30 %
Steam: Take consumption at 2·5–3·5 kg/kg water evaporated
Electricity: 1 kW/m^2
Labour: $\frac{1}{2}$ man/shift to operate a dryer of any size.

8.6.3 Conduction solid-band dryer

There is relatively little basic cost data here. Some information based on one form of multi-band continuous dryer is summarised below.

Capital: £550–1 100/m^2
Steam: 1·5–2 kg/kg water evaporated
Water: 0·1 m^3/kg water evaporated for vacuum operation
Electricity: 20–30 kW largely independent of size of dryer
Labour: 1 man/shift to operate a dryer of any size

8.6.4 Continuous-tray dryer

Cost information on through-circulation dryers with stacking and on tunnel dryers for chemical products is too scarce to quote.

Horizontal continuous-tray dryers can be assumed to have similar costs to the corresponding band dryers.

Comparison of costs

The performance data given in Section 8.5 is for very different drying conditions and costs worked out from them and the cost figures in the present sections cannot be compared. Each particular case must be worked out on its own merits and the information in the detailed description (Section 8.4) and on performance (Section 8.5) can be used for guidance. As a general rule the through-circulation dryer is the most efficient and is to be preferred where it can be used.

APPENDIX 8.1 PREFORMING MACHINES

Many different types of preformer are possible; the main types of machine are described below with indication of their performance and field of use. Further information will be found in References 23 and 24.

A.8.1.1 Preformers dependent on partial drying

Fluid pastes have to be concentrated before a preform can be made. Partial drying can in some cases be combined with the preforming operation itself and two examples are:

The grooved drum preformer: The machine shown in Fig. 8.13 is a hollow

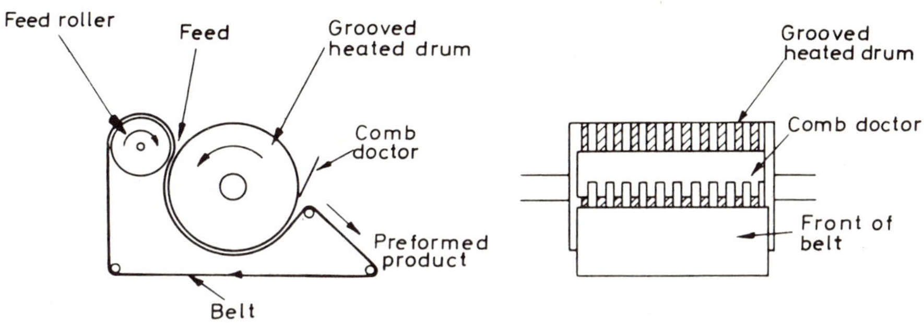

Fig. 8.13. Grooved drum preformer

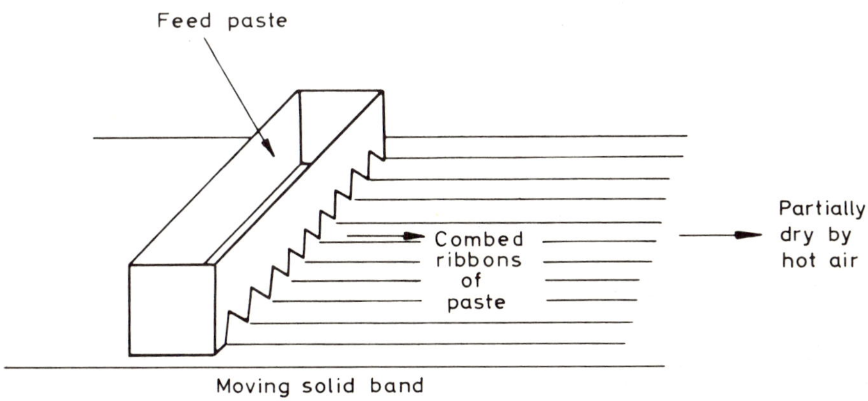

Fig. 8.14. Comb feed to band preformer

steam heated drum the surface of which contains a series of peripheral grooves about 6×6 mm section. The fluid paste is fed on to the drum and squeezed into the grooves by a plain feed roller. It is held in the grooves by means of a belt and partially dried. It is then pulled out of the grooves, by a comb-like doctor blade, in the form of rectangular chips of 6×6 mm section which break naturally into lengths of 10–25 mm. Operation is critical as many materials expand slightly and cannot be removed from the grooves without breaking them down to powder.

A variant comprising two grooved drums arranged after the fashion of a double drum dryer has also been used. The wet paste is fed to the nip between the two drums which revolve towards each other. The field of use of this type of design is probably limited to adhesive fibrous materials.

The band preformer: In the design shown in Fig. 8.14, the fluid paste is fed into a bottomless box which sits on top of a solid moving band. The leading edge of the box is fitted with a serrated edge or comb so that the paste is spread on the band as a series of parallel ribbons about 6 mm thick, 6–12 mm wide. The ribbons are partially dried by blowing hot air on to or across the band and then taken off against a doctor blade when they break up to form chips about 6 × 6 mm cross-section and 10–25 mm long.

Feed to a band preformer can also be achieved by extrusion of the feed paste in various ways. The system is in general cumbersome and expensive and it may be preferable to condition the fluid paste by back mixing of dry solid where this is permissible.

A.8.1.2 Preformers dependent on extrusion of soft pastes

The mincing machine: The domestic mincing machine and the larger motorised versions can be used as paste preformers. The field of application is somewhat limited since if the paste is too soft the extrudate tends to coalesce at the head of the machine and, if it is too hard, it slips against the scroll and no extrusion takes place. A further disadvantage for large-sacle application

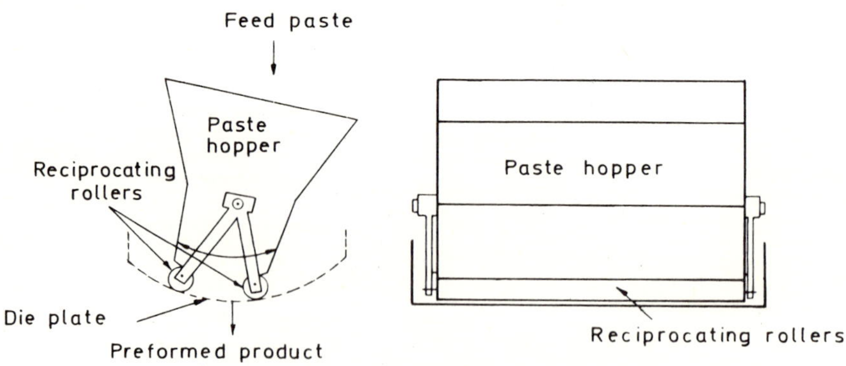

Fig. 8.15. Reciprocating extruder preformer

is that both feeding and discharging tend to be at points rather than over areas and hence the machine is not particularly convenient for accepting feed from a continuous filter or for delivery of preformed material to a continuous band.

Nevertheless, in practice, the mincing machine has been found to be a useful laboratory machine for determining whether material can be preformed by the soft paste extrusion technique and for preparing small samples of a few pounds for laboratory drying trials.

Reciprocating extruders: Fig. 8.15 shows a trough-like die plate containing many holes, over which a pair of heavy rubber rollers reciprocate. A hopper containing the paste is attached to the roller assembly. The machine is simple and massive in construction. Further details are given in Reference 3.

Trials have established that this machine has a great range of applicability, particularly as regards soft paste into which it introduces a minimum of work-softening. It will handle any paste that will form an extrudate which does not coalesce on itself (it will in fact handle much softer pastes but the exercise is academic since the extrudate collapses on the drying band). At the other end of the scale it will handle fairly stiff pastes. A further advantage is that it will act as a feed hopper. The hopper can be loaded with paste to a depth of 0·5–1·0 m and extrusion will start when the reciprocating rollers are started, and stop when they are stopped.

Direct pressure extruders: Another method of preforming is to extrude directly through a die plate by feeding the paste to it under pressure. Thus in one design the paste is pumped from a hopper by positive ram pump (with retractable cylinder liner) into a tee-shaped extrusion head containing a line of 5 mm holes through which the material is extruded directly by pressure. In order to obtain even extrusion across the length of the die, it is desirable to extend the orifices at the ends to provide a greater pressure, for example by 6 mm copper pipes of varying lengths.

Direct pressure extrusion is limited to pastes which flow reasonably easily under pressure and which remain as a distinct phase. This limits the application considerably since many pastes which are stiff enough to give a stable preform will separate water under pressure. Again blockages of the die can be a serious problem and it is usually necessary to fit a coarse filter in the line from the pump to remove such materials.

Direct pressure feed to an agitated preformer is possible but in such cases it is usually better to feed from the pump line at atmospheric pressure and allow the mixing action of the preformer to carry out the actual extrusion.

A.8.1.3 Preformers dependent on extrusion of stiff pastes

A number of proprietary machines are available in this field most of which have been specifically developed for pelleting fibrous material such as animal foodstuffs.

Pelleting machines: Figs 8.16 and 8.17 illustrate the principles of two of the many designs found to be suitable for relatively stiff pastes containing 40–80% solids. Fig. 8.16 shows a horizontal circular die plate like the rim of a wheel which is rotated and has a pair of rollers free to rotate on the inner periphery. Material is fed into the die chamber, extruded through the die holes into an outer annular trough and discharged from this trough at a single point.

Fig. 8.17 shows a pair of rollers arranged so that they rotate towards each

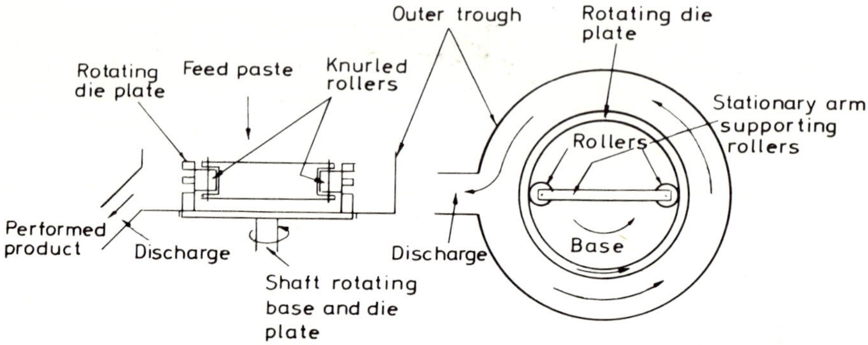

Fig. 8.16. Pelleting machine for stiff pastes (Type S)

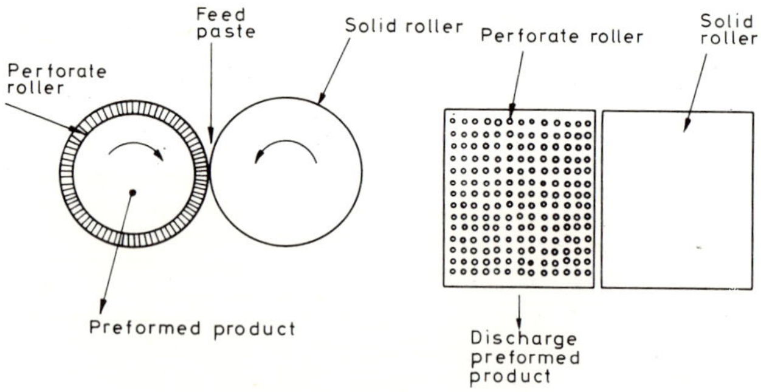

Fig. 8.17. Pelleting machine for stiff pastes (Type A)

other after the fashion of a double drum dryer. One roll is perforated and one is plain (in earlier models both rolls were perforated). The perforate roll rotates at about twice the speed of the plain roll. The paste is fed into the nip between the rollers, extruded through the holes in the perforate roller and discharged along the axis of the perforate roller.

A.8.1.4 Preformers dependent on size reduction

In addition to true extrusion, preforming of stiff pastes can be carried out by what amounts to a rough grinding or screen classification. Machines which grind the paste through a sieve fall into this class rather than that of extrusion. Various coarse grinding mills such as rotating disc crushers and so-called granulating machines can be used in this way. This type of technique tends to produce a rather variable size and irregular shaped preform and is not so generally useful as true extrusion. It can, however, be simple and cheap on suitable products of high solids content.

APPENDIX 8.2 WORKED EXAMPLE OF DESIGN OF THROUGH-CIRCULATION DRYER

Note: The figures in this appendix have been converted to metric units from the British units used in Reference 2.

A.8.2.1 Problem

A continuous band through-circulation dryer is required to dry 180 kg/h (50 g/s) of paste of 50% solids content to 0·5% moisture. The dry weight of material is therefore 25 g/s.

The paste has been preformed by extrusion. The maximum permissible air temperature has been found to be 70°C. The material does not case harden in an experimental dryer. Other operating conditions have been decided from economic considerations to be:

Mass velocity of drying air = 1 360 g/s m^2
Diameter of extruded paste rods = 4·5 mm
Bed loading (dry) = 7 kg/m^2
Approach air humidity 0·005 kg/kg dry air.

The dryer is to have two zones of equal area with counter-current air circulation and is to be of the general design shown in Fig. 8.5. It is required to calculate the area of drying band from the results of laboratory experiments with an experimental dryer of the type shown in Fig. 8.4.

A.8.2.2 Results from laboratory experiments

In a typical experiment with the laboratory equipment described in Section 8.3, the apparatus was first warmed by passing air through it at 70°C at a rate equivalent to 1 360 g/s m^2 (referred to the area of basket containing

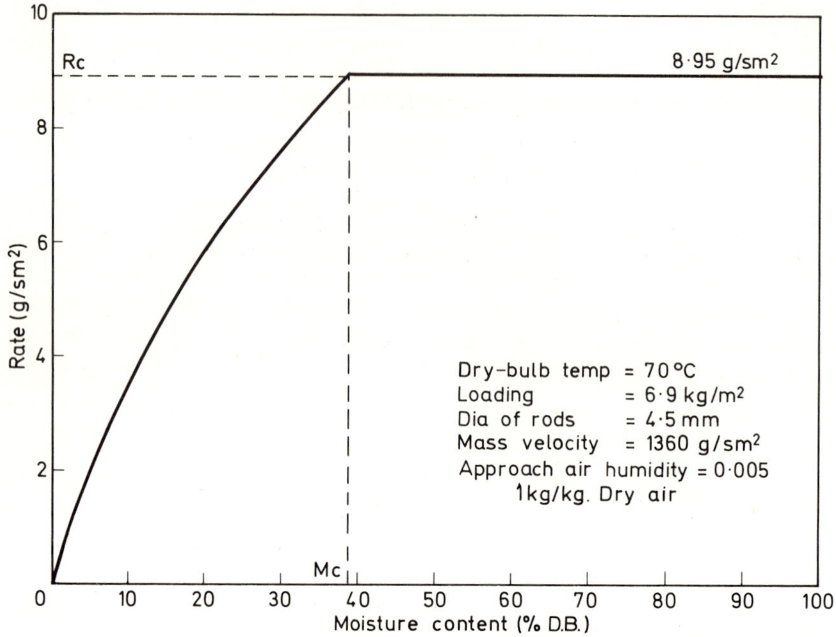

Fig. 8.18. Laboratory drying curve under standard conditions

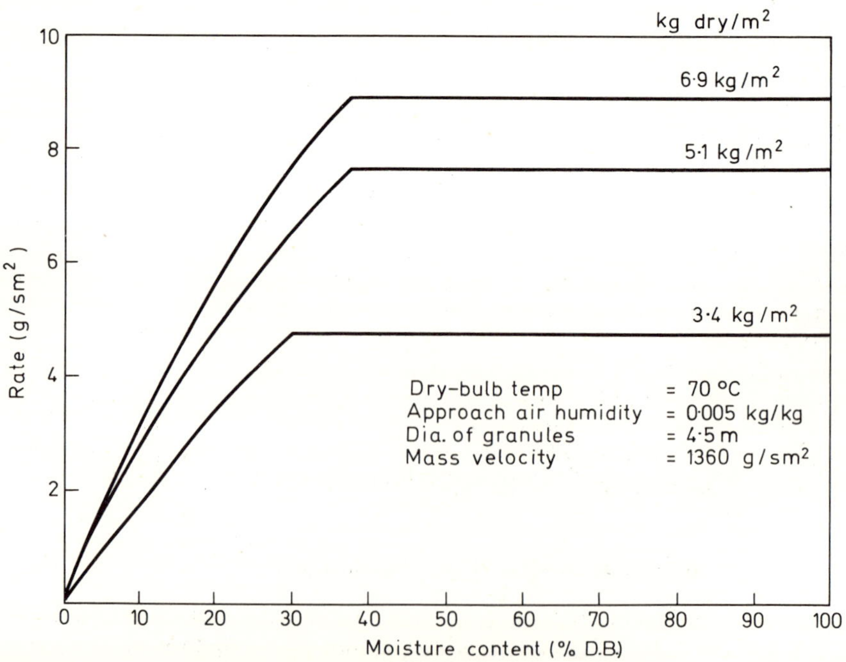

Fig. 8.19. Variation of drying rates with bed loadings. Dry-bulb temperature = 70°C. Diameter of rods = 4·5 mm. Approach air humidity 0·005 kg/kg dry air. Mass velocity of air = 1 360 g/s m²

the rods of material). The air humidity was adjusted to 0·005 kg/kg dry air. The air stream was then diverted through the by-pass and the weighed basket filled with rods of material to a depth of 50 mm (corresponding to a loading of dry material of 7 kg/m^2). The hot air was then diverted through the basket to determine the drying rate. At 2 min intervals the air was by-passed whilst

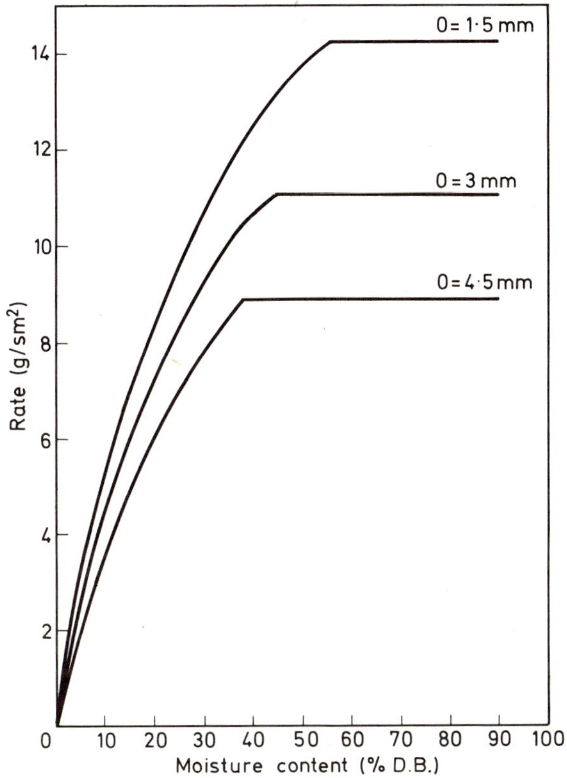

Fig. 8.20. Variation of drying rates with rod diameter (t). Dry-bulb temperature = 70°C. Bed loading = 7 kg/m^2. Approach air humidity = 0·005 kg/kg dry air. Mass velocity of air = 1 360 g/s m^2

the basket was weighed, an operation taking about 5 s. When drying was completed the basket was removed and its weight checked. The pressure drop across the bed remained substantially constant at 0·25 mb. The experimental data are shown in Fig. 8.18 (which is similar to Fig. 8.1). The results of this and a few runs under other conditions are given in Table 8.6 and Figs. 8.18–8.23. The graphs show clearly the sharp reduction in drying rate after the critical moisture content. It follows that since the mechanism of drying is quite different during the constant-rate and falling-rate periods, each period has to be considered separately when scaling-up the laboratory data.

Effect of depth of bed: Fig. 8.19 shows drying-rate curves with bed depths

Table 8.6 Calculations from laboratory experiments of drying constant K'_H for several conditions of bed loading and approach air humidity

Approach air humidity H_g (kg/kg)	*Bed loading* L (kg/m²)	*Constant drying rate* R'_c (g/s m²)	*Log mean humidity difference* $(\Delta H)_{lm}$ (kg/kg)	*Drying constant* (*equation* 8.5) (K'_H/L) (g/s m²/g/m² = s⁻¹) *for unit humidity difference*
0·005	7·0	9·0	0·018	72
0·005	5·2	7·6	0·019	77
0·005	3·5	4·8	0·020	69
0·043	7·0	6·4	0·013	70
0·081	7·0	5·2	0·009	82

Drying bed temperature = 70°C. Mass velocity of drying air = 1 360 g/s m². Diameter of preformed rods = 0·45 mm.

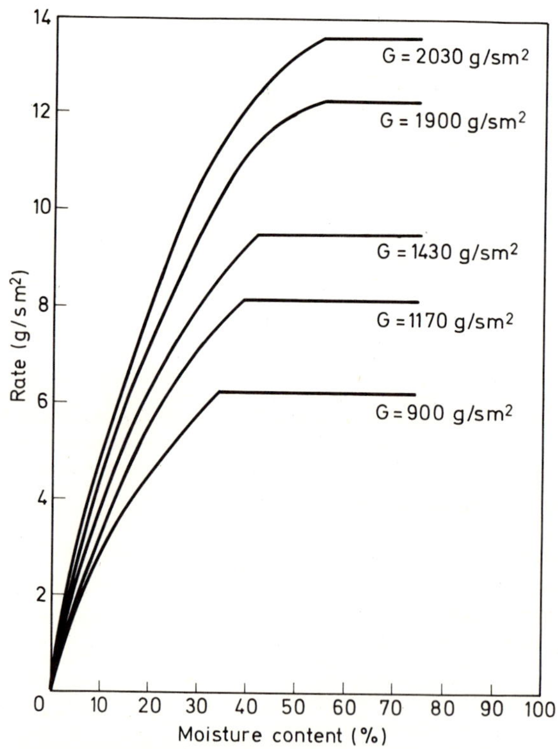

Fig. 8.21. Variation of drying rates with mass velocity of air stream, G—see equation 8.7

of 2·5, 3·8 and 5 cm, corresponding with bed loadings of 3·5, 5·2 and 7 kg/m^2. Over the range covered, the constant drying rate is almost proportional to the bed depth. Hence the time taken for constant-rate drying is almost independent of the bed depth. Effect of diameter of extruded rods: Fig. 8.20 shows drying rate curves for rods of 1·5, 3 and 4·5 mm. The relationship

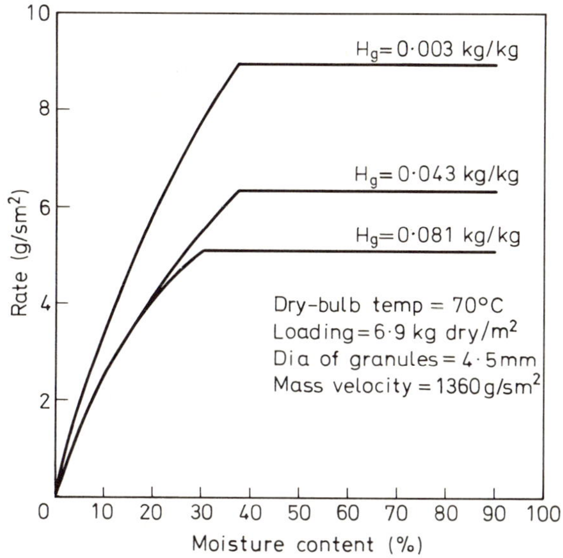

Fig. 8.22. Variation of drying rates with approach air velocity (H_g)

between the constant drying rate and the rod diameter within the range of diameters studied is given in equation 8.6.

Effect of air velocity: Fig. 8.21 shows drying rate curves for a few mass air velocities within the range 900–2 000 g/s m^2; the relationship is given in equation 8.7.

Effect of humidity of inlet drying air: Fig. 8.22 shows the drying rate curves with the air controlled at approach humidities of 0·005, 0·043 and 0·081 kg/kg dry air.

A.8.2.3 Basis of design method

Referring now to the equations given in Section 8.2, the data from laboratory through-drying experiments is applied in the following manner. Equation 8.5 is used to calculate the constant drying rate for a required approach air humidity (H_g), temperature and bed loading in kg dry material/m^2. Equation 8.6 can be used for calculating the effect of change in the state of sub-division of the material and equation 8.7 for the effect of change in air velocity.

The falling drying rate can then be calculated from the constant drying rate and critical-moisture content for the required conditions by graphical solution of equations 8.8–8.13 as indicated in Fig. 8.23. It is, however, implied that the critical moisture content and the shape of the falling-rate curve are unaffected by change in the drying conditions. These assumptions are not strictly true, but do not create serious error over the limited range of conditions involved.

When designing a full-scale through-circulation dryer it frequently happens that factors such as the loading, air velocity, and size of the material (also very often the drying temperature) are decided beforehand by economic

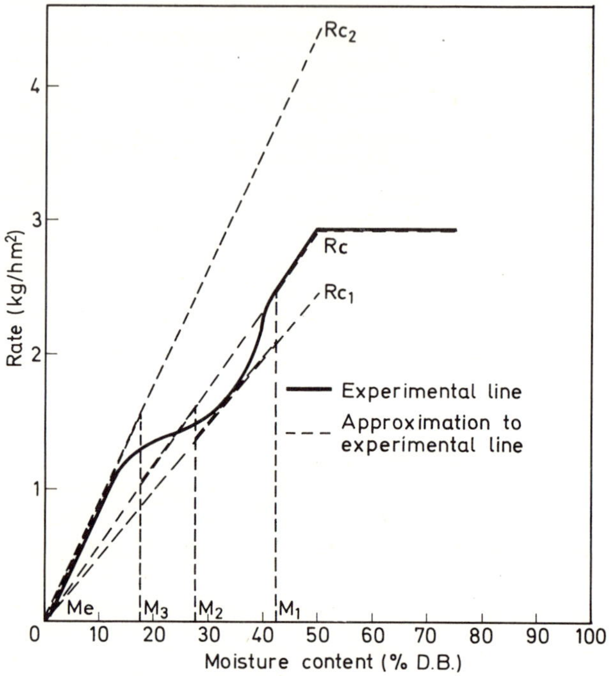

Fig. 8.23. Graphical construction to calculate time of drying in falling-rate period

considerations and the mechanical and thermal stability of the material. The laboratory experiment can be made so that these conditions are identical with those to be used and the design becomes essentially a matter of calculating the effect of the different humidities which will be used in a multi-zone dryer with air recirculation. Since humidity calculations are necessary anyway, the laboratory experiment can conveniently be carried out at room humidity, thereby simplifying the apparatus (cf. Fig. 8.4). If the available laboratory data is for different conditions of loading, material size, temperature or air velocity it should first be converted to the required basis by the methods outlined above.

A.8.2.4 Calculation of area of dryer

The method is one of successive approximation. The first step is to draw up an air/water balances on the lines of Fig. 8.24, assuming values for the areas of each zone of the dryer and also a probable distribution of water driven off in the air inlet stage, zone 1, and the final stage, zone 2, where the wet material enters. A few other factors must also be decided. In this worked example the assumptions made are:

(a) That each zone has an area of 4·2 m^2.
(b) That the make-up air is saturated at 20°C.
(c) That the air exhausted from the recirculating system in the dryer has a moisture content corresponding to a dew point of 50°C (this

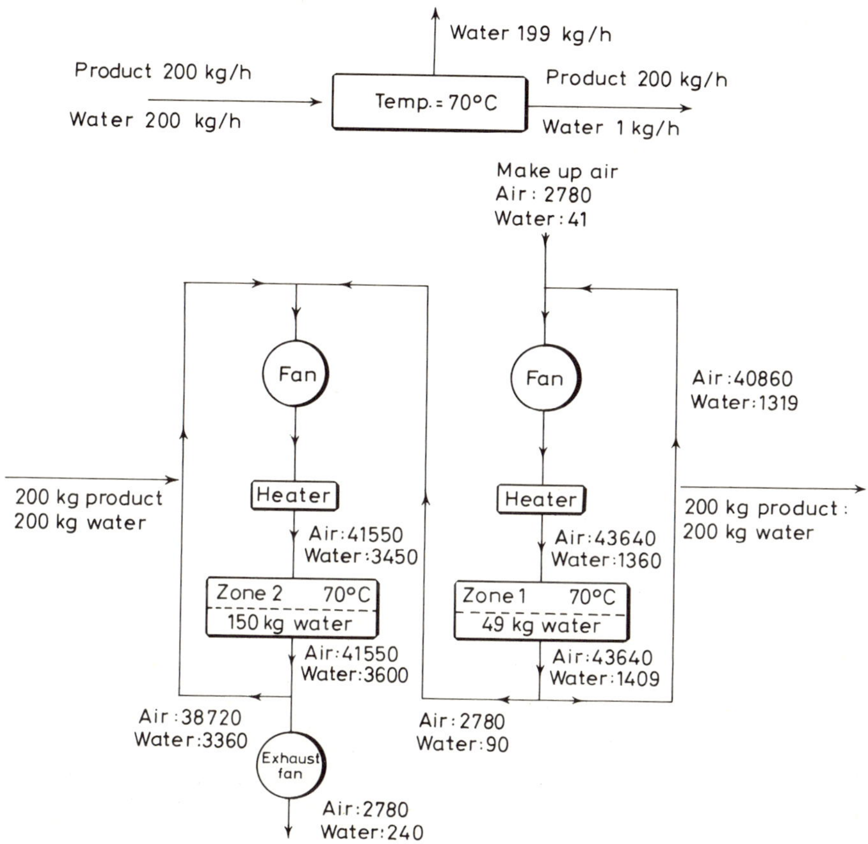

Fig. 8.24. Solids/water and air/moisture balances for dryer

is a reasonable temperature difference from the drying temperature of 70°C in order to ensure that condensation does not occur at any cold spot in the system).

(d) That 75% of the moisture is removed after half the drying time. (This has been estimated by inspection of the laboratory determined time/moisture content curve, Fig. 8.25, allowing for the increase in relative drying rate as the material enters zone 2.)

These assumptions are made as the result of experience. A less experienced designer would need to make several successive approximations to arrive at the quantities given below.

Humidity driving force in each zone: the mean can be calculated on the assumption that the material is at wet-bulb temperature.

Zone 2: The approach air humidity (H_1) = 0·083 kg/kg is derived from

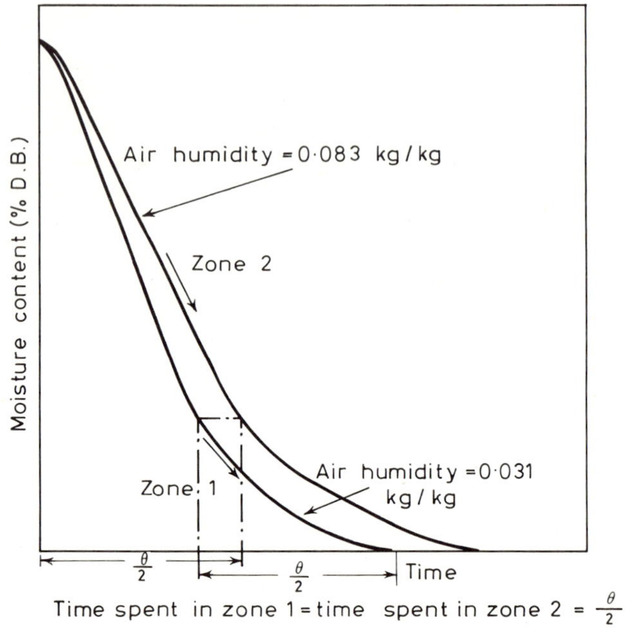

Fig. 8.25. Moisture content/time curves to determine experimentally the rate of drying for two-zone conditions

the preliminary air/moisture balance given in Fig. 8.24. From the experimental figures plotted in Fig. 8.18, which is for an approach air humidity of 0·005 kg/kg, the constant drying rate is 8·15 g/s m^2. This rate will be substantially lower when H_1 = 0·083 kg/kg and a figure of 4·1 g/s m^2 will be assumed as a first approximation; from this the humidity H_2 of the air leaving zone 2 is calculated to be H_2 = 0·087 kg/kg, see Fig. 8.23. Pyschrometric tables give the corresponding wet-bulb humidity H_{sw} = 0·094 kg/kg. Hence for zone 2

$$(\Delta H)_{\mathrm{lm}} = \frac{(H_{sw}-H_1)-(H_{sw}-H_2)}{\ln(H_{sw}-H_1)/(H_{sw}-H_2)} = 0{\cdot}009 \text{ kg/kg} \qquad (8.20)$$

The drying constant K'/L calculated by equation 8.5 from the laboratory measurements for the constant-rate drying period is 72 s^{-1} for unit humidity difference (see Table 8.6) and the humidity constant $K'_H = 72 \times 7 = 504$ g/s m^2. From this we can determine the constant drying rate which would occur in zone 2, namely $R'_c = 72 \times 7{\cdot}0 \times 0{\cdot}009 = 4{\cdot}5$ g/s m^2. The estimate of 4·1 g/s m^2 made above is sufficiently close to make little difference to the value of $(\Delta H)_{\mathrm{lm}}$.

Zone 1: Using the same method for an approach air humidity of 0·031 kg/kg (see Figs 8.23, 8.24), $(\Delta H)_{\text{lm}} = 0{\cdot}015$ kg/kg. Then for zone 1

$$R_c' = 72 \times 7{\cdot}0 \times 0{\cdot}015 = 7{\cdot}6 \text{ g/s m}^2$$

These constant-rate values are now plotted alongside the curve for the laboratory experimental results as in Fig. 8.26. The falling-rate curve for each zone condition is then constructed from the laboratory curve by simple proportion on the assumption that the shape and critical moisture content are unaffected by the change in drying conditions, thus giving complete drying curves for each of the zone conditions. Figs 8.18 and 8.26 show that the critical moisture content is 38% dry basis. The water evaporated in constant-rate drying is therefore 56 kg (155 g/s) and the required dryer area under zone 2 conditions is 155/4·5 = 3·5 m^2.

The remaining 0·7 m^2 of zone 2 and the whole of the 4·2 m^2 of zone 1 will thus be taken up with falling-rate drying. The path of the material

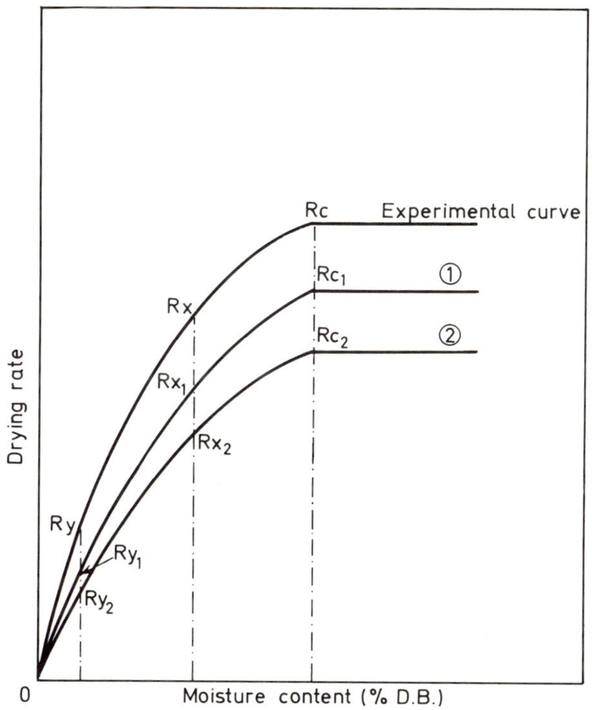

Fig. 8.26. Construction of drying rate curves for design purposes

going through the dryer can now be drawn on the rate curves as shown in Fig. 8.27. The material will pass along the line R_{c1} under zone 2 drying conditions and then follow the falling-rate curve under zone 2 conditions until, at a moisture content M_x, it passes into zone 1 and follows the falling rate curve under zone 1 conditions of lower circulating air humidity. This path is represented on Fig. 8.27 as the series of straight lines *h-b-d-s-r-q-p-n-o'*.

At this stage we can check the value of M_x which was assumed to be 25% from the laboratory data. Since a single band at constant dry loading is passing at a constant speed through both zones, we have

$$\frac{\theta_{M_c-M_x}}{\theta_{M_x-M_f}} = \frac{0{\cdot}7}{4{\cdot}2} = \frac{1}{6} \tag{8.21}$$

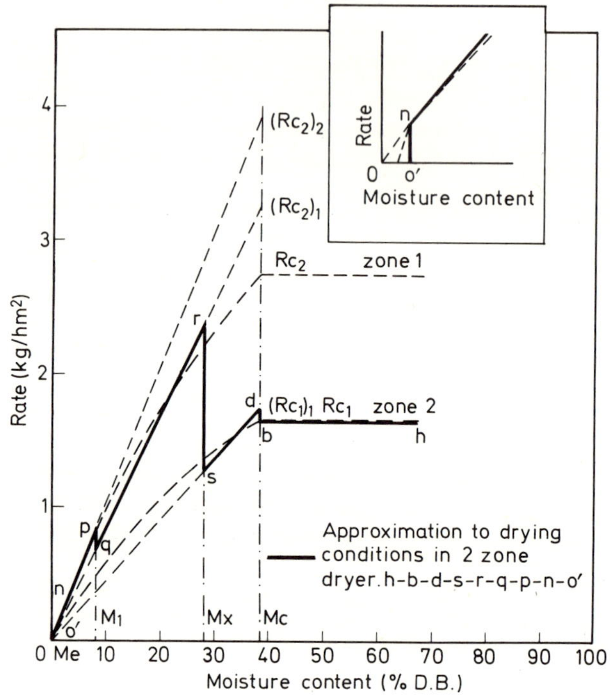

Fig. 8.27. Graphical calculation for 2-zone continuous dryer

Hence from equation 8.11 and values of R'_c read from Fig. 8.27, we have

$$\frac{6}{(R'_{c1})_1}\ln\frac{M_c}{M_x} = \frac{1}{(R'_{c2})_1}\ln\frac{M_x}{M_1}+\frac{1}{(R'_{c2})_2}\ln\frac{M_1}{M_f} \tag{8.22}$$

$$\frac{6}{4{\cdot}75}\log\frac{0{\cdot}38}{M_x} = \frac{1}{8{\cdot}8}\log\frac{M_x}{0{\cdot}08}+\frac{1}{10{\cdot}7}\log\frac{0{\cdot}08}{0{\cdot}005}$$

$$1{\cdot}26\log 0{\cdot}38 - 1{\cdot}26\log M_x = 0{\cdot}114\log M_x - 0{\cdot}114\log 0{\cdot}08 + 0{\cdot}0935\log 0{\cdot}6$$

$$1{\cdot}374\log M_x = -1{\cdot}26\times 0{\cdot}42 - 0{\cdot}114\times 1{\cdot}097 - 0{\cdot}113 = -0{\cdot}767$$

$$\log M_x = -0{\cdot}557 = \bar{1}{\cdot}443$$

$$M_x = 0{\cdot}277$$

Hence $M_x = 28\%$, which is in satisfactory agreement with the assumed value of 25%. We can therefore complete Fig. 8.26 by addition of points *s* and *r* at 28% moisture content.

Equation 8.13 is now used to calculate the area required in zone 2 to dry from the critical moisture content of 38% down to 28%

$$A'_{M_c - M_x} = \frac{W_d M_c}{(R'_{c1})_1} \ln \frac{M_c}{M_x} \tag{8.13}$$

$$= 25 \times 0{\cdot}38/4{\cdot}75 \times 2{\cdot}303 \log 0{\cdot}38/0{\cdot}28 = 0{\cdot}61 \text{ m}^2$$

Similarly for the area of zone 1.

$$A'_{M_x - M_o} = TM_c \left(\frac{1}{(R'_{c2})_1} \ln \frac{M_x}{M_1} + \frac{1}{(R'_{c2})_2} \ln \frac{M_1}{M_f} \right)$$

$$= 25 \times 0{\cdot}38 \times 2{\cdot}303 \left(\frac{1}{8{\cdot}8} \log \frac{0{\cdot}28}{0{\cdot}08} + \frac{1}{10{\cdot}7} \log \frac{0{\cdot}08}{0{\cdot}005} \right)$$

$$= 25 \times 0{\cdot}38 \times 2{\cdot}303\,(0{\cdot}0618 + 0{\cdot}113)$$

$$= 3{\cdot}84 \text{ m}^2$$

The area of dryer required is thus

	m^2
Zone 1 – falling rate	3·84
Zone 2 – falling rate	0·61
Zone 2 – constant rate	3·50
Total	7·95, say 8·0 m^2

This is in close agreement with the assumption of 4·2 m^2 per zone made at the beginning of this design calculation. The total area of dryer for design purposes might therefore be taken as 8·0 m^2. In practice a dryer of 9 or 10 m^2 is likely to be purchased in order to cover variation in the drying characteristics of the wet material (unless the laboratory experiments were also repeated with a 'bad average' sample).

This design example shows that there is a reasonably straightforward procedure for scaling-up laboratory data despite lack of knowledge of the fundamental mechanisms occurring during the falling-rate period. The main assumptions are that there are not large changes in the shape of the falling-rate curve nor in the value of the critical moisture content on changing from the laboratory apparatus to one of the full-scale machines available, and some indication of this should show up in a series of laboratory tests with several samples. The experimental errors with the laboratory apparatus are about 5%. Because of wall effects in the small laboratory equipment, it is to be expected that it will overestimate the size of a full-scale dryer.

CHAPTER NINE

Continuous drum or film dryers

9.1 DESCRIPTION

The process of drum drying is essentially the drying of a thin film of solution or suspension on a heated drum and subsequent removal of the film of dry solids by means of a doctor knife. The process is sometimes known as roller drying; the term cylinder drying is usually confined to the drying of thin continuous sheets of material, e.g. paper, film etc. [16,17,18,20] The drum dryer consists basically of one or two hollow heated drums to which the solution (or suspension) is applied by various feed methods. Steam is the normal heating medium for drying under atmospheric conditions, although hot gases have been infrequently used; for drum drying under vacuum the drums may be heated by hot water or steam at sub-atmospheric pressure.

Materials of construction of the drums are usually fine grain cast iron (semi-steel), gunmetal, and very occasionally stainless steel. Semi-steel may also be chrome surface hardened to give increased resistance to corrosion and wear. Auxiliary fittings such as end frames, conveyors, knife carriers, etc, may be fabricated in cast iron, mild steel or stainless steel.

Cleaning of the drum dryer is simple. Maintenance costs of the twin drum dryer are low; for the double drum dryer and the vacuum dryer they may be appreciably higher.

For safe stress reasons the maximum size of drum is limited to approximately 12 m^2 drying area and it may be necessary to use a multiplicity of machines for high outputs. In these circumstances, the considerable capital expenditure and increased labour requirements involved may tend to reduce the advantages of the relatively cheap drum drying process.

Ancillary equipment normally provided includes a draught hood for removal of water vapour and dust, conveyors and chutes for dry powder, feed tank, feed tray and recirculation tank and pump or special arrangements such as splash feed fittings for drying slurries. In vacuum drum drying a condenser and vacuum pump or ejector are necessary.

9.2 APPLICATIONS AND LIMITATIONS

The drum dryer can be used to dry many materials at a relatively low operating cost since supervision charges are low and there is a minimum of materials handling. It is particularly suitable for heat-sensitive products where short-time high-temperature drying is permissible. A wide range of drying conditions is obtainable by variation of steam pressure and drum speed. The dry product is quite often powdery and may not require grinding. It is sometimes possible to control drying to give flake or chip material where a particular physical form of material is required.

Nevertheless all types of drum dryer are unsuitable for drying materials which, in the dry form, exhibit lack of adhesion to metal. Materials which decrepitate on heating may be also unsuitable for drum drying unless special precautions are taken to recover the powder lost from the surface of the drum. A method of drum drying powders between two layers of cloth has, however, been described.[22]

Solutions of soluble silicates invariably cause knifing difficulties; also suspensions of some insoluble silicates. Formation of a glazed film of solids on the drum results in incomplete removal of the solids film and excessive knife wear. Where 'glazing' does occur it must be generally concluded that the material is unsuitable for drum drying.

Materials which are thermoplastic often dry unsatisfactorily and cannot be knifed cleanly. Use of a cold air jet at the knifing point has shown some improvement, also the addition of 10% of sodium sulphate or sodium chloride to the material where this is permissible.

Material losses are low on well designed machines and comparable to closed drying processes. Hence the drum dryer can be used where a number of products is handled although the relative size of each production is small. Its ideal application however, is in the continuous drying of a single product, or a group of closely related chemical compounds.

9.3 GENERAL THEORY

9.3.1 Formation of liquid films on heated drums

The output rate of any drum dryer depends on the thickness of the liquid film on the drum and the drying characteristics of the film.

In double drum drying the thickness of the film is limited by the fixed clearance between the drums and is thus independent of the physical properties of the solution or slurry being dried.

On single and twin drum dryers, the physical properties of the feed solution control the liquid film thickness and may considerably affect the output rate. Statistical analysis of experimental data from plant and semi-technical drum dryers established that the weight and thickness of the liquid film picked up by a rotating heated drum can be expressed as a function of adhesion tension and peripheral drum velocity. In the range of drum temperatures normally employed in practice, 150–160°C, the relationship is satisfied by the equation:

$$\rho t_l = \frac{a\gamma_a^{\frac{3}{2}}}{v_d^{\frac{1}{3}}} = \frac{a(\gamma \cos \phi)^{\frac{3}{2}}}{v_d^{\frac{1}{3}}} \tag{9.1}$$

Viscosity is shown to have no significant effect on film thickness in the viscosity range 1–20 cP; most feed solutions have viscosities within this range.

In practice, employing semi-steel heated drums, the thickness of the liquid film varies between 25 and 125 μm.

There is no immediate application of this data. It does indicate, however, that the maximum film thickness is obtained from solutions having adhesion tensions approaching 75×10^{-3} J/m^2 (75 dyne/cm^2), the highest value normally attainable for aqueous solutions. Conversely solutions of wetting agents, or addition of these to feed solutions, will correspondingly decrease the film thickness on the drum.

9.3.2 Drying of thin films on heated drums

Drying on heated drums is a process of heat transfer through the drum wall to the film of material being dried. In the evaporation and constant drying rate zones heat transfer may be expressed by the general equation (cf. Section 2.4.1.)

$$Q = \frac{10^{-3} A\theta_h (T_h - T_b)}{1/h_h + 1/h_w + 1/h_l} \tag{9.2}$$

In the falling-rate zone, which normally occupies 80–98 % of the total drying time, the temperature of the solids film commences to rise and the heat-transfer rate decreases. Because of difficulties in measuring the temperature of the solids film during the falling-rate period, individual film coefficients are rarely calculated and it is only practicable to calculate heat transfer over the entire drying period using the general equation:

$$\frac{Q}{\theta} = 10^{-3} U A \Delta T_\theta \tag{9.3}$$

Assuming that the solids temperature at the knifing point equals the steam temperature, calculated overall heat-transfer coefficients range from 60 W/m^2°C for a material which is dried with difficulty, to 400 W/m^2°C for an easily dried material.

Approximate values of individual thermal conductivities and heat resistances likely to be obtained in practice are

Condensing steam film $\qquad R_{st} = \frac{1}{11} \cdot \frac{°\text{C}}{\text{kW/m}^2}$

10 mm metal wall $\quad k_w = 42$ W/m° C $\qquad R_w = \frac{1}{4{\cdot}2} \cdot \frac{°\text{C}}{\text{kW/m}^2}$

75 μm liquid film $\quad k_l = 550 \times 10^{-3}$ W/m°C $\qquad R_l = \frac{1}{7{\cdot}3} \cdot \frac{°\text{C}}{\text{kW/m}^2}$

75μm solids film $k_m = 65 \times 10^{-3}$ W/m°C $\qquad R_s = \frac{1}{0{\cdot}86} \cdot \frac{°C}{kW/m^2}$

9.3.3 Correlation of drying data

Calculation and comparison of overall heat-transfer coefficients give some indication of the relative drying properties of material. To date, there is no satisfactory method of scaling laboratory tests to plant operation; satisfactory data can only be obtained by trials on a small-scale drum dryer, say 0·5 m diameter.

Correlation of some drum drying data has been possible by plotting the logarithm of unaccomplished moisture content change, defined as the ratio of the free water still to be removed at time θ to the total free water originally present, vs. logarithm of the drying time.

$$\text{i.e. } \log \frac{W_\theta - W_q}{W_o - W_q} \text{ vs. } \log \theta$$

Straight line relationships have been obtained for certain solutions and slurries; these could be interpolated to give moisture contents at intermediate drying times though further work is required to substantiate this approach.

9.4 DETAILED DESCRIPTION

9.4.1 Classification

Drum dryers may be classified according to the number of drums, the direction of rotation of the drums, and the feed arrangements, as shown in Fig. 9.1.

A Single drum,	dip feed (a) splash feed (b) transfer roll (c)	Atmospheric or vacuum operation
B Twin drum,	dip feed (d) splash feed (e)	Atmospheric or vacuum operation
C Double drum,	nip feed (f)	Atmospheric or vacuum operation

All three types are in general use in chemical industry but the vacuum dryer (usually double drum arrangement) is mainly used in the foodstuffs industry where products are particularly heat sensitive. Hence available information is largely confined to atmospheric single, twin, or double drum dryers. Also only two feed systems are normally employed (1) dip feed in which the drum rotates partially immersed in the feed solution and (2) nip feed in which the solution is introduced into the nip formed by two drums.

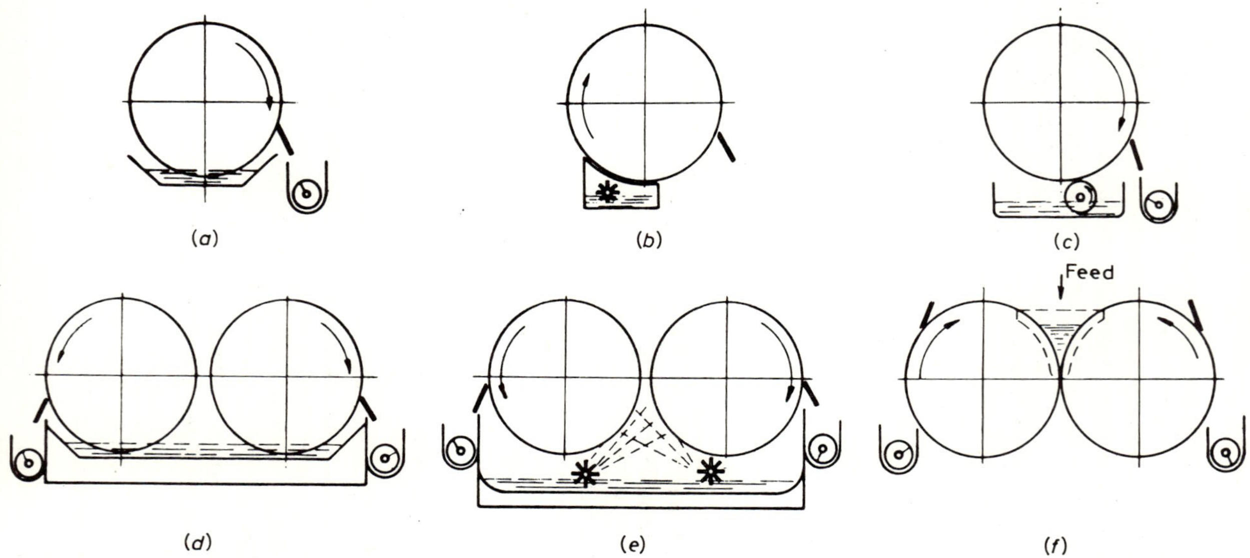

Fig. 9.1. Types of drum dryer

Splash feed and feed by transfer roll or weir box are not in general use and are confined to specific applications.

9.4.2 The atmospheric single drum dryer

This is the simplest form of dryer, consisting of a single (steam) heated drum of dimensions up to 1·25 m diameter × 3 m wide depending on working pressure. The drum rotates with an immersion of $1-2\frac{1}{2}$ cm into feed solution contained in a shallow trough. Alternatively the solution may be splashed or sprayed on to the drum or applied from a side trough. As the drum rotates, the film of solution, usually 25–125 μm, is dried and the dry product removed by a doctor knife pressed against the surface of the drum. Typical arrangements are shown diagrammatically in Fig. 9.1 a–c.

9.4.3 The atmospheric twin drum dryer

Designed to eliminate mechanical difficulties associated with the nip and the end plates of the double drum machine, the twin drum dryer consists essentially of two (steam) heated drums rotating outwards, Fig. 9.1 d–e. The feed solution to be dried is contained in a shallow trough or tray and each drum is immersed in the solution to a depth of $1-2\frac{1}{2}$ cm. Constant level in the feed tray is maintained. The alternative arrangements of splash feed or spray feed may be used. This type of dryer is also suitable for the single or double transfer roll type of feed where the feed solution is very viscous. The maximum size of drum employed is usually 1·25 m diameter × 3 m wide for operation at 5·5 pressure, or correspondingly smaller at higher pressures.

9.4.4 The atmospheric double drum dryer

In this type of dryer the two drums rotate towards each other, i.e. downwards into the nip, Fig. 9.1 f. The drums are set to maintain a small clearance (typically 125 μm) when hot and the feed liquid is introduced into the V-shaped trough formed by the nip of the two drums and spring-loaded end plates. One feature of this type of dryer is the concentration of the feed liquor by boiling that takes place in the nip thus augmenting the normal drying process. Since the gap between the drums is fixed, however, a film of definite thickness is applied to each drum. The knifing point is positioned so that the dry powder falls down the drum surface, countercurrent to a stream of air and water vapour.

9.4.5 Vacuum drum dryer

Normally used for drying heat-sensitive materials, to minimise oxidation, or for solvent recovery, the vacuum dryer was originally an atmospheric single drum dryer totally encased for operation at sub-atmospheric pressures. Present practice is to utilise the double drum machine under vacuum. The heating medium is usually hot water but sub-atmospheric or pressure steam may be used. Auxiliary equipment is necessary and usually consists of two receiving vessels for dried solids, wet dust collector and vacuum pump. Operation is complicated by vacuum; any adjustment of knives etc, must be made from outside the housing and involves vacuum-tight stuffing boxes, sight glasses etc.

Its use is confined chiefly to food products.[1, 9]

9.5 APPLICATION

9.5.1 The atmospheric single drum dryer

The major application of this type is in the drying of sticky materials or products forming hard lumps during drying; the latter accumulate between the end plates of the double drum machine and often force the drums apart. Although attempts have been made to improvise a nip feed to the single drum, the dip or splash feed is normally used. For feed solutions which are extremely viscous or which do not wet the hot drum surface particularly well, single or double transfer rolls may be used but they are difficult to operate.

9.5.2 The atmospheric twin drum dryers

This machine, employing dip feed, is suitable for drying solutions of viscosity 20 cP (0·02 Ns/m^2) at 80°C, which will not readily deposit crystals in the feed tray. Alternatively it may be used with a splash feed arrangement for drying slurries. It is the preferred machine for large scale production where batch-to-batch carry-over of material (tray residues) is not detrimental. It should be adopted as a general-purpose machine wherever possible on grounds of reliability, low maintenance and low operating costs.

Although solutions of 50% w/w solids contents may be handled on this machine, difficulty in forming a uniform film on the drums often arises at this concentration; 40% w/w solids is normally the highest concentration used.

The thickness of a liquid film picked up by a heated drum is roughly proportional to the surface tension of the liquid. Solutions having surface tensions approaching that of water (0·072 J/m) usually give the highest output rates. When drying solutions of low surface tension it is desirable to operate at the highest possible solids concentration to minimise the reduction in output of dried solids which would otherwise occur.

The twin drum dryer is not recommended for drying solutions containing $<12\%$ w/w solids since it is difficult to remove the resulting thin film of dry solids unless excessive knife pressure is used. In addition heat loss to the feed liquor is high and boiling or foaming in the feed tray may cause increased material loss.

9.5.3 The atmospheric double drum dryer

Although the double drum dryer is said to have the widest range of application its success depends largely on careful maintenance. It may be used for drying dilute solutions; boiling of the solution in the nip of the drums adds to the capacity of the dryer. Concentrated solutions, slurries, and even sludges which can only be pumped with difficulty may be handled in this type of machine. It is not advisable to use the double drum dryer, however, if the solution contains a solute of limited solubility or when abrasive solids or lumps may be thrown out of solution. Since the feed solution is pumped directly into the nip of the drums there is no need for a recirculating system, and the fixed clearance between the drums ensures uniformity of film and eliminates the need for spreaders. The design is particularly useful for materials whose final drying stages are difficult, for example materials which form stable hydrates; concentration of solution in the nip produces the effect of extended drying time without reduction in output rate. This advantage particularly applies when handling solutions of low surface tension, e.g. wetting agents, which would normally give thin films and low output rates on the single and twin drum machines. There is no heat loss from the drums to the bulk of the feed solution when feeding directly into the nip, cf. the twin drum dryer where heat loss from the partly immersed drum raises the temperature of the circulating feed liquor to 80–90°C. Hence particularly heat-sensitive solutions may be dried with the minimum amount of decomposition.

The absence of a heel or residue after drying makes the double drum dryer attractive for small-scale production of a large number of products where a high proportion of residual material could be an embarrassment, e.g. expensive chemicals. Nevertheless, whilst the double drum dryer can be applied to a wide range of products, its mechanical disadvantages tend to give lower output rates than would normally be expected; the proportion of lost time can be high unless intensive maintenance is carried out.

9.6 PERFORMANCE

9.6.1 Methods of operation

The basic methods of operation are typified by the particular machines, e.g. the twin drum dryer requires dip or splash feed operation. In general the required moisture content of the dried product fixes the operating conditions of drum speed and steam pressure for a feed solution of given strength. To achieve the maximum output of dried solids, however, control of feed rate and solution strength is necessary for all types of machine. The temperature of the feed solution should be as high as possible consistent with chemical stability of solution, deposition of solids from solution, tendency to frothing etc.

In both splash-feed and dip-feed operation there should be sufficient overflow from the feed tray to prevent stagnation and deposition of solids at the corners of the tray. In nip-feed operations, the feed is preferably metered to the dryer. Control by instrumentation is not normally practised since the drum dryer, once set up correctly, requires little attention and manual control is sufficient. Pressure gauges are the only essential instruments but a steam flowmeter and a rotameter for feed solution are sometimes used.

Where uneven film formation on the drums occurs, reducing the gap between the drums (until contacting if necessary) is preferable to the use of pilot rollers. Conversely films which are too thick may be controlled by the use of auxiliary scrapers fitted, in the case of single or twin drum machines, 5–15 cm after emergence of the drum from the feed solution.

Knifing conditions vary according to material of construction of the drums. For tempered steel blades against semi-steel drums it is permissible to use a high knife pressure, hence relatively hard thin films may be removed. With stainless steel or gunmetal drums there must be little or no pressure of knife on drum; hence only relatively thick films can be knifed satisfactorily.

Each type of drum dryer is flexible in operation; steam pressure can normally be varied over the range 3–5·5 bar and drum speed 5–15 rev/min. Conversion from one type to another cannot be conveniently carried out, however, mainly because of differing end frame and knife carrier designs. Multi-purpose machines are usually small and their use is mainly confined to experimental work.

9.6.2 Performance data

Criteria for assessing the performance of a drum dryer are output rate of dried solids, evaporation rate, residual moisture content, and thermal efficiency. Output rates of dried solids vary widely, not only according to type of dryer; they are usually dependent on variations in the physical properties of the solutions being dried. The output rate is fixed within fairly narrow limits for a pre-determined residual moisture content and more often than not the output rate cannot be advantageously altered by control of solution properties.

Table 9.1 indicates typical performance data for a variety of products and can form a basis for estimating output rates by analogy; for an accurate assessment, a small-scale drying trial must be carried out.

Output rates for most inorganic salts are high on any type of drum dryer. Conversely wetting agents give low rates unless highly concentrated solutions are fed to a double drum machine. Organic chemical products vary considerably in output rate. The presence of additives such as wetting agents, sodium bicarbonate etc., tends to make the drying process more difficult. Products of high solubility in water often exhibit poor heat transfer properties

Table 9.1 Performances of drum dryers (size range between 0·4 × 0·4 and 0·8 × 2·25 m)

Type of dryer and feed size by letter, A, B or C	*Drum speed* (rev/min)	*Steam press* (bar, g)	*Type of material*	*Physical form of feed*	*Solids in feed* (%)	*H_2O in product* (%)	*Output of dried product* (g/s m²)	*Evaporation rate of water* (g/s m²)
			Inorganic salts					
Single (dip)	4·4	3·5†	Alk. carbs	—	50	8–12	5·5	4·9
Single (splash)	1	3·0	$Mg(OH)_2$	Thick slurry	35	0·5	1·9	1·5
Twin (splash) A	3	3·0	$Fe(OH)_3$	Thin slurry	22	3·0	4·3	1·3
Double	3–8	5·0	Na Acetate	Solution	20	0·4–10	2·0–7·0	8–24
Double and twin	7–9	2–3	Na_2SO_4	Solution	24	0·15–5·5	4·7–6·1	11–12
Double and twin	5–9	4–6	Na_2HPO_4	Solution	44	0·8–0·9	8·2–11·1	9–14
Twin (dip) A	5	5·5	Organic salts	Solution	27	2·8	1·9	5·2
Twin A	3	5·5	Organic salts	Solution	33	13·0	1·4	2·6
Twin B	2*	3·5	Organic salts	Solution	20	1·0	1·0	3·8
Twin C	5	5·5	Organic salts	Solution	39	0·4	3·9	6·1
Twin C	5½	5·5	Organic salts	Solution	42	1·0	2·1	4·6
Twin C	6	5·5	Organic salts	Solution	35	5·0	4·1	7·2
Twin (splash) A	3–5	5·0	Organic salts	Thin slurry	20	1·7–3·1	1·0–1·9	3·7–7·3
Double A	5½	6·0	Organic salts	Solution	11	—	1·1	9
Double B	6½	5–6	Organic salts	Solution	40	3	3·4	4·9
Twin (dip) A	5	5·5		Thin slurry	30	1·2	2·4	5·5
Twin A	5	5·0	Organic	Viscous Soln.	28	10·5	1·9	4·2
Double	2	3·0	Compounds	Viscous Soln.	—	6·0	0·7	—
Double	4½	3·5		Thin slurry	25	1·0	0·4–1·9	3·5–5·0
Twin (dip)	5	5·0	Organic	(a) Solution	25	0·5	0·3	0·8
Twin	10	5·5	compounds of	(b) Thick slurry	30	2·5	2·0	4·6
Twin	10	5·5	low surface tension,	(c) Thick slurry	35	—	3·1	—
Double	11	5·5	similar letters	(b) Thin paste	46	—	6·4	7·3
Double	12	5·5	for same	(c) Thick paste	58	—	6·0	4·3
Double	11	5·5	compound	(a) Solution	20	0·5	0·24	1·0

†Plus external hot air flow *Stainless steel drum Dryer dia. and width, metres A – 0·457 × 0·457 B – 0·71 × 1·52 C – 0·91 × 2·54

during the intermediate and final stages of drying. The low output rates of such products are a direct result of the extended contact time required for complete drying.

Evaporative rates also vary considerably; whilst values as high as 25 g water/s m^2 drying surface have been quoted in literature, 15 g/s m^2 is rarely exceeded in practice and 7–11 g/s m^2 is the normal range. The thermal efficiency of the drum dryer may vary from 40% in the case of a very dilute solution (twin drum) to 70% for a feed solution of high solids content, i.e. the steam usage varies from 1·4–2·5 kg/kg water evaporated. Radiation and

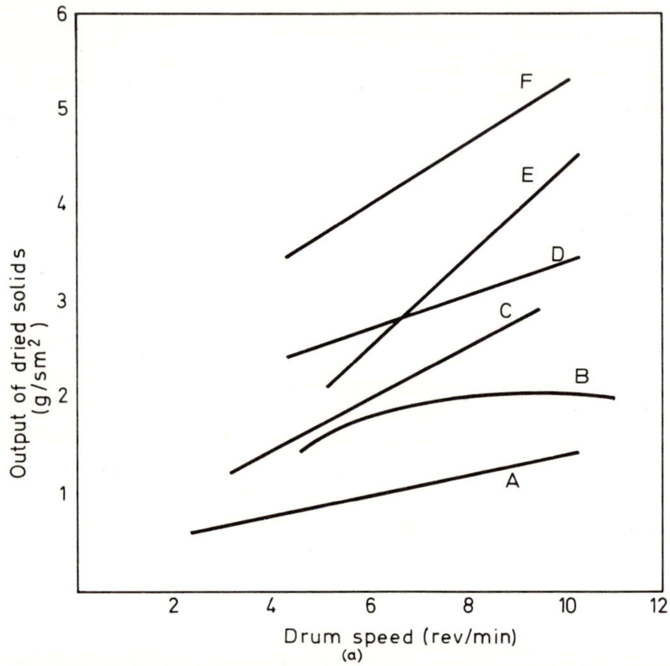

Fig. 9.2. Variation in output rate and moisture content of dried product with drum speed. (a) Single and twin (dip feed). (A) wetting agent, (B) an organic compound, (C) a sodium salt of an organic sulphonic acid, (D) a wetting agent, (E) a sodium salt of an organic sulphonic acid; (b) Double (nip feed). (A) wetting agent, (B) sodium acetate, (C) a water soluble dyestuff; (c) (A–D) sodium salts of organic sulphonic acids, (E) an organic compound

convection losses from the drying surface and the ends of the drums are fairly high and in the case of organic materials have been determined as 2·5 kW/m^2 °C on the total drum surface, e.g. a 15 m^2 drum dryer will consume 65 kg/h of steam under non-drying conditions. The influence of steam pressure in the drums on drying rate is usually not so marked as one would expect.[2] The highest practicable drum pressure is always advantageous but increased radiation losses at high steam pressures lower the overall thermal efficiency. The normal range of steam pressures is 3·5–5·5 bar but pressures as high as 12·5 bar may be used in stainless steel drums.

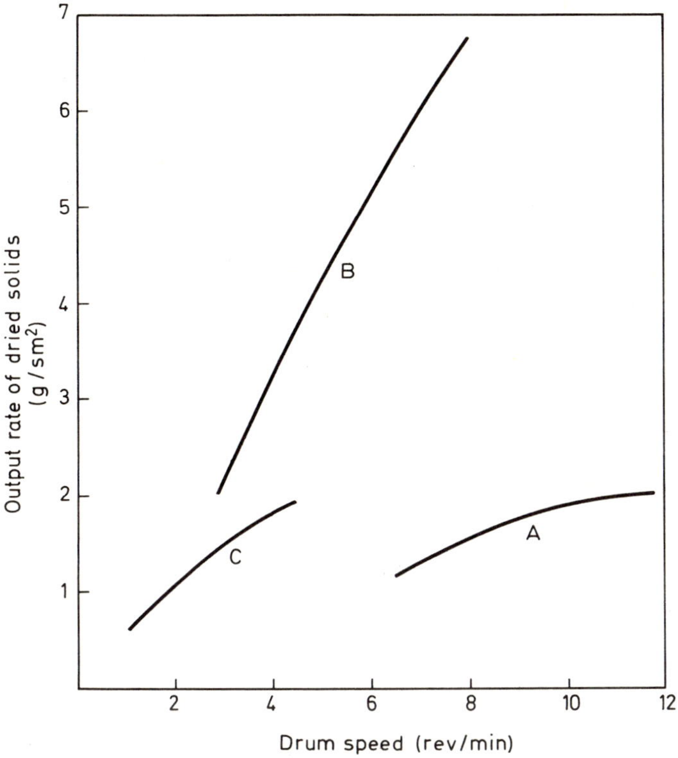

(b)

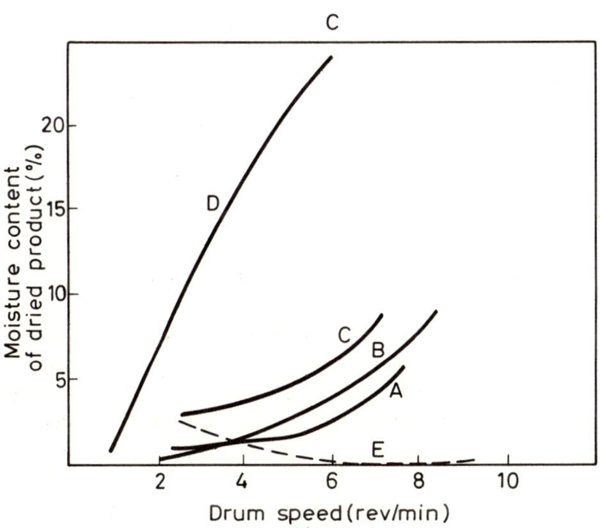

(c)

Fig. 9.2 (continued)

The effects of variables such as steam pressure, drum speed etc. are summarised in Table 9.2.

Typical output vs drum speed and moisture content vs drum speed curves are shown in Fig. 9.2. For most purposes it may be assumed that output rate varies linearly with drum speed. Some drying curves are shown in Fig. 9.3.

Table 9.2 Effects of variables on double and twin drum dryer performance

Double-drum dryer:

Effect	*Variable increased*			
	Steam pressure	*Feed temperature*	*Drum speed*	*Drum clearance*
Increased	Film thickness. Evaporation on drum	Film thickness. Evaporation on drum		Film thickness. Evaporation on drum
			Moisture content of powder	Moisture content of powder
Decreased	Moisture content of powder	Moisture content of powder	Evaporation on on drum	

Single- and twin-drum dryers:

Effect	*Variable increased*			
	Steam pressure	*Feed temperature*	*Drum speed*	*Drum clearance*
Increased	Evaporation on drum	Evaporation on drum	Evaporation in tray	Evaporation in tray
	Evaporation in tray	Evaporation in tray	Moisture content of powder	Moisture content of powder
Decreased	Film thickness. Moisture content of powder	Film thickness. Moisture content of powder	Film thickness. Evaporation on drum	

9.7 DESIGN BASIS

9.7.1 Assessment of material and type of dryer

The suitability of a solution or slurry for drum drying can only be determined by a small-scale test, preferably on a multi-purpose machine with drums about $\frac{1}{2}$ m diameter and $\frac{1}{2}$ m long. The test machine should have drums of the same material of construction as the proposed production machine otherwise results may be misleading. Output rates (g dried material/s m^2) may be scaled up directly from a machine of this size but it is doubtful whether the same reliance could be placed on tests with 15 cm diameter drums. Optimum drying conditions (drum speed and steam pressure) should be established, first using the twin drum arrangement with dip feed, to give the required moisture content or physical form of product. If both drum speed and steam pressure still limit the drying rate, parallel tests on the double drum dryer using nip feed will show whether evaporation of feed liquor in the nip can be used to assist drying. Slurries will require tests by splash

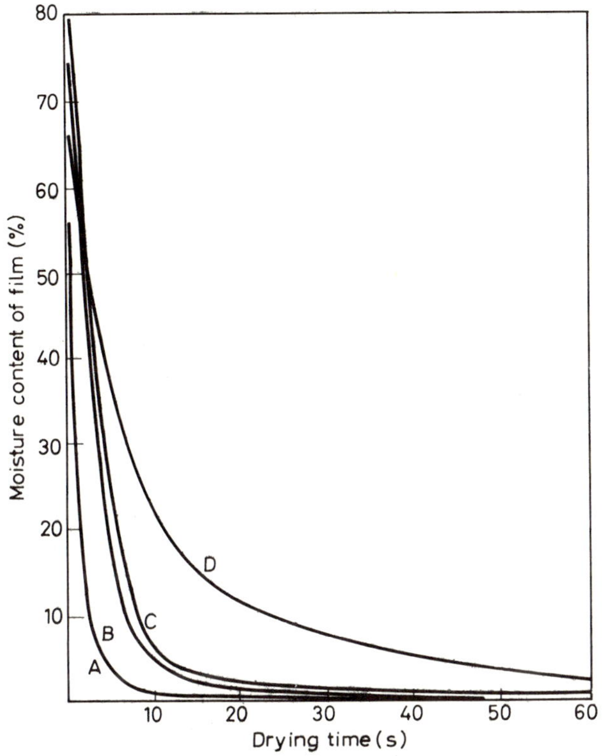

Fig. 9.3. Typical drying curves for drum dryers. (A) dyestuff, (B) wetting agent, (C) iron oxide, (D) sodium salt of an organic sulphonic acid

feed (single or twin drum). Tests with a double drum (nip feed) are not recommended; separation of lumpy material may force the drums apart and cause frequent shear pin failure, and leakage from the end plates is a constant problem.

Tests should preferably be carried out with different concentrations of feed liquor or slurry. The highest output rates can normally be obtained when the solids concentration in the feed solution is approaching that at saturation point. A typical drying test procedure is outlined in Appendix 9.1.

9.7.2 Size of dryer

The output rate of dried product per unit area per hour will have been determined by preliminary tests and, with a moderate allowance (around

25%) for maloperation, this figure can be used to determine the required drying area for large-scale production. The number of drums (or machines) employed (according to type) depends largely on the operating steam pressure since this limits the size of the drum. As a rough guide Table 9.3 indicates the maximum steam pressure and length of drum for a given diameter.

Table 9.3 Dimensions and normal working pressure for various diameter drums

Diameter (mm)	*Length* (mm)	*Working pressure* (*gauge*) (bar)
700	1 500	7 (semi-steel)
800	1 300	12 (stainless steel)
900	2 500	5·5 (semi-steel)
1 200	3 000	3·5 (semi-steel)

Individual makers' designs may vary from the figures quoted as these refer to general practice rather than absolute values. Wherever possible a twin drum machine should be used in preference to a single drum of the same area so that higher steam pressures may be used on the smaller diameter drums of the twin machine.

9.7.3 Drive

The drive to the drum dryer is normally obtained through a variable gear box with a ratio change of 3:1 giving a final drum speed range of 4–12 rev/min for general purpose drying. For multi-product drying, however, a 5:1 ratio gear and drum speed range 2–10 rev/min would be preferable. Final drive from the gearbox to the drums may vary according to make, e.g. chain drive, worm wheel and pinion or standard gearing. Size of driving motor corresponds to type of machine and size of drum(s). Table 9.4 indicates sizes of motors fitted to some typical drum dryers.

Table 9.4 H.p. requirements for drum dryers

Type of dryer	*Surface area of drums* (m^2)	*Diameter and width of drums* (mm)	*Motor* (hp)
Multi-purpose (twin)	1·25	450 × 450	5
Single	3·3	700 × 1 500	$7\frac{1}{2}$
Double	5·7	700 × 1 300	10
	6·6	700 × 1 500	
Twin	6·6	700 × 1 500	15
Twin	14·0	900 × 2 500	20

Auxiliaries: Add 50% to the dryer H.P.

9.7.4 Dust elimination

Major dust losses occur at the knifing points, particularly when the moisture content of the dried product is <2%. The loss may be minimised by one of three methods.

(a) Total enclosure of the machine, which may be applied to all types of dryer. In this method it is essential to prevent condensation inside the casing by effective draught control or by secondary heating. Because of this tendency towards condensation this method is not normally preferred. A typical installation is, however, shown diagrammatically in Fig. 9.4a.
(b) Removal of product at the doctor knife by means of a pneumatic extraction system. A cyclone separator and bag filters effect separation of the entrained solids. 1% loss of total solids in the feed is reported.

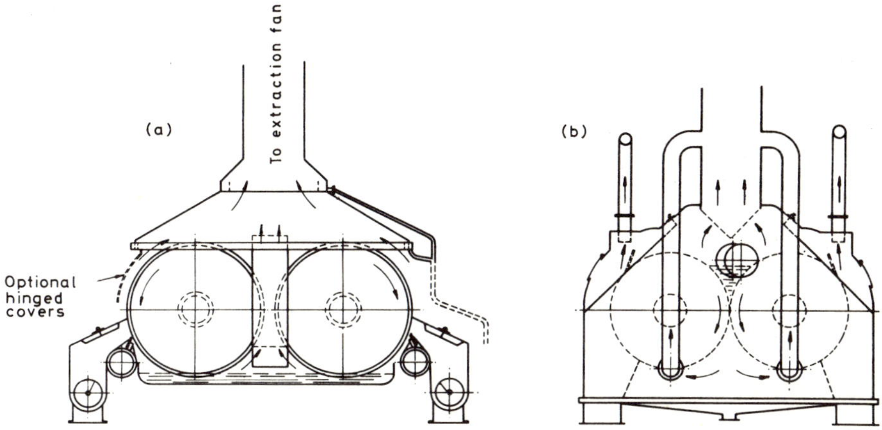

Fig. 9.4. Typical draughting systems. (a) Twin drum, (b) double drum

Care is essential when using this method otherwise condensation in the separator will occur and wetting of the filter bags (which would choke them and cause failure).
(c) By partial enclosure of the knife assemblies, as the greatest proportion of dust loss occurs at these points. This method is applicable to single and twin drum machines and reduces dust loss at the knives to approximately $\frac{1}{2}$% of the total solids dried. The arrangement is outlined in Fig. 9.4b.

9.7.5 Draughting

For partial hooding, in which the vapour hood covers the top of the machine only, the air velocity through the clearance between the periphery of the hood and the drums must be at least 0·8 m/s. A 'duplex' hood, in which a high velocity air stream is drawn through the annulus between two spaced canopies, has also been used to obtain high air velocities above the knifing points. In the case of twin drum dryers, vapour chutes should be provided at each end of the dryer to carry water vapour from the feed tray into the extraction hood. A typical extraction system is shown in Fig. 9.4b. It will be noted that provision is made for drainage of the vertical stack and also for collecting condensation on the inner surface of the hood.

In the case of total enclosure of the machine, one of two conditions must be satisfied: (a) the casing must be heated or effectively lagged to prevent condensation, or (b) there must be sufficient ingress of air to take up the evaporated water without condensation occurring.

9.7.6 Doctor knives

Correct setting and maintenance of the cutting edge are of primary importance, particularly when the dry material forms a hard thin film on the drum surface. The recommended cutting edge shown in Fig. 9.5a is stronger and retains its cutting edge for a longer period of time than the plain 'knife-edge'.

Hardness of the doctor blade is of primary importance if gunmetal or stainless steel drums are employed. Gunmetal and stainless steel normally have Brinell Nos. of 180–200 and the doctor blades should not exceed 260–280 Brinell. For semi-steel, cast iron or chrome-hardened drums the doctor knives may be tempered to a Brinell No. of 450–480. The above conditions apply to both tempered steel and stainless steel knives. General practice is to employ stainless steel (on grounds of durability and corrosion resistance) or tempered steel knives – there is little information on other materials for knife fabrication. The normal positions of the knife in relation to the drum are shown in Fig. 9.5b.

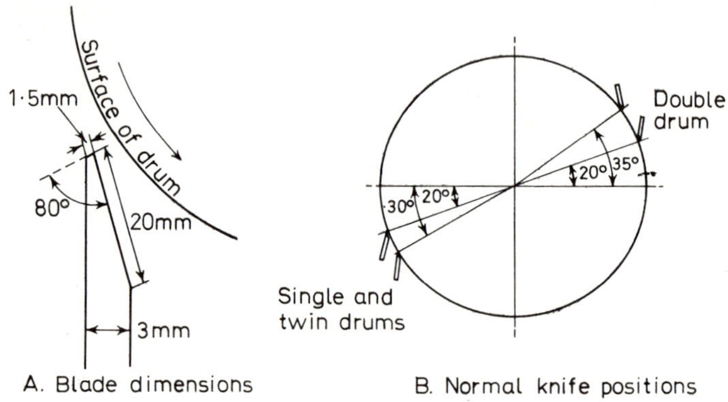

Fig. 9.5. (a) Blade dimensions and (b) knife positions

When stainless steel drums are being operated the doctor blade should project not less than 25 mm out of the blade carrier to give additional flexibility; the knifing point should be nearer the centre line of the drum than in standard practice in order to present a shallow knifing angle and minimise the risk of 'turning' the drum.

9.7.7 Nip and end-plate settings (double drum)

On the double drum dryer reliable means of setting the (hot) drums to a gap of 25–125 μm are essential. Sliding trunnion brackets adjusted by means of vernier screwed rods and captive handwheels are normally used. End-plates are preferably spring loaded to ensure uniform contact against the end surfaces of the drums. A marked reduction in leakage can be obtained by facing the contact surface of the end-plate with Tufnol sheet.

9.7.8 Material of construction of drums

For normal production work involving no corrosion or particular quality aspects, drums are fabricated in semi-steel, i.e. cast iron to which a proportion of steel has been added. This type of drum presents a slightly porous surface which assists film pick-up but after prolonged use 'pitting' occurs and the drums require skimming. Chrome hardening of the drum surface gives increased resistance to pitting but 'plated' drums are not satisfactory.

Gunmetal drums are employed when the product is very slightly acid or when contamination of the product must be kept to a minimum. Careful knife setting and maintenance are required for this relatively soft alloy.

Stainless steel may be employed where corrosion resistance is of importance but the manufacturers do not recommend this material for normal use. Experience suggests that the 'mirror' finish of stainless steel tends to hinder film pick-up in the case of twin drum dryers. It has been shown in practice that measurable distortion of the stainless steel drums may occur when steam at 12 bar gauge pressure is applied and extreme care is required in knife setting to prevent scoring of the drum surface.

9.7.9 Discharge arrangements

Provision for discharge of the dried powder varies considerably. In general, however, the dryer is elevated and the screw conveyors, situated immediately beneath the doctor knives, discharge directly into tared drums. Discharge of dried product into bulk hoppers is not normally practised, although the product may be transferred to bulk storage by elevators or Redler type conveyors.

Pneumatic extraction of the product from the knifing points may also be used.

9.8 COSTS

9.8.1 Capital cost

Approximate capital costs of various sizes of drum dryers are given in Table 9.5. Each cost represents the machine only at 1968 values. Additional expenditure for driving motor and installation of machines is about 70% of the cost of the bare dryer.

Table 9.5 Approximate capital costs of drum dryers

Machine (mm)	*Material of construction*	*Total drying surface* (ft^2)	*Cost* (*1968*) (£/m^2 *drying surface*)
450 × 450 Twin	Semi-steel	1·25	1 150
700 × 1 500 Single	Semi-steel	3·3	950
700 × 1 500 Twin	Stainless steel	6·6	1 300
900 × 2 500 Twin	Semi-steel	14·0	500

9.8.2 Labour

Major labour costs are for preparing the feed solution or slurry, starting, final cleaning of the dryer, and handling the drums of dried material. Preparation of the feed solution or slurry is not necessarily an intrinsic part of the drying operation however, as the product may be received batchwise from the preceding process as solution or slurry. For estimation purposes it is usual to take one man as fully-occupied during preparation of feed solution, start-up of machine (2 h) and cleaning of machine (1 h). During the actual drying operation 30–50% time allocation per machine is usually sufficient.

9.8.3 Services

Electricity requirements may be estimated from the hp requirements shown in Table 9.4 assuming that one hph consumes 1 kWh at factory meter. Steam requirements will vary according to product, strength of solution, etc. but for estimating purposes a figure of 1·8 kg steam per kg water evaporated can be used. Cooling water, additional steam and electricity will be assessed from consideration of ancillary equipment.

9.8.4 Maintenance

Maintenance costs vary according to type of dryer and nature of product. For double drum dryers or when the material is such that excessive knife wear occurs (e.g. drying of solids of an abrasive nature), maintenance costs may be considerably higher.

For semi-steel drum dryers a maintenance figure of $7\frac{1}{2}$% of the capital cost may be used. The figure should be increased to 10–15% for stainless steel machines.

9.9 WORKED EXAMPLE OF CAPITAL AND OPERATING COST

The following example is not necessarily representative of actual drying costs.

9.9.1 Problem

Calculate the direct expenses (i.e. excluding capital charges and works overheads) for drying 200 t/y of a 25% solution of an organic compound to a final moisture content of 2%. The batch size is limited to 1 100 kg. The dryer's availability is 46 weeks/y.

Tests on a 25% feed solution indicate an output dry solids rate of 1·3 g/s m^2 with a drum speed of 5 rev/min using steam at a gauge pressure of 5·5 bar.

9.9.2 Size of dryer

A working rate of 4 batches per week is 202 t/y. Four batches per week on continuous shift = 42 h per batch. The time taken for the 'loading' of each batch and the adjustment and cleaning of the dryer between batches is 5 h, hence actual duration of drying each batch is 37 h. The size of dryer required is, therefore:

$$\frac{1100 \times 1000}{1{\cdot}3 \times 37 \times 3600} = 6{\cdot}4 \text{ m}^2$$

A standard dryer 700 mm diameter, 1 500 mm wide has an area of 6·6 m^2 and will just suffice.

9.9.3 Cost of dryer

Dryer alone 6·6 × 1 300	=	£8 500.
Ancillaries at 70%	=	£6 000.
Total capital cost	=	£14 500.

9.9.4 Operating costs per batch

Labour: 3 h at 100% allocation + 37 h at 50% allocation = $21\frac{1}{2}$ man-h.

Taking the cost of labour as £0·7/man-h for wages, insurance, pension, fringe benefits etc., labour cost per batch = ... £15

To this is added up to 33% for overlooking including foreman etc.

Hence total labour cost per batch = ... £20 (46%)

Steam: Water evaporated from 25% solution = 3 250 kg

Steam required at 1·8 kg/kg water evaporated = 5 850 kg

Adding 25% for heating vessels, steam main losses etc., total steam requirement = 7 300 kg

With steam at £1/t, steam cost = ... £7 (16%)

Electricity: Total hp for dryer and ancillaries = 22·5 hp

Taking into account motor efficiency and distribution losses,

the electricity usage from factory meter = 22·5 kW
Hence per batch run of 37 h, kW/batch = 835

With electricity at £6/MWh, cost/batch = ...	£5	(8%)
Maintenance: Total per year at $7\frac{1}{2}\%$ of total capital is per batch = ...	£6	(14%)
Sundry direct expenses: These include analytical services, consumable stores, haulage etc. and amount per batch to about	£5	(12%)
Hence total cost per batch	£43	

or cost per 1 000 kg = £3·9

Cost for small outputs: The above cost is for large batches on a large machine. It will be 4 or 5 times higher for small batches on small machines.

APPENDIX 9.1 PROCEDURE FOR A DRUM DRYING TRIAL

A.9.1.1 Plant

(a) Multi-purpose 0·45 m × 0·45 m drum dryer, adaptable for use as twin drum, double drum or splash feed (twin) dryer and supplied with variable ratio gear to give 1–10 rev/min drum speed.
(b) 100 litre feed vessel with effective agitator.
(c) 20 litre overflow return vessel.
(d) Slurry pump, about 150 litre/h.
(e) Dip and splash feed trays for twin drum operation.
(f) Steam supply through reducing valve to give 3–6 bar gauge pressure in drum.

A.9.1.2 Procedure

Preliminary information will have indicated whether the physical properties of the feed liquor or its heat sensitivity require double or twin drum operation, and the machine will be set up accordingly. A typical sequence of operations for each method of drying is as follows:

A.9.1.3 Twin drum drying

(a) Set steam pressure to 5·5 bar (gauge) (or less if product is heat sensitive) and drum speed to 3 rev/min.
(b) Adjust overflow in dip tray to give 12 mm immersion.
(c) Heat 70–100 litre of feed solution to 80°C (or less if heat sensitive) and feed to tray at such a rate that an overflow of approximately 100 litre/h is returned to the feed tank via the pump.

(d) Operate dryer for $\frac{1}{2}$–1 h, collect and weigh dried product and determine moisture content.
(e) Repeat drying procedure at 5 and 7 rev/min if moisture content requirements will allow, or repeat at 3 rev/min (or less) if the moisture content must be reduced.
(f) Plot output rate and moisture content curves against contact time.
(g) Repeat drying procedures with various strengths of feed solution, say 20%, 30% and 40% w/w, and determine optimum drying conditions.

A similar procedure will be required for splash feeding.

A.9.1.4 Double drum drying

(a) Set steam pressure to 5·5 bar (or lower for heat sensitive products), clearance of (hot) drums to 75 μm and drum speed to 3 rev/min.
(b) Heat 50–75 litre of feed liquor to 80°C (lower for a heat sensitive product) and feed to dryer so that a liquor level of approximately 5 cm is maintained in the nip.
(c) Operate dryer under these conditions for $\frac{1}{2}$–1 h, collect and weigh dried product and determine moisture content.
(d) Decrease drum speed to 1 or 3 rev/min or increase to 5, 7 or 9 rev/min as moisture content requirements permit.
(e) Plot output rate and moisture content curves against contact time.
(f) Repeat drying procedures with different strengths of feed liquor and determine optimum drying conditions.

CHAPTER TEN

Continuous rotary dryers

10.1 PRINCIPLES AND APPLICATION

A continuous rotary dryer is one in which the wet material is continuously lifted by the rotation of the dryer so that it falls through a current of hot air passing down the length of the dryer shell. The dryer is sloped so that the material gradually flows along the length of the shell. The air flow may be either counter-current or co-current. Large dryers are followed by a product cooler working on the same principle as a counter-current direct dryer, or by a fluidised dryer cooler (Chapter 11).

The rotary dryer is one of the most widely used forms of continuous dryer. It is simple and reasonably versatile and, in consequence, it is suitable for drying a wide range of materials rapidly and at a low unit cost when quantities are large. Thus dryers can be designed to provide drying times ranging 5 to 60 min and solid handling capacities range from a few hundred kilograms up to 200 t/h.

A major use in the chemical industry is for drying fertiliser salts such as ammonium sulphate, nitrate and phosphate, potassium salts and compound N.P.K. fertilisers in world quantities of more than a hundred million tons per year.

Rotary dryers are ideally suitable for drying free-flowing granular products requiring moderate drying times. They can be adapted to handle fairly sticky materials by special devices to dislodge caked solids—in fact the tumbling action is beneficial within limits for all products since any semi-permeable crust on the surface of the particles (case-hardening) is disrupted, thus allowing easier escape of moisture from their inside. Rotary dryers are quite suitable for heat-sensitive materials provided that the restriction of drying temperature does not result in any excessive drying time. They are not, however, suitable for the following applications; slurries, sludges and very sticky materials; and slow-drying materials requiring more than one hour for drying. In addition there are limitations in handling friable, dusty or light materials which could easily be carried away in the air stream: indirect dryers rather than direct dryers should be given consideration for such cases.

10.2 TYPES OF ROTARY DRYERS

Rotary dryers can be classified into three types—the cascading direct rotary dryer, the direct dryer with peripheral louvres for the hot air and the indirect dryer. The louvred dryer is really a continuous agitated through-circulation dryer (cf. Chapter 8).

Except with small dryers it is essential to cool nearly every product with a stream of air in a rotary cooler or in a fluidised bed cooler (Chapter 11), since the heat content of a continuous discharge of 20–200 t/h leaving a dryer at 100°C would be excessive, especially if the material is to be packed into plastic or fibre sacks without previous storage in a bunker or silo. Moreover the particles of many types of product are likely to cake together if put hot into a bunker or silo. In general cooling to a temperature 20–30°C above ambient is desirable. In humid areas it is desirable to dry the inlet cooling air partially, for example by scrubbing in an absorption tower with calcium chloride solution, which is subsequently re-concentrated.

Other ancillaries include direct- or indirect-fired air heaters and air tempering chambers, efficient dust collectors and fans; also a system of (belt) conveyors for the incoming and dried product. For description and performance of dust collectors see Reference 1.

A system of automatically operated rappers is necessary on some plants to dislodge material caking in the lifters. These are noisy and sometimes troublesome to maintain in effective condition.

10.3 THE CASCADING DIRECT ROTARY DRYER

The arrangement of the most widely used form of rotary dryer is indicated in Fig. 10.1. A direct oil-fired heater is shown, but any fuel may be used and

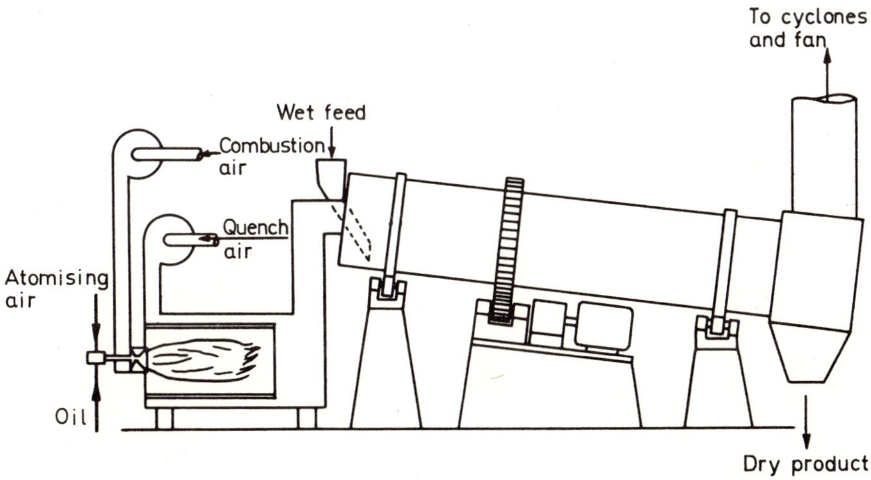

Fig. 10.1. Typical cascading direct rotary dryer arranged for counter-current operation

indirect heaters are used if necessary. The air flow through the shell may be either counter-current as shown, or co-current.

The rotating shell may have a diameter up to 3 m according to the output required; it is usually between four and fifteen diameters long. To promote longitudinal travel of the drying solid, the shell is mounted on a slope of about

1 in 40 to 1 in 20 (though occasionally co-current dryers may be nearly horizontal). Attached to the inside wall of the dryer cylinder are continuous rows of lifters (flights) – see Fig. 10.2. As the dryer rotates, these lifters pick up the material and it then cascades through the air stream. The dryer is rotated at 4–5 rev/min. The air velocity is within the range 1·5–2·5 m/s, depending on the size of the particles of material to be dried and on the amount of fine material formed in the dryer.

Direction of solid/air flow: Co-current flow of solid and air is essential when a large proportion of the moisture is to be evaporated during the early stages of drying, as it permits the use of high air temperatures without unduly

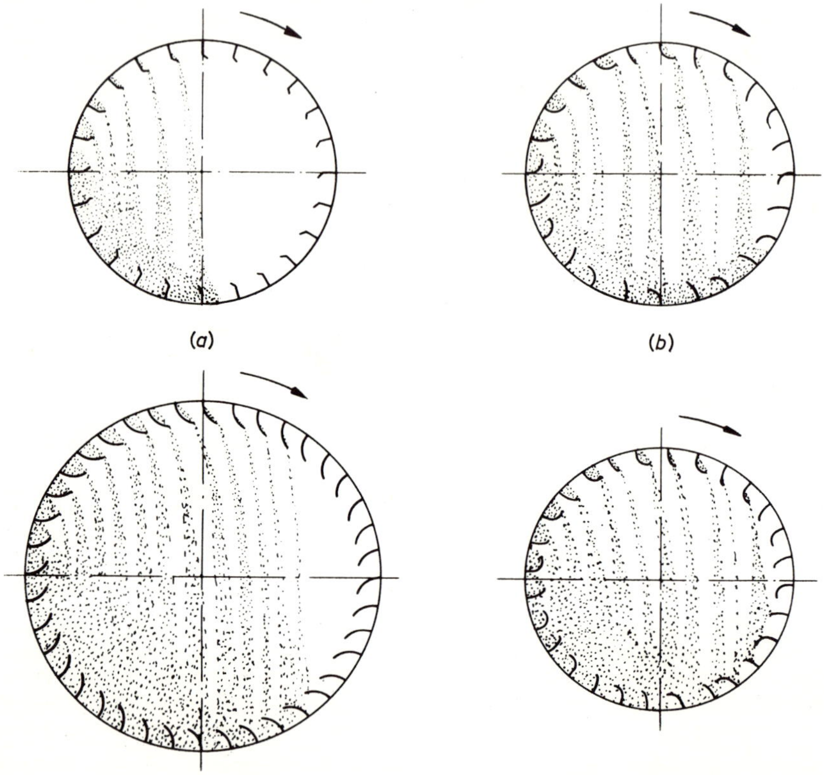

Fig. 10.2. Types of lifter

high product temperatures. It is also indicated for drying temperature-sensitive materials, including those which contain volatile components such as the ammonium salts in fertilisers. The latter should not be heated above 90°C in order to avoid fume formation. Since the air and solid temperatures converge towards each other at the exit, the temperature of the dried solid leaving the plant is easily controlled to a maximum whilst still retaining the advantage of a high temperature difference in the early stages of drying. In the counter-current dryer the dry solid temperature tends towards that of the inlet air. Consequently counter-current operation is indicated whenever it is necessary to heat the solid to a relatively high temperature to complete the drying process.

The air exit temperature in a co-current dryer is 10–20°C higher than the solid exit temperature. The air may then be so near its dew point that dust deposits are likely to be formed in poorly lagged dry dust collectors: the trouble can be prevented by adding to this air the dusty air leaving the cooler or by using a wet scrubber such as a venturi scrubber.

The air exit temperature of a counter-current dryer is less clearly defined but 100°C is a reasonable working value for design purposes.

Chimney plumes containing condensed water mist droop seriously and can cause a nuisance.

10.4 LIFTERS FOR CASCADING DRYERS

It is not always appreciated that any material not held up in the lifters must remain as a rolling bed in the bottom of the dryer shell and thus effectively by-pass the drying operation. Maximum heat and moisture transfer is obtained when all the material is in the process of falling in cascades. Under these conditions the transfer coefficients per unit length of the dryer are proportional to the 'hold-up'. It is also important that the cascades or curtains of falling material should be carried well across the dryer in order to achieve the best possible contact with the air stream. It is for this reason that lifters should be used which give the curtain pattern shown (d) in Fig. 10.2. With this particular lifter design the maximum hold-up is 14% of the shell volume. A disadvantage is the tendency for sticky material to build up in the curve of the lifter. This can be overcome to some extent, though with some loss of effectiveness by making the curvature less pronounced. If, however, this has to be taken to the stage where the lifter approaches a simple radial design, the loss of effectiveness is so great that research should be conducted to improve the feed by modifications to the process producing it or by some form of pretreatment such as back-mixing: alternatively consideration must be given to the use of a different type of dryer.

Many special designs of lifters have been devised by dryer manufacturers. Whilst a few may be successful, it is important to remember that a complicated lifter system is likely to be more difficult to clean and more costly to maintain particularly as some abrasion is inevitable with the high throughput of rotary dryers. Hence simple systems are to be preferred whatever the theoretical attractions of a more complicated arrangement. Moreover short term experimental tests cannot cover day to day variations in material to be dried, or perhaps, more important, variations produced by changes in raw materials to the production process.

10.5 THE LOUVRED DRYER

Fig. 10.3 shows a form of rotary dryer in which the drying air flows along channels formed by special shaped louvre plates and escapes through the louvres into the bed of the solid. A valve plate ensures that the air flows only to those louvres which are actually beneath the bed of solid, and the downward pointing, overlapping plates which form the louvres ensure that no solid material enters the air channels. The louvres thus form an inner drum which is slightly tapered; the axis of the drum is horizontal but the solid moves down the slope by kiln action. A dam ring is formed at the exit end by a system of overlapping hinged blades. In normal operation a relatively deep bed, occupying about 40% of the dryer volume, is used but the

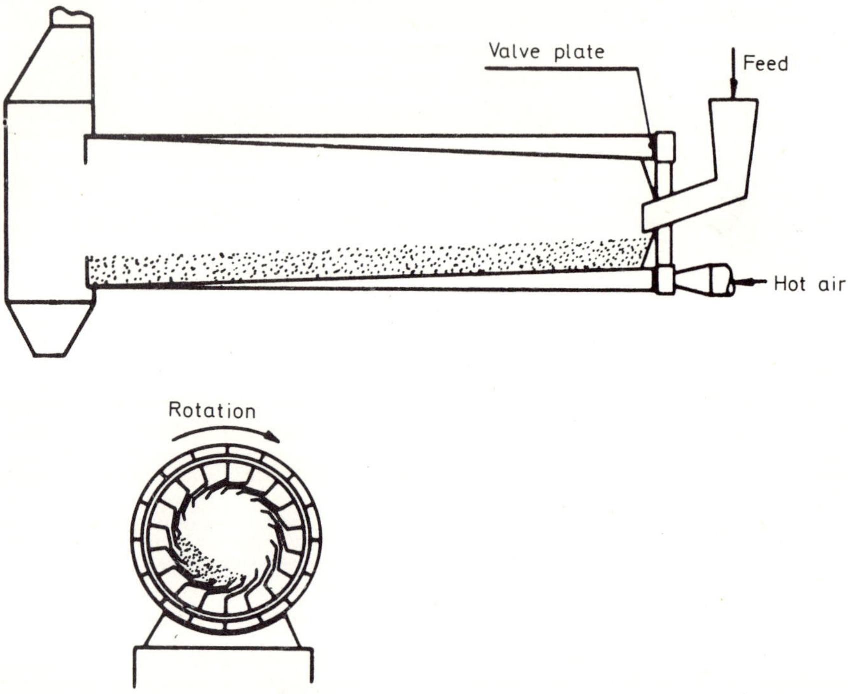

Fig. 10.3. Principle of louvred dryers

bed depth can be modified by altering the angle of the dam ring blades; in the extreme, the dryer can be completely emptied very rapidly. The air is drawn off through a wind box which usually incorporates a nest of high efficiency cyclones.

The gentle rolling action of the bed prevents caking and minimises attrition of friable materials. The relatively deep bed of material may allow a satisfactory retention time in a smaller machine compared with the cascading dryer and the power consumption is lower as the speed is low. The high heat-transfer coefficient which can be achieved also results in a compact machine for materials whose drying is mainly in the falling-rate period.

This dryer is not suitable for sticky materials which tend to choke the louvres or to form large balls as a result of the rolling motion. Because it is a cross-flow heat exchanger, the exit solids temperature approaches the inlet air temperature, which may limit the maximum value of the latter. Also the air temperature drop in the falling-rate region is comparatively small, so that the thermal efficiency tends to be low. The thermal efficiency is highest in the constant-rate region where the temperature difference is greatest.

Since the air passes through the bed, the maximum air rate per unit area may be limited by fines carry-over and, in general, this dryer does not appear to be suitable for drying dusty material. Its best application appears to be to the drying of coarse dust-free material, where drying is mainly in the constant-rate region and with no severe restriction on the exit solids temperature.

Because the air enters the bed at a constant temperature along its length a small parcel of feed material goes through the same drying history as it travels through a louvred dryer as it would if it were exposed to similar conditions in a batch dryer. It is therefore possible to carry out tests in a small laboratory-size dryer in order to find an acceptable air rate per unit area and air temperature and the corresponding drying time. The full-scale dryer is then sized to give the same conditions. Some manufacturers have laboratory test apparatus and are also prepared to loan self-contained laboratory-scale, batch and continuous units to potential purchasers.

10.6 THE INDIRECT DRYER

One form of indirect dryer uses an inclined rotating shell as in the direct dryer, but the lifters are replaced by tubes containing steam or hot water. In small machines there is one row of tubes, but in most sizes two rows are employed, the inner row being of smaller diameter than the outer (see Fig.

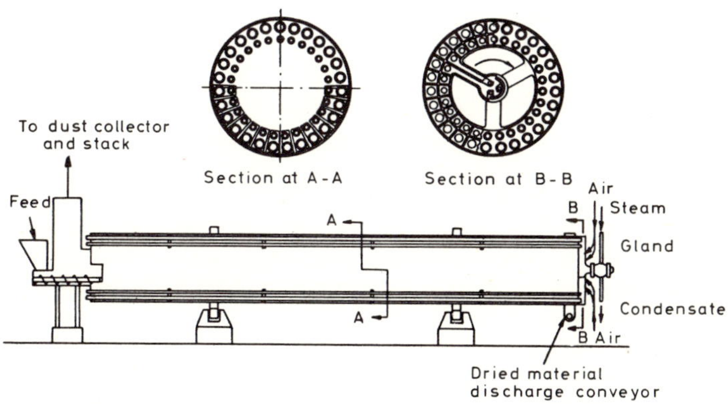

Fig. 10.4. Principle of indirect dryer, arranged for counter-current flow

10.4). The feed is heated by direct contact with the tubes over which it flows in a thin layer. Air is drawn through the dryer in sufficient quantity to remove the water vapour. The air leaves the dryer almost saturated so that the quantity required is usually much less than in a direct rotary dryer. The problem of dust removal from the exit air is simplified and in many cases natural circulation without the use of fans is adequate. The air velocity is usually about 0·3 m/s.

The heating steam is introduced into the tubes, and condensate is removed, through a rotary gland at the solid discharge end of the dryer. The dry solid is discharged through ports in the shell, which have dams to maintain a sufficient depth of bed.

This dryer is especially suitable for materials with a long falling-rate period as it can hold them at a substantially constant temperature for a relatively long period, while the rolling action prevents caking and, by mixing the bed, facilitates escape of the water vapour. Because of the low heat losses in the exit air stream the efficiency is high. It is suitable for heat-sensitive materials as the maximum temperature can be accurately controlled by control of the temperature of the heating medium. It is not suitable for any materials that cake strongly on the heating tubes.

Indirect dryers cannot, for mechanical reasons, be as large as the largest direct dryers required for high throughputs but fortunately the outputs of material for which they are particularly suitable are relatively small. Plants have been supplied of diameters 1·5–3·0 m, 10–20 m long. Typical diameters of the tubes are: outer row 110 mm, inner row 75 mm and a typical heat transfer rate is 2 kW/m°C.

10.7 AIR HEATERS FOR DIRECT DRYERS

A typical modern gas or oil-fired direct air heater is illustrated in Fig. 10.1. The combustion chamber is cylindrical and has a length of about 2 diameters. The hot products of combustion discharge into a mixing chamber where they are mixed with the remainder of the total air flow (termed the 'quench' or 'dilution' air). With oil fuel it is essential that combustion should be complete before the quench air is mixed with the products of combustion otherwise the combustion of the carbon particles in the flame will be arrested and smoky dirty gases will be obtained. To obtain complete combustion in a combustion chamber of reasonable dimensions a forced draught burner is essential; because of the unavoidably leaky nature of a dryer, with its chutes for the solid material, the induced draught of the main fan is insufficient for all oil burners and low-pressure gas burners. If fuel gas is available at a fairly high pressure, however, self-inspirating burners making use of the energy of the gas may be employed. The possible effects of the sulphur in fuel oil must also be considered; see Section 10.10.

Nearly all the heat released in the flame appears as sensible heat of the gases and, as the heat loss through the wall is comparatively small, the wall face temperature is only about 100–200°C below flame temperature. If the burner were operated with the low excess air ratios usual in boiler practice this would result in a face temperature of about 1 700°C and expensive super-refractories would be necessary. This can be overcome by operating with about 100% excess air, admitting 30% through the burner and 70% through a ring of ports near the circumference of the combustion chamber. This will keep the refractory temperature down to about 1 400°C and give a flame temperature of about 1 500–1 600°C; these temperatures are high enough to ensure complete combustion, but low enough to permit the use of cheaper grades of refractory. (The quenching temperature for carbon formation is about 1 000°C.) This additional air should always be supplied by a fan as the induced draught on the combustion chamber is rarely adequate. Since the products of combustion are to be further diluted with air there is no merit in operating the combustion chamber with less than 100% excess air. It is in this, together with the low draught and the need to limit the flame temperature that the dryer combustion chamber differs from that in a boiler. Even manufacturers of good repute who claim to specialise in this type of equipment do not always fully appreciate the significance of these differences. These remarks apply especially to oil-fired heaters. The lower emissivity of non-luminous gas flames results in a larger temperature difference between flame and refractory and there is usually no difficulty due to overheating of the refractory.

It is common practice in large units to circulate the quench-air through an annular space surrounding the combustion chamber. This cools the refractory lining but at the same time significantly reduces the heat loss, since most of the heat picked up by the air is utilised. It is necessary to use a fan for the quench air in these circumstances and this refinement is sometimes omitted in smaller installations to save the additional fan and ducting.

The quench-air mixing chamber ideally should be about 10–12 diameters long to ensure satisfactory mixing. This is not likely to be practicable but the introduction of one or more right angle bends or an annular mixing baffle (Reference 1, p. 583) improves the mixing in three diameters and is often easy to arrange. A bend also acts as a radiation shield which is especially important with oil firing as the combustion chamber is effectively a black body at 1 500°C.

10.8 FIRING EQUIPMENT FOR DIRECT AIR HEATERS

10.8.1 Oil burners

Complete combustion is essential in many applications as even a slight trace of soot can discolour light coloured material. Three conditions must be met to achieve this:

(a) The atomisation of the fuel must be very good so that the non-volatile residues from the droplets are small.
(b) A turbulent fast mixing flame is necessary to burn all the secondary carbon particles, formed by thermal reactions in the flame, before they grow to a large size.
(c) The flame gases must not be cooled by dilution until combustion is complete.

Forced draught is necessary to achieve condition (b) and in many installations the 'low pressure air atomising' (L.P.A.) burner is suitable. In these burners all or at least a large proportion of the total combustion air is supplied to the burner at a pressure of about 75 mb and also serves to atomise the fuel. Secondary combustion air should be supplied by a fan, which should also supply the additional excess air required to limit the flame temperature.

A better standard of atomisation is achieved by the 'medium pressure air' atomising burner (M.P.A.). This burner employs compressed air at 1–2 bar for atomisation; the quantity required is about 2% of the total combustion air. For an even higher standard of cleanliness it is necessary to use 'high turbulence combustors' and M.P.A. atomisers. Pressure jet atomisers are unsuitable as the standard of atomisation is inferior to that of the L.P.A. and M.P.A. atomisers, and the operating range is too small. Both the L.P.A. and M.P.A. burners have a range of about 4:1. The pressure jet is virtually a constant-rate atomiser; some special designs have been evolved for which a 3:1 turn-down is claimed but they still have the disadvantage of relatively poor atomisation for this exacting duty.

10.8.2 Gas burners

Gas burners may be divided into two main categories; the low pressure gas burner in which the energy of mixing is supplied by the combustion air at a pressure of about 15–25 mb, and the high pressure gas burner, in which the energy of mixing is supplied by the gas stream; pressures up to 1·5 bar are commonly used. The flame is very stable in these burners and it is possible to obtain a 3:1 turn-down ratio at full air rate, and about 10:1 if the air and fuel are turned down in the same proportion. A high proportion of the combustion air is entrained as primary air in the ejector, but some secondary air is necessary to complete combustion. As there is no problem of flame quenching and the flame is stable in quite high air velocities, it is usual to mount the burners in the quench-air stream.

Advice on suitable gas burners for the supply available from the local Gas Board in Britain should be obtained from their Industrial Burner Department. This is particularly necessary at the present time when high pressure natural gas is becoming available and new types of burner for it are being developed.

10.8.3 The use of solid fuel

Solid fuel may be used for direct fired dryers if there is no objection to contamination of the product by particles of ash. Automatic stokers should be used to give a steady operation and facilitate temperature control. Pulverised coal may be used on large dryers. As with oil firing, the combustion chamber must be large enough to ensure complete combustion before the hot gases meet the dilution air.

10.9 INDIRECT AIR HEATERS

Indirect air heaters are often preferred when the highest standard of cleanliness is necessary, although they are more expensive than direct heaters.

For low temperatures, steam may be used as the source of heat and the design of the heater follows conventional heat exchanger practice. The relative values of the inside and outside transfer coefficients indicate the use of finned tubes. A fan is necessary to overcome the pressure drop across the exchanger, and it is usual to install an inlet air filter. Dryers are often in dusty places and the use of a filter maintains the efficiency of the heat exchanger.

For high temperatures fuel-fired heaters must be used. The design of these heaters closely resembles that of the tubular process heaters extensively used in the chemical industry. Radiant tubes surround the combustion chamber and this is followed by a convection section. All-radiant heaters are available but they are relatively inefficient and are only justified when fuel is exceptionally cheap. Provided that the tube temperatures do not require the use of high nickel-chrome alloys, heavy fuel oils may be used; these heaters may then provide an economic alternative to the use of expensive clean fuels in a direct heater where clean gases are required. If high nickel-chrome alloys are necessary, however, the heater becomes very expensive and there is a danger that the tubes will be attacked by vanadium compounds from residual fuel oils. Gas and oil burners are similar to those which are suitable for the direct air heaters and the firing technique is similar to that required by a boiler. Solid fuel may be used also; some form of mechanical stoker is essential to obtain consistent performance and close control of the air temperature.

10.10 EFFECT OF SULPHUR IN FUEL OIL

Heavy fuel oil contains 2–4% sulphur unless derived from low sulphur crudes. The oxidation of the SO_2 in the combustion gases to SO_3 with consequent raising of the dew point of the flue gases and thereby risk of corrosion in the cooler parts of the system (e.g. dust arrestors) is reduced in boiler furnaces by operation with only 1–2% of oxygen in the flue gases above about 400°C. This is not, of course, possible in systems where tempering air is added as for direct heaters for dryers. Nevertheless the oil companies should be consulted for the latest practical experience in reducing SO_3 formation in this type of system. Apparatus for continuous measurement of SO_3 is described in Reference 5.

If contamination of the product through chemical reaction with SO_2 and SO_3 in the flue gases cannot be tolerated, consideration must be given to the use of an indirect heater or better still, use of a low sulphur oil. The relative price of low sulphur oils is expected to fall during the next few years.

10.11 PRINCIPLES OF DESIGN

Most drying problems concern what may be termed simple drying in which the object is to obtain a product having a stated maximum final moisture content. For the large tonnages of salts and fertiliser compounds there is, however, a somewhat rigid specification, not only for the total moisture content of the finished product, but also for bound moisture, temperature (leaving cooler), shape, size analysis and uniformity of chemical analysis of relatively small samples, all of which can be affected during passage through a rotary dryer and cooler. Hence, as well as knowledge of the capability of a dryer as a heat- and matter-transfer apparatus it is necessary to consider the influence of a particular design of rotary dryer on the physical characteristics of the materiai to be processed. Oversize and undersize material can be sieved out and recycled but this expensive necessity should be minimised.

Design is, in fact, greatly handicapped by the paucity of research studies of this type of dryer, particularly on full-scale plant, largely because variations in plant conditions to obtain information might lead to temporary rejection of plant product and costly handling and pre-processing. Most published investigations (cf. References 2–4) have been on idealised systems in which material characteristics have been deliberately suppressed by study of virtually non-porous materials; also relatively low temperatures have been used. Moreover hold-up times and the distance of particle falls are short in small-scale plant so that even quite detailed investigations do not provide sufficiently accurate information for scaling-up to large dryers. Little is known of the influence of flight design and of the effect of moisture content, particle shape and size distribution on the angle of repose even in a straightforward convection dryer. Moreover in the design method given below conduction and radiation heat transfer are ignored. The latter may be sometimes appreciable in co-current dryers. Significant increases in capacity of even direct dryers could be achieved, though not necessarily economically, by the use of heated centre tubes and heated hollow flights.

On coolers more information is required on the mechanism of the process and on the optimum internal design for what should be free flowing particles; also on the economies of cooling the cooler drum by water sprays with partial immersion in a trough of circulating cooled water.

The basic design of any dryer is a problem in simultaneous heat and mass transfer. Four equations are required:

(a) A heat balance.
(b) A mass balance.
(c) A heat-transfer rate equation.
(d) A mass-transfer rate equation.

The design must satisfy these four equations. In general there will not be a unique solution and the final design may be selected on economic grounds or in the light of the considerations indicated at the beginning of this section.

The following data are required:

(a) Throughput of dry solid.
(b) Initial moisture content and temperature of the feed.
(c) Required moisture content of dry product.
(d) An estimate of an acceptable maximum temperature for the dry product.
(e) Vapour pressure data for the material to be dried over a sufficient range of temperatures and moisture content (see Appendix 10.1).
(f) Effective latent heat of vaporisation of water from the solid.
(g) The specific heat of the dry solid.

(h) An estimate of the humidity of the drying air at the dryer inlet.
(i) A heat-transfer coefficient in the dryer.
(j) Rate of drying data for the feed material.

The latent heat is included because the heat of crystallisation of very soluble materials causes the effective value of the latent heat to differ considerably from that of pure water. The effective value can be obtained from the vapour pressure data by means of the Clausius-Clapeyron equation as described in Appendix 10.1. When estimating the humidity of the drying air, it must be remembered that in direct fuel-fired heaters the products of combustion may contribute about 100 mb of water vapour in addition to that originally present in the atmosphere.

It must first be decided whether the dryer is to be co-current, or counter-current. A water vapour partial pressure at the dryer exit is then chosen. This must be less than the vapour pressure of water over the solid and 50–80% of the latter value is a reasonable choice in most cases. Thereafter it is possible to make an overall matter balance and determine the air rate, from which the dryer diameter can be calculated. The length of the dryer is next calculated using the transfer data and finally the slope and speed are decided to give the correct time. There is usually some freedom of choice of conditions and some limitations to be taken into account, so that the calculation may have to be repeated until a satisfactory combination of conditions is achieved. Formulae for the complex calculations required are given in Appendix 10.2.

10.12 CALCULATION OF DRYER DIAMETER

Having chosen the mode of operation and decided on a reasonable humidity and temperature for the exit air, the quantity of air required and its inlet temperature can be calculated by a heat and matter balance. The air velocity must not be too great or excessive carry-over of solids will result. For coarse materials the practical limit seems to be a mean velocity of 2·5 m/s at the air exit; for fine materials a much lower velocity must be adopted. A useful empirical rule is to specify an air velocity equal to half the free-falling terminal velocity* of the smallest particles present in a substantial quantity. The cross-sectional area and the diameter can then be calculated, assuming that the area available for air flow is 85% of the total cross-section.

10.13 THE CALCULATION OF DRYER LENGTH

10.13.1 Simple method

If the drying time can be determined, either on an existing plant or in the laboratory and under identical transfer conditions to those which will be used in the proposed new dryer, then all that is required is to design

*See Appendix II of Chapter 12

the dryer mechanically to have a residence time a little greater than the drying time. Correlations which can be used for this are given in Section 10.17. The important factor is the identity of the transfer conditions in the experimental and proposed dryers; in general, this means that the drying time must be determined on a full scale or large pilot plant rotary dryer.

10.13.2 Formal method

The calculation of dryer length is a problem in simultaneous heat and mass transfer. The procedure to be followed is stated below and details are discussed in subsequent sections.

The overall change in moisture content is divided into a convenient number of stages. These could be a preheating stage to heat the feed to the wet-bulb temperature; a constant-rate drying stage, and then from six to ten stages during the falling-rate drying period. This is the procedure for a desk calculation; if the calculation is to be done on a computer then many more and smaller stages may be used with a consequent increase in accuracy. The designer must treat each problem individually and, while other people's work may be a source of inspiration, he cannot expect to find a ready-made, fully automatic design procedure.

A heat-transfer coefficient referred to unit length of the dryer is required (see Section 10.14) and as the feed, air and wet-bulb drying temperatures are known, then the lengths of the preheating and constant-rate drying sections can be calculated by reference to heat transfer only. (It is convenient to assume that no drying occurs in the preheating section.) For the falling-rate period, mass-transfer coefficients are required – referred to unit length and expressed as a function of the moisture content of the material; the heat-transfer coefficient may be assumed constant throughout the dryer. It may be preferable to estimate the drying time required and to calculate the dryer length to give this residence time. The method of calculation is basically the same although the arithmetical route is different.

All conditions at the beginning of the first section are known; at the end the moisture content of the solid has been arbitrarily chosen but the temperature is not known. Assume a temperature, and ignoring heat losses at this stage, calculate the air temperature and humidity by heat and mass balances. Take an average value for the mass-transfer coefficient over the section and calculate the length by both the mass-transfer and heat-transfer equations. Repeat the process, assuming different terminal temperatures until acceptable agreement between the two estimates of the length is achieved. The exit conditions for the section then constitute the inlet conditions for the next section and the calculation is repeated section by section until the dryer terminal conditions are reached. The total dryer length is then the sum of the sections. This is clearly a laborious procedure with ample room for errors to have a substantial cumulative effect and there is a very strong case for employing a computer. To assist the preparation of a programme a schedule of the data required and a design procedure are summarised in Appendix 10.2.

Whilst most rotary dryers are only purchased after discussion with and pilot plant trials by experienced manufacturers, this design method is of considerable value in analysing the performance of existing plant.

10.14 THE HEAT-TRANSFER COEFFICIENT

At first sight it would be thought that heat transfer would be excellent in a rotary dryer where the solids are cascading freely through the air space. In fact the solid is concentrated into a number of curtains with relatively clear spaces between them, and the resistance to air flow in the curtains is large compared with that in the spaces. Most of the air flows along the clear spaces and the solid only comes into contact with a relatively small quantity of air drawn into the curtains as a result of momentum transfer from the particles to the air. The rate of heat transfer within the curtains is high because of the large area exposed and the air and solid probably approach equilibrium, but only a small fraction of the total air flow is involved. Consequently the overall rate of heat transfer is much less, by a factor of 30–40, than, for example, in a pneumatic conveying dryer where the solid is well dispersed in the air stream (Chapter 12).

Heat may also be transferred indirectly via the shell and lifters but this is negligible except possibly in very small dryers; this is one reason why the results of measurements on very small laboratory-scale dryers cannot safely be scaled up. It has also been suggested that the curtains may be treated as though they were solid surfaces and that the usual correlations for heat transfer in ducts may be used. This is a fallacious argument because these correlations are based on the assumption of a stagnant boundary layer and zero flow normal to the surface whereas the opposite conditions actually exist.

By assuming that the solids and the entrained air reach thermal equilibrium the heat-transfer coefficient can be expressed in terms of the air entrainment rate. This concept is useful for analysing the performance of existing dryers and coolers and for designing similar machines, but as there are no correlations between the entrainment rate and other parameters it is not possible to predict the entrainment rate under completely new conditions.

For design purposes an empirical value for the heat-transfer coefficient must be assumed, or an empirical value of entrainment rate. There is evidence that the heat-transfer coefficient expressed per unit mass of solid hold-up in the dryer is reasonably constant. Experience with material of $-5+10$ B.S. mesh (1·7 – 3·3 mm) suggests a value of 2 kcal/kg h°C (0·12 W/kg°C); some experimental data published by Saeman and Mitchell[2] indicate that for a finer material (ammonium nitrate of $-16+35$ B.S. mesh $\equiv$ 0·4–1·0 mm), this value may rise to about 6 kcal/kg h°C (0·36 W/kg°C).

10.15 VARIABLES AFFECTING HEAT TRANSFER

The rate of entrainment of air into the falling curtains will depend on the rate of transfer of momentum to the air. The theory of entrainment of gases by free jets shows that the momentum flux of the jet, that is the product of mass flow rate and velocity, is the main parameter determining the rate of entrainment of gas into the jet; the rate of entrainment per unit length of jet is proportional to the square root of the momentum flux. The mass flow rate in a dryer 'curtain' is proportional to the product of the speed of rotation (r) and the hold-up per unit length (H_l) and the velocity will be proportional to the square root of the dryer diameter (D_s). The entrainment, and therefore the heat-transfer coefficient per unit length of dryer is thus approximately proportional to

$$(N_r \,.\, H_l)^{\frac{1}{2}} \,.\, D_s^{\frac{1}{4}} \,.\, D_s \tag{10.1}$$

At a constant fraction filled, H_l is proportional to D_s^2, so that the heat transfer coefficient per unit mass of solid is proportional to

$$\frac{(N_r D_s^2)^{\frac{1}{2}} \,.\, D_s^{\frac{1}{4}} \,.\, D_s}{D_s^2} = N_r^{\frac{1}{2}} D_s^{\frac{1}{4}} \tag{10.2}$$

Thus the heat-transfer coefficient per unit mass of solid is little affected by the diameter but depends on $N_r^{\frac{1}{2}}$. Observation that the heat-transfer coefficients per unit mass in coolers of 2·7 and 3·6 m diameter but running at nearly equal speeds, were approximately equal supports this. The results of experiments in a small dryer reported by Saeman and Mitchell[2] are also consistent with this though there are insufficient measurements to establish it beyond doubt.

The effect of the air velocity has been the subject of much controversy. In the heat transfer mechanism considered here it is postulated that, because of the high resistance of the curtains, nearly all longitudinal air flow takes place between them; consequently the effect of the air velocity should be small. However, since the air velocity does have some effect on the motion of the particles (as shown by its influence on the residence time), it is also likely to have some effect on the heat-transfer coefficient. Saeman[3] reviewing published work on rotary dryers, pointed out that many investigators had assumed a log-mean temperature difference, although this is incorrect for the temperature distribution usually found in dryers, and had also taken insufficient precautions to obtain accurate air temperatures; he considered that the large influence of air velocity reported by these investigators was a false effect due to these errors. In his opinion, the work of Friedman and Marshall[4] was the most reliable and they found that the heat-transfer coefficient varied as $V^{0 \cdot 16}$.

For design purposes (see Appendix 10.2) it is suggested that:

(a) The heat-transfer coefficient should be taken as 0·12 W/kg°C (2 kcal/kg h°C) for coarse materials of about $-5+10$ mesh and may be increased to 0·36 W/kg°C (6 kcal/kg h°C) for materials of $-16+35$ mesh, but higher values must not be used because of lack of evidence to support them.

(b) The values recommended in (a) apply to dryers running at about 4–5 rev/min; for higher speeds they may be increased in proportion to the square root of the speed.

(c) The above values of the heat-transfer coefficient refer to air velocities of about 1·5–2·5 m/s. In general it is probably safe to ignore the effect of air velocity.

10.16 THE MASS-TRANSFER COEFFICIENT

The mass-transfer coefficient must be defined in terms consistent with those used for the best heat transfer and the practical units to be used in this chapter are kg/kg h atm. As in the case of the heat-transfer coefficient this value is multiplied by hold-up per unit length of dryer to give a mass-transfer coefficient per unit length.

The mass-transfer parameter must be determined experimentally in suitable apparatus. Direct through-circulation dryers or small rotary dryers have been recommended for this purpose. Either is satisfactory for materials which do not 'case-harden', i.e. the pores of the granules do not become blocked by crystals deposited as the water evaporates. In practice the rough handling of the granules in a rotary dryer often prevents case hardening by breaking up the new crystal growths as they form, so that it is essential to simulate this in the laboratory.

Under constant-rate drying conditions the mass-transfer coefficient will be related to the heat-transfer coefficient and will therefore be influenced by the same variables as the heat-transfer coefficient. In the falling-rate period however the mass-transfer coefficient is determined almost entirely by the properties of the solid so that it will not be significantly affected by other variables.

10.17 RESIDENCE TIME AND HOLD-UP

The residence time and the hold-up are related thus

$$H_l L_d = 10^{-3} W_d \theta \qquad (10.3)$$

The residence time is determined by the speed, diameter and slope of the dryer and by the air velocity. A formula published by Friedman and Marshall[4] is widely used for predicting residence times but an alternative formula published by Saeman and Mitchell[2] has been found more satisfactory. This is

$$\theta = \frac{L_d}{a N_r D_s (s_d + b v_m)} \text{ minutes} \qquad (10.4)$$

where a is a constant, dependent on lifter design and b is a constant, dependent on particle size. Note that v_m is negative for counter-current flow.

According to Saeman and Mitchell the value of a may lie between 2·0 and 3·14 (π). With lifters designed for the best cascading action the value of a may approach 3·0 but 2·5 is probably a more typical value. The value of b applicable to coarse inorganic materials is $9{\cdot}1 \times 10^{-4}$; Saeman and Mitchell found a value of $1{\cdot}52 \times 10^{-3}$ for fine material.

The term bv_m represents the equivalent slope due to the displacement of the falling solid by the air stream. In counter-current machines bv_m should not exceed half the actual slope or irregular flow of the solids will result.

It is usual for large machines to rotate at 3 or 4 rev/min and smaller ones at higher speeds. Although higher speeds give better heat-transfer coefficients this advantage is not usually worth the large power consumption involved, particularly if the material has a long falling-rate period when the heat-transfer coefficient is not important.

10.18 POWER REQUIREMENTS

The mechanical design of rotary dryers is a major problem in mechanical engineering and it is not proposed to discuss this at length. The only mechanical matter which will be discussed is the estimation of the power required to rotate the shell. Since a large dryer may require as much as 300 kW it is important to be able to estimate the power with reasonable accuracy. The following formula has been found satisfactory:

$$\text{Power (kW)} = 4{\cdot}5 \times 10^{-4}\ W_t v_r + 1{\cdot}2 \times 10^{-4} B D_s f_r N_r \qquad (10.5)$$

f_r is the average number of lifts (flights) per revolution of the shell (a reasonable value is 1–2).

The power required by the dryer fan is largely determined by the pressure drop through the gas cleaning equipment (see Reference 1 for information).

10.19 INSTRUMENTS

Recording thermometers are necessary for inlet and outlet gas temperatures and for the discharged product; also for control of the dust arrestors. A continuous moisture meter based on physical principles, such as dielectric constant, is desirable on the product stream. In addition an automatic product sampler is necessary.

10.20 COSTS

The 1968 cost of large dryers in mild steel was fairly well represented by the expression

$$\pounds(\text{cost}) = 600\ L_d D_s \qquad (10.6)$$

when the length and diameter of the dryer are in metres.

This includes the cost of shell and lifters, tyres, girth wheel, gear box and motor. The largest dryers, more than 3 m in diameter will cost about 10% more than this, because of the relatively high cost of the girth wheel and motor. A girth wheel for a 3·65 m diameter dryer costs £15 000.

The costs of foundations, buildings and dedusting equipment will vary widely with local circumstances and are not included. They are likely to be at least as large again depending on the efficiency of the deduster. The costs of the air heater and product cooler are not included.

The use of stainless steel may increase the cost by 100% or more. Only small dryers are commonly made wholly in stainless steel; in larger dryers it is usual to line a mild steel shell with stainless steel. As an example 3 × 12 m dryer, according to the above formula costs about £22 000 in mild steel; lining with stainless steel costs an extra £11 000.

Small dryers cost rather less than the above formula indicates because of the lighter construction and the use of relatively cheap standard units for the drives. On the other hand a small dryer is more likely to be made entirely in stainless steel.

10.20.1 Labour

Two men per shift should be sufficient for operation of a large system (one for a small system) checking inlet and outlet gas temperatures and flames in heaters; also supervising conveyors and dryer rappers. In addition at least one labourer may be required to shovel up spillage from conveyors and empty dust collector hoppers.

10.20.2 Maintenance

Meticulous alignment of a dryer is essential, first cold and then hot, when new and after renewal of supporting runners. General maintenance includes renewal of bearing and runners, rappers and lifters; also of refractories in combustion chambers.

The total cost of maintenance should not exceed 10% of the capital cost of plant (excluding buildings etc).

APPENDIX 10.1 MEASUREMENT OF VAPOUR PRESSURES OF MOIST SOLIDS

The vapour pressure of water over a moist solid can be measured in the apparatus illustrated in Fig. 10.5. A small quantity of the solid is enclosed in the bulb; both this bulb and the balancing mercury manometer are then frozen in a solid CO_2—methanol bath. The bulb is then evacuated to an absolute pressure of less than 1 mb. The cock above the bulb is closed and the bulb and manometer are transferred to a water bath and slowly heated, allowing ample time for the solid to reach the same temperature as the bath at successive stages. As the pressure in the bulb rises, the mercury manometer is balanced by admitting air to the vacuum system. The pressure and temperature are read at intervals.

Vapour pressure measurements should be made on several samples covering the full range of moisture content of interest. It will be found that at low moisture contents the vapour pressure is reduced by surface absorption. The vapour pressure p is related to the absolute temperature T and the latent heat by the Clausius-Clapeyron equation.

$$\frac{dp}{dT^*} = -\frac{\lambda}{RT^{*2}} \tag{10.7}$$

If ln$p$ is plotted against $1/T^*$ a straight, or nearly straight, line is usually obtained. The equation to this line is the integrated form of the Clausius-Clapeyron equation

$$\ln\frac{p_2}{p_1} = \frac{\lambda}{R}\left(\frac{1}{T_1^*} - \frac{1}{T_2^*}\right) \tag{10.8}$$

The latent heat λ is easily found from this equation. When drying soluble materials, a saturated solution is being evaporated, so that crystallisation

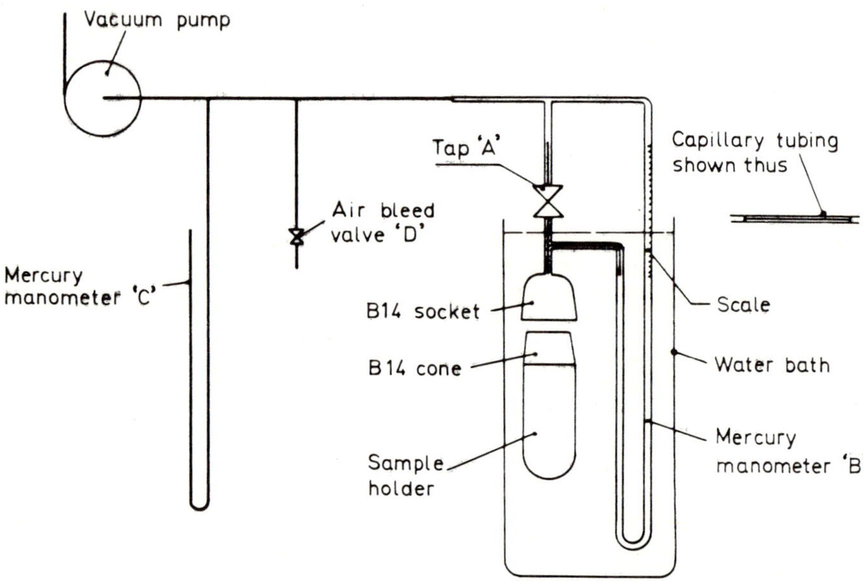

Fig. 10.5. Vapour pressure apparatus

takes place at the same time. The heat of crystallisation supplies part of the latent heat of evaporation and the value of λ derived from the vapour pressure is the net value corresponding to the overall energy change. With very soluble materials the effect can be large; the material for which curves are given in Fig. 10.6 has a latent heat of 2 135 kJ/kg compared with 2 387 kJ/kg for pure water at about 100 mb vapour pressure. Ammonium sulphate is an example of a less soluble material; the latent heat of its solutions is 2 320 kJ/kg at a similar vapour pressure.

The curves for this particular material given in Fig. 10.6 show the decrease in vapour pressure with decreasing moisture content. There are also signs of

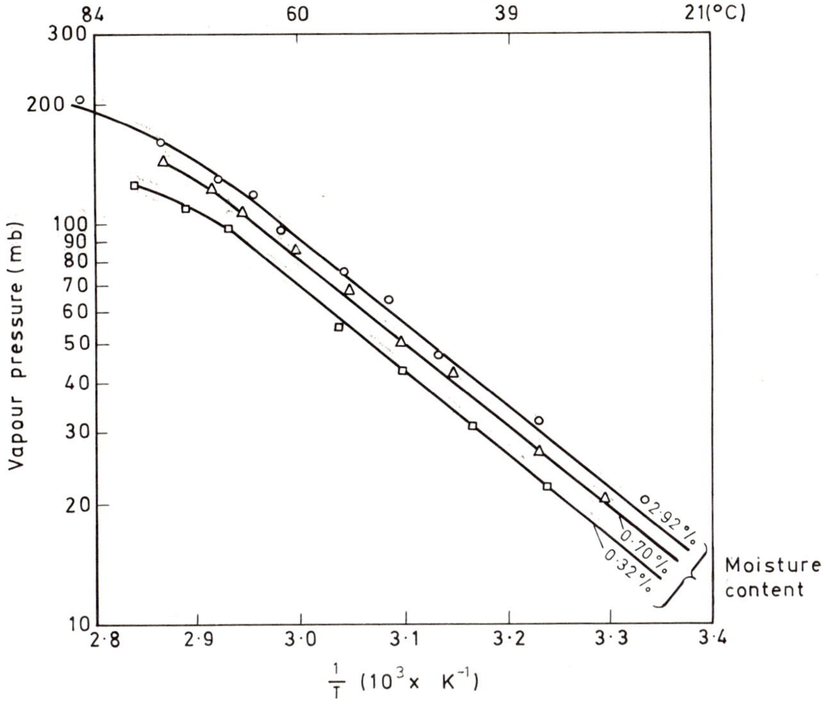

Fig. 10.6. Vapour pressure of an inorganic material

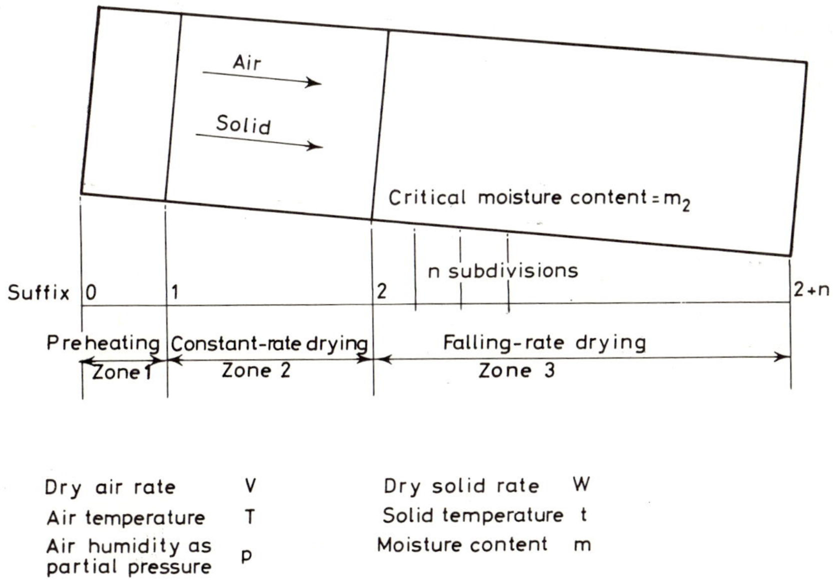

Fig. 10.7. Key to nomenclature for calculation

a maximum at about 100°C. This latter is characteristic of very soluble materials which have a continuous transition from solutions to the fused anhydrous material.

APPENDIX 10.2 DESIGN PROGRAMME FOR ROTARY DRYERS

The following programme for calculating the dimensions of a rotary dryer has been prepared in a form suitable for desk calculation or preparation of a computer programme. Definitions of symbols are given in kg, kcal, h, °C units. The definitions in the general list of symbols in SI units given at the beginning of this book do not apply. The key to the nomenclature is given in Fig. 10.7—this is for co-current drying. For counter current drying the air velocity is negative.

Appendix 10.2 Design of rotary dryers

Nomenclature (See Fig. 10.7)

Solid stream		
Dry solid rate	W	kg/h
Moisture content (dry basis)	m	kg/kg
Temperature	t	°C
Maximum permissible temperature of dry solid	t_{max}	°C
Specific heat of dry solid	C_d	kcal/kg °C
Specific heat of liquid water	C_m	kcal/kg °C
Effective latent heat of vaporisation	λ	kcal/kg
Vapour pressure of water in equilibrium with moist solid (as a function of moisture content)	P	atm
Bulk density	B	kg/m³
Gas stream		
Total humid air rate	Q	m³/h
Dry air rate	V	m³/h
Humidity as partial pressure	p	atm
Air temperature	T	°C
Ambient air temperature	T_a	°C
Specific heat of dry air	C_p	kcal/kg °C
Specific heat of water vapour	C_w	kcal/kg °C
Density of dry air	ρ_a	kg/m³
Density of water vapour	ρ_w	kg/m³
Air velocity	u	m/h
Ratio p_{n+2}/P_{n+2}	α	—
Heat and mass transfer		
Sensible heat transfer to solid	H	kcal/h
Heat-transfer coefficient per unit mass dry solid	h_m	kcal/kg h °C
Heat-transfer coefficient per unit length of dryer	h_L	kcal/m h °C
Mean temperature difference	Δt_m	°C
Mass-transfer coefficient per unit mass dry solid (as a function of moisture content)	k_m	kg/kg h atm
Mass-transfer coefficient per unit length of dryer	k_L	kg/m h atm
Mean partial pressure difference	Δp_m	atm

Dryer

Length	L	m
Diameter	D	m
Speed	N	rev/min
Slope	s	m/m
Residence time	θ	min
Hold-up per unit length	H_L	kg/m
Fraction filled	f	—
Coefficients in Saeman and Mitchell equation		
Filling coefficient } values in Section 10.17	a	—
Slope coefficient } values in Section 10.17	b	h/m

Design procedure

Overall heat and mass balances

Assume an exit temperature for the product: $t_{n+2} = t_{max} - 10$ (say)

Assume an exit air temperature: $T_{n+2} = t_{n+2} + 10$ in a co-current dryer, or $T_{n+2} = 100°C$ in a countercurrent dryer.

Find the equilibrium vapour pressure over the solid at exit: P_{n+2}

Choose an exit air humidity: $p_{n+2} = \alpha P_{n+2}$

Determine the dry air rate by an overall water mass balance

$$Wm_1 + V\rho_w\left(\frac{p_1}{(1-p_1)}\right) = Wm_{n+2} + V\rho_w\left(\frac{p_{n+2}}{(1-p_{n+2})}\right)$$

From which, and since $p_1 = p_0$ as no drying occurs in the preheating zone

$$V = \frac{W(m_1 - m_{n+2})}{\rho_w\left(\frac{p_{n+2}}{(1-p_{n+2})} - \frac{p_0}{(1-p_0)}\right)}$$

Determine the inlet air temperature by a heat balance, neglecting losses

$$T_0V\left(\rho_aC_p + \frac{p_0}{(1-p_0)}\rho_wC_w\right) + t_0W(C_d + m_1C_m) = \lambda W(m_0 - m_{n+2}) + t_{n+2}$$

$$(C_d + m_{n+2}C_m) + T_{n+2}V\left(\rho_dC_d + \frac{p_{n+2}}{(1-p_{n+2})}\rho_wC_w\right)$$

Dryer diameter

Choose a fractional filling: f

Calculate the total air flow at the exit conditions: Q_{n+2}

Choose an exit air velocity appropriate to the particle size: u

Calculate the dryer diameter

$$D = \left(\frac{4 \cdot Q_{n+2}}{\pi u(1-f)}\right)^{\frac{1}{2}}$$

The preheating zone (*Zone 1*)	
Calculate the heat-transfer coefficient per unit length	$h_L = \frac{\pi}{4} D^2 f B h_m$
Estimate the wet-bulb temperature	$t_1 = t_2$ since the wet-bulb temperature is constant over the constant-rate drying period for the air/water system
Calculate the sensible heat transfer in preheat zone	H_1
Calculate the air temperature at end of preheat zone	$T_1 = T_0 - \frac{H_1}{V\rho_a C_p}$
Check wet-bulb temperature and recalculate if necessary	
Calculate log mean temperature difference and length of preheat zone	$L_1 = \frac{H_1}{\Delta t_{lm} h_L}$
The constant-rate drying zone (*Zone 2*)	
Calculate the sensible heat transfer to the solid	$H_2 = \lambda W (m_0 - m_2)$ Note $m_0 = m_1$
Calculate exit gas temperature T_2 by heat balance	$T_1 V \left(\rho_a C_p + \frac{\rho_w p_1 C_w}{(1-p_1)} \right) + W m_1 C_m t_1 = H_2 + T_2 V \left(\rho_a C_p + \frac{\rho_w p_2 C_w}{(1-p_2)} \right) + W m_2 C_m t_2$
	$p_1 = p_0;\ t_1 = t_2;\ W C_d t_1 = W C_d t_2$
Calculate the log mean temperature difference	Δt_{m2}
Calculate the length of the constant-rate zone by heat transfer	$L_2 = \frac{H_2}{\Delta t_{m2} h_L}$
The falling-rate drying zone (*Zone 3*)	
Divide the zone into sections numbered 3, 4, …, n having arbitrary final moisture contents $m_3, m_4, \ldots, m_{n+2}$ (m_{n+2} is the required final moisture content)	
Calculate the exit air humidity by a mass balance	$\rho_w V \left(\frac{p_2}{1-p_2} \right) + W m_2 = \rho_w V \left(\frac{p_3}{1-p_3} \right) + W m_3$
Assume a solid exit temperature t_3 and calculate the air exit temperature T_3 by a heat balance	$V T_2 \left(C_p \rho_a + \frac{p_2}{(1-p_2)} C_w \rho_w \right) + W t_2 (C_d + m_2 C_m)$
	$= V T_3 \left(C_p \rho_a + \frac{p_3}{(1-p_3)} \right) + W t_3 (C_d + m_3 C_m) + \lambda W (m_2 - m_3)$

Calculate the sensible heat transfer H_3 between gas and solid	$H_3 = VT_2\left(C_p\rho_a+\frac{p_2C_w\rho_w}{(1-p_2)}\right) - VT_3\left(C_p\rho_a+\frac{p_3C_w}{(1-p_3)}\right)$ $= Wt_3(C_d+m_3C_m)-Wt_2(C_d+m_2C_m)+\lambda W(m_2-m_3)$
Calculate log mean temperature difference and length of section from heat-transfer equation	$L_3 = \frac{H_3}{\Delta t_{m3}h_L}$
Find the vapour pressure of water over the solid at the assumed exit temperature. Calculate the mean vapour pressure driving force Δp_m	Δp_m
Find the drying coefficient corresponding to the mean moisture content in the section from the experimental drying data. Calculate the mass-transfer coefficient per unit length of dryer*	$k_{L3} = \frac{\pi}{4}D^2 f B\, k_{m3}$
Calculate the length L'_3 from the mass-transfer equation	$L'_3 = \frac{W(m_2-m_3)}{k_{L3}\Delta p_m}$
Repeat the calculation with different assumed values of t_3 until acceptable agreement between L_3 and L'_3 is obtained	
The exit conditions for section $(n-1)$ are the inlet conditions for section (n), and the calculation is repeated progressively through the dryer. All values of L_n are then summed to give the total active length.	

The slope and speed

Calculate the residence time (cf. equation 10.3)	$\theta = \frac{60\pi D^2 f B}{4W}$
Choose a reasonable speed and calculate the slope from the Saeman and Mitchell equation (from equation 10.4). If necessary repeat with a different speed	

*Strictly, the gas phase resistance should be taken into account when calculating the mass-transfer coefficient, but as it is usually small compared with the solid drying resistance in the falling-rate period, little error is introduced by neglecting it.

CHAPTER ELEVEN

Fluidised bed dryers

11.1 FLUIDISED SYSTEMS

The general properties of fluidised systems are well known. They are described in References 1, 2, 3, 15 and will not be described here. Most of the applications described in these books are, however, in respect of catalytic systems, calcining, or bulk drying of materials such as sand and coal, rather than conventional chemicals. Some of these applications are discussed in Reference 4. Fluidised bed drying as a special subject is described in Reference 5 (which should be used to supplement the information in this chapter).

The theory and practice of fluidised bed dryers for chemical products of the type described in preceding chapters of this book is not yet, however, well established. This chapter describes the use of fluidised bed dryers in which closely sized material is gently roused by passage of a stream of hot gas upwards through the bed. The bed of material is not then 'boiling' with the passage of discrete gas bubbles but is in the form of dense-phase fluidisation.

For the drying of most chemicals, however, the air quantities required for the usual maximum temperature permitted are so large that uneconomically large cross-sectional areas would be required to ensure placid fluidisation. It then becomes necessary to use vigorously 'bubbling' or 'spouting' beds with the disadvantage that complete mixing of the bed occurs and uniform thorough drying is not achieved. Fully dispersed systems in which the material to be dried is conveyed pneumatically are described in Chapter 12.

11.2 DESCRIPTION

The principle of fluidised bed drying is shown in Fig. 11.1. Heated gas is blown upwards through a bed of the particulate material to be dried. The gas is distributed evenly through small orifices supporting the material at a rate sufficiently high to cause incipient fluidisation of the particles but not so high as to give the appearance of a vigorously boiling bed. Above the bed is a disengagement space 1–2 m high in which all but the smallest entrained particles can settle out. In some designs of dryer this upper section of the dryer is expanded to a greater cross-sectional area than that of the fluidised bed as shown on the right of the dryer section in Fig. 11.1. The drying gas leaves through some form of dust arrestor such as the cyclone shown. The collected fines can either be returned to the bed or incorporated in the wet feed. Large plants with deep fluidised beds are often also equipped with an induced draught fan after the dust arrestor and the system is run under balanced atmospheric pressure in the disengagement space: this arrangement facilitates the introduction and withdrawal of the material.

Fig. 11.2 shows some of the possible variations in the dryer chamber of continuous plant. In *a* the material flows horizontally across a vessel which is divided by baffles to give a zig-zag course. Alternatively the baffles may extend right across the vessel and be arranged to give an up and down zig-zag flow as shown in *b*. The section *c* shows a simple two-stage vertical arrangement in which the air flows through both stages in series. As shown, dusty air passes from the lower vessel through the distributor of the upper vessel without dedusting, and chokage of this distributor may occur. In

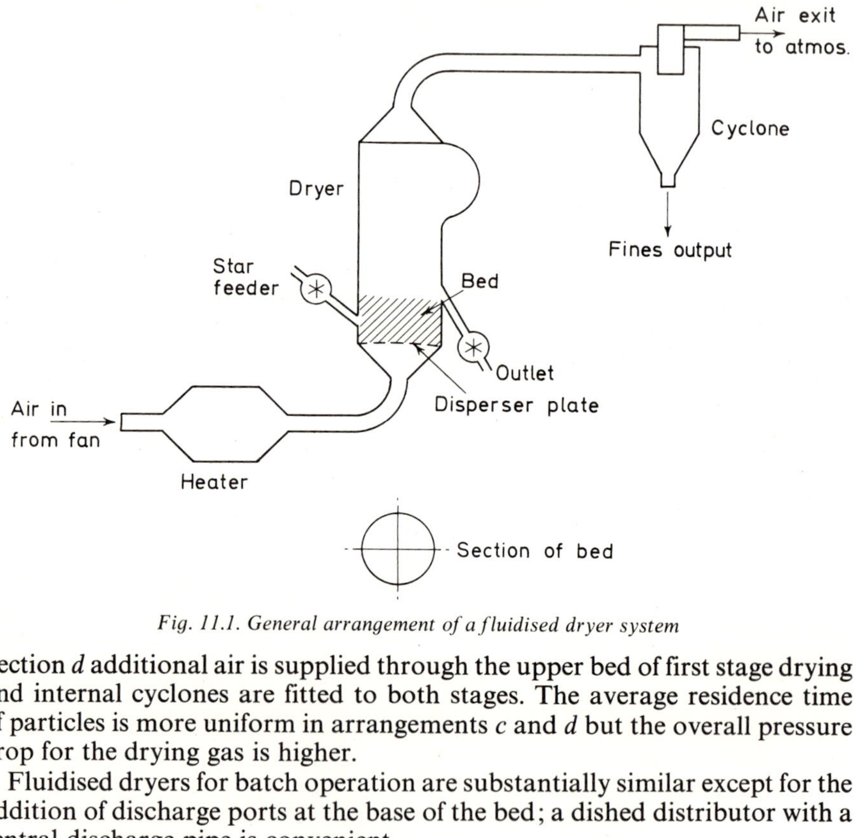

Fig. 11.1. General arrangement of a fluidised dryer system

Section *d* additional air is supplied through the upper bed of first stage drying and internal cyclones are fitted to both stages. The average residence time of particles is more uniform in arrangements *c* and *d* but the overall pressure drop for the drying gas is higher.

Fluidised dryers for batch operation are substantially similar except for the addition of discharge ports at the base of the bed; a dished distributor with a central discharge pipe is convenient.

11.3 GAS DISTRIBUTORS

For efficient fluidisation it is essential that the drying gas is distributed uniformly under and throughout the whole bed. Distribution will be uniform

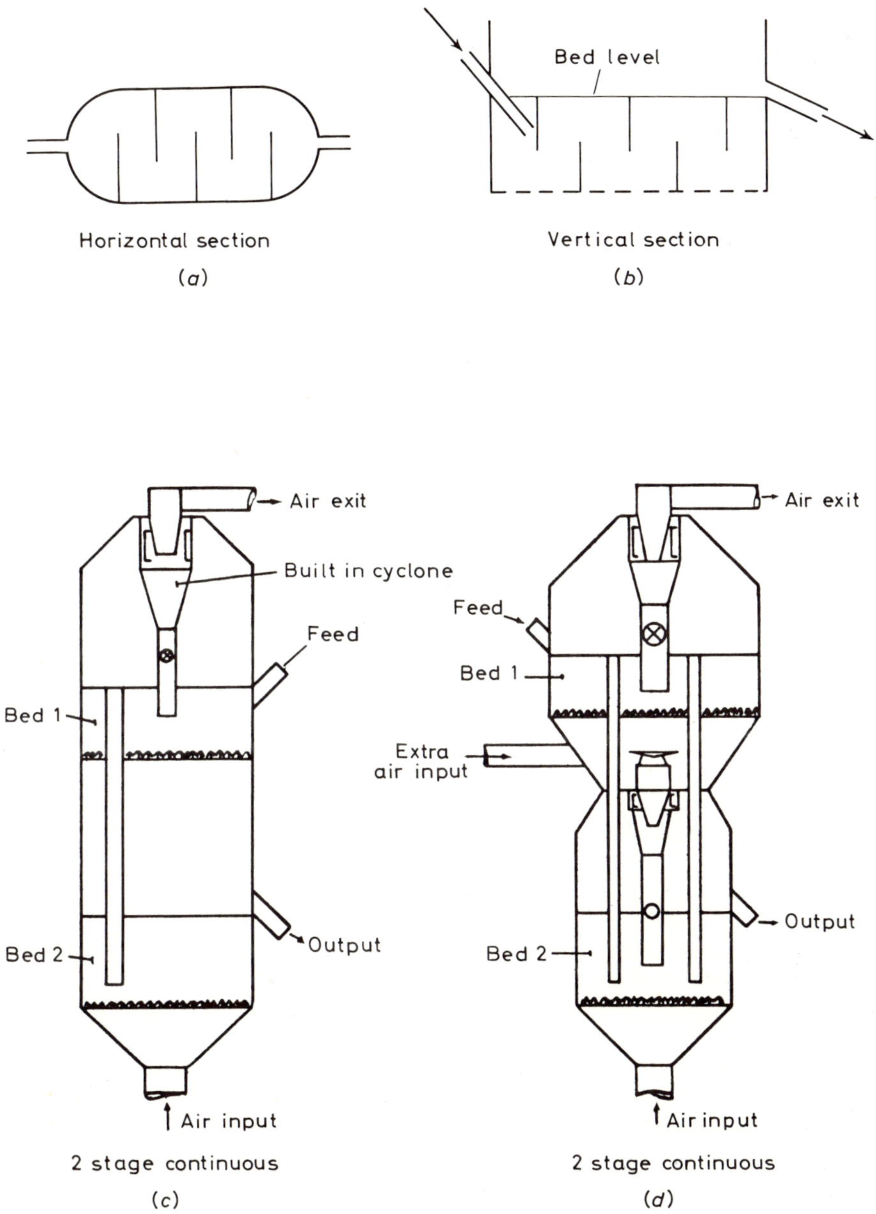

Fig. 11.2. Examples of fluidised dryer vessels

if the pressure drop across the distributor is sufficiently high, generally at least equal to the pressure drop through the bed.[6] Some plant suppliers employ distribution plates perforated with holes about 1 mm in diameter, others use circular slots and yet others use bubble plate type distributors of the general form shown in Fig. 11.3. Outlet holes of 2 mm diameter in the latter can be arranged to impart a circular motion to the fluidised bed; up to

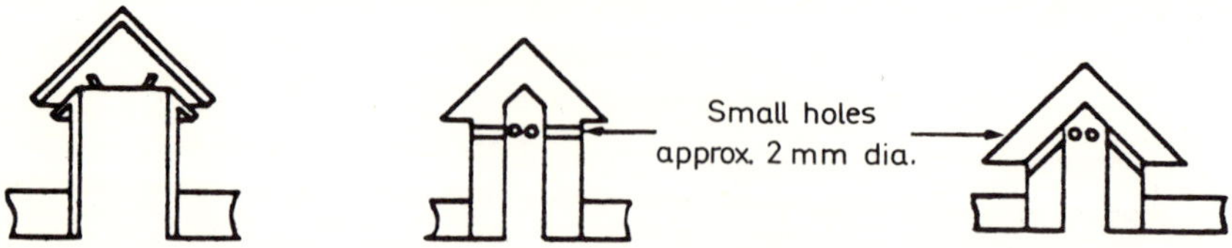

Fig. 11.3. Capped gas distributors

1 000 nozzles per square metre of distributor plate can be accommodated. When perforated plates are used, it is preferable to dish them (centre down) in order to increase the flow of gases near the walls of the vessel, since the normal tendency is for higher flows in the centre of the bed. This arrangement also facilitates emptying the vessel.

Porous porcelain tiles also serve as excellent gas distributors when the required gas velocity is low.

In order to reduce back flow of solids when a plant is shut down, the gas flow should be stopped by a valve after the blower and the pressure relieved from the bottom upward through the bed; if the blower is just tripped electrically, dusty gas flows down through the distributor and out through the blower.

11.4 ANCILLARY EQUIPMENT

11.4.1 Methods of feeding[13]

The feed is introduced into the drying chamber at the top of the fluidised bed. An essential feature of any arrangement for feeding material is that the material is in a dispersed form. Although some muddy wet powders can be made suitable for fluidisation drying by mixing with some dried product followed by suitable preforming, it must be remembered that agglomerates are large particles which will sink to the bottom of the bed of smaller particles unless they break up immediately they arrive on the bed and begin to dry. Alternatively, since attrition in a gently roused fluidised bed is usually negligible, it may prove possible to dry firmly bound porous granules without excessive carry over of fines. When drying mixtures, e.g. NPK fertilisers, it is, however, necessary to check that the fines have the same composition as the granules. Suitable feed mechanisms are screw-type feeders, star feeders and rotary valves but only the first can deliver material into a region of higher pressure. When the moist material is free flowing it can be fed from a steeply-sided supply bunker down a wide pipe terminating near the base of the bed. The bunker can then be supplied by any form of conveyor suitable for the moist material. The supply leg from the bunker acts as a gas seal. It is necessary also to use a feed system which enables the rate of feed to be metered in order to match the feed rate with the evaporation rate of the system. Any temporary chokes can readily be freed by compressed air injected above the slide valve.

11.4.2 Discharge of dried solids

In most continuous units, the dried material overflows from the surface of the bed into an outlet receiver and the material in the downcomer acts as a gas seal. When a rotary air lock is employed as a sealing device, it may be necessary to measure the rate of solids discharge as well as the rate of input of wet material in order to keep the bed level and the pressure drop through it constant. Batch plant is frequently supplied with discharge ports at several levels, thus enabling several bed depths to be used.

11.4.3 Gas moving equipment

The pressure drop through a fluidised bed may be as high as 25 mb and hence high efficiency fans are necessary for the forced draught and induced draught (when installed). Automatic control is desirable to maintain the gas pressure balanced – usually slightly below atmospheric pressure above the fluidised bed so as to prevent escape of dust through small leaks in the equipment.

11.4.4 Dust arrestors

High-efficiency cyclones operating at pressure drops of 8–12 mb are the most suitable equipment for cleaning the outlet gases. Filter bags are not suitable because they blind very easily if operated near the dew point of the gases. If the cyclones are installed outside instead of within the fluidising chamber, it is essential to lag them and the gas ducts to a high standard and to make provision for preheating them before starting up the dryer. When the collected dust from the cyclone is fed directly back into the fluidised bed, the cyclone must be mounted sufficiently high to ensure that there is an effective seal of dust to prevent backflow of gas into the cyclone base.

Further information on choice of cyclones will be found in Reference 7.

11.5 MATERIALS SUITABLE FOR FLUIDISED DRYING

An overriding material property in deciding whether to consider fluidised bed drying is the drying curve coupled with the specification for the dried product. The particles in a fluidised bed are thoroughly stirred by the fluidising gas and hence the residence time in any one compartment of a continuous dryer may be negligible for some of the particles, and the material leaving a compartment will contain some only partly dried particles despite the intimate contact of drying gas with every particle. (This does not, of course, apply to batch drying.) The superficial water will be evaporated in a few seconds. This process corresponds to the constant-rate drying period. Any material which has a very long falling-rate drying period is therefore likely to be unsuitable for continuous fluidised drying and batch drying of such material is seldom likely to be economic.

Another point to consider is the uniformity of residual moisture in the dried product. For many purposes it is essential that the range of moisture distribution in small portions of dried material is small since otherwise caking of the product could occur in storage. With fertilisers this could also affect the free flow of particles through application drills. Thus difficulties might occur if a material of 0·2% average moisture content contained 2% of particles containing 2% of moisture and 98% of particles containing 0·16% of moisture.

Once this point has been settled, the suitability of a particulate solid for fluidised drying is best determined by small-scale experiment using a transparent chamber of at least 120 mm diameter. In general the maximum size of particle which can be used in a dense phase bed is about 4 mm. The lower limit is about 0·1 mm, e.g. synthetic cracking catalyst.

The range of particle size is important. When a reasonably closely sized mixture is fluidised, the incipient fluidisation velocity (Section 11.10) is some mean of the individual values for each size and observation confirms that, as soon as the actual gas velocity in the bed is reduced below some critical value, the coarse particles settle out whilst the rest remain fluidised. On the other hand, when the gas velocity through a widely sized material is slowly increased, entrainment of the smaller particles commences as soon as the gas velocity above the bed exceeds their terminal falling velocity. In practice, it is not possible to select a gas velocity which will avoid both segregation and entrainment of particles of the same density if the ratio of the largest to smallest diameters exceeds 8.[8] Naturally this ratio is smaller if the particles become even slightly porous on drying. Moreover, a check must be made that particle size distribution does not change during the drying process; thus some particles tend to agglomerate when fluidised (e.g. wet and sticky powders) and some to disintegrate (e.g. preformed shapes).

It follows also from the description of the plant given in Section 11.2 that it is necessary to check on a fairly large scale that wet material can be fed to the drying chamber without choking and that the dried material will run out smoothly.

11.6 SOME APPLICATIONS OF FLUIDISED BED DRYERS

Information is scanty on the design and performance of full-scale plants for the thorough drying of conventional chemicals of the types described in this book.

A list of materials which have been dried in semi-technical plants is given in Table 11.1. It has also been demonstrated[5] that solutions of salts can be evaporated to crystals in fluidised dryers at a lower capital cost but higher running cost than in conventional multiple-effect evaporators.

Several plants have been installed for the drying of PVC and of polymer chips. Reference 9 gives a brief description of a batch plant for rapid drying of hot, dripping wet polymer chips, size 3 × 7 × 7 mm, with a residence time of

Table 11.1 Materials dried in small scale fluidised dryers

Material	*Size* (mm)	*Moisture content* (%) *In*	*Out*
Na Cl	0·15–0·6	3	0·03
KCl		8	0·2–0·5
$Na_2CO_3 1OH_2O$		62	0·05
$Na_2SO_4H_2O$		24–28	0·05
$(NH_4)_2SO_4$	1·2–4·2		0·2–0·09
Mixture of Na Cl and NH_4NO_3			0·01
Urea	<1	1·2–4·2	0·5

Note: $FeSO_4{\cdot}7H_2O$ can be dehydrated to $FeSO_4{\cdot}4H_2O$ in 3 min at 50–65°C and to $FeSO_4{\cdot}H_2O$ in 18 min at 120–160°C.

20 min. Sufficient air is drawn through the bed to cool this heat-sensitive material in the dryer. The machine occupies one-quarter of the floor space which would have been occupied by the previous type of dryer consisting of vertical cylindrical vessels through which air is gently blown and its capital cost was one-third of that of the old type.

Batch type fluidised dryers are used extensively for heat-sensitive pharmaceutical products.

On the tonnage scale, continuous fluidised dryers are used for drying of sand[10] and coal.[11] One U.S.A. plant for coal consists of two dryers, each 4·3 m diameter, each removing 35 t/h of water from 350 t/h coal of size 0–25 mm. The bed depth is only about 30 cm. The hot gas velocity is so high that about 50% of the coal is carried out of the bed into the cyclone. With a time of contact of only 0·1 s, the drying current of hot flue gas is cooled from 750°C to 65°C. The thermal efficiency is 96%, excluding radiation loss of $5\frac{1}{2}$%. The dried coal and sand have, however, a much higher moisture content than would be acceptable for most chemical plant products

The continuous fluidised bed process is suitable for cooling granular materials such as NPK fertilisers leaving rotary dryers. A 10 cm deep fluidised bed in a horizontal trough, 10 m long, has been described.[5] The material flows smoothly along the trough and overflows at a weir at the end remote from the feed. Tubes through which cooling water is passed are installed in the last third of the trough for use in hot weather. The air rate is 1 500 m^3 and the power consumption of 1·5 kWh/t is less than that of rotary coolers. The plant is far more compact and cheaper in capital cost than a rotary cooler.

11.7 ADVANTAGES OF FLUIDISED DRYERS

Fluidised dryers are suitable for materials of size 0·05–15 mm provided the size range of the particular product to be dried is small since otherwise the bed is not uniformly fluidised and large or heavy particles tend to accumulate at the base of the bed. As a corollary, fluidised bed dryers will perform the function of removing unwanted fines from the product being treated.

The contact between solid and drying gas in a fluidised bed is far better than in any other type of dryer because the agitation of the particles almost eliminates the stagnant gas film around them and the gas itself is mixed thoroughly during its passage through the bed. Consequently, the mass- and heat-transfer coefficients are so high that the gas leaves the bed at or only just above the bed temperature and may be almost at the wet-bulb temperature. The intrinsic thermal efficiency is therefore nearly 100%. The extremely rapid heat transfer enables higher inlet gas temperatures to be used than in other types of dryer because of the rapid fall in gas temperature.

Although agitation is violent within the bed, abrasion and breakage of the particles is small because they are cushioned from each other by the gas stream – in fact, it has been reported that the sharp edges of an abrasive grain have been maintained without degradation.

Hold-up and residence time are small for constant-rate drying and quick and easy discharge is possible, a particular advantage with heat-sensitive materials dried in batches (but a disadvantage in continuous plant). Extended residence times required when the falling-rate period is long can readily be provided by multi-deck dryers in cascade of the type shown in Fig. 11.2. The last stage of the cascade arrangement can be supplied with cooling air with the advantage that cooling is provided within the same enclosed vessel and, moreover, that the dew-point of the mixed air streams leaving the vessel for the dust arrestor is raised.

A fluidised dryer occupies less floor area than nearly any other type of dryer handling the same quantity of material or evaporating the same mass of water, but it may require more headroom. It is considerably cheaper in capital cost, particularly when corrosion-resistant materials of construction are required such as stainless steel. It has no moving parts, except for necessary ancillaries such as fans and rotary feeders and extractors. Maintenance costs are therefore low.

The control of moisture content in the dried product is simple compared with that required with transport or pneumatic dryers. Labour for operation is likely to be low because of the mechanical simplicity of the system.

11.8 DISADVANTAGES OF FLUIDISED DRYERS

It might be thought from the above list of advantages that fluidised dryers could be used for almost any batch or continuous drying requirement. This is not so. In the first place, they are difficult to use when the falling-rate period is long; also many compartments in series may be necessary because there is such a wide spread of residence time in each compartment and there is the chance that some partly dried particles may short circuit many compartments. Secondly, the material to be dried must be suitable for fluidisation. Some muddy wet powders can be dried by suitable preforming with dried product though dust losses may be high. Alternatively, a bed of heavy uniform inert particulate material may be used as the main fluidised medium and the dried product separated from the medium by screening or by carry-over in the gas stream. There is then a risk of product contamination. Again a fluid can be injected into the base of a fluidised bed but there is a danger of blockage of the feed nozzles. There is, also, great uncertainty that a full-scale plant using these devices will perform as satisfactorily as a pilot plant since the characteristics of the wet mud to be dried may change appreciably from time to time.

In batch fluidised dryers difficulty may occur when changing from one product to another because of the different fluidising velocities required.

The heavy power costs for fan power required may outweigh the substantially lower capital cost; it will be recalled that the pressure drop through the distributor is about equal to that through the bed and that the pressure drop through an efficient cyclone dust arrestor is also appreciable.

Finally, no design method has been established sufficiently firmly to avoid some risk even after trial with a 0·75–1·0 m diameter pilot plant such as is available at the works of some manufacturers of this type of dryer.

11.9 SOME FACTORS AFFECTING PERFORMANCE AND DESIGN

(For further information see Reference 14 and associated papers.)

11.9.1 Gas flow within the bed

A bed may still appear to be uniformly fluidised with uneven distribution of gas within the bed. Relatively little drying gas may penetrate the bed near the walls of the containing vessel.[13] Most of the information published on gas flow has, however, been in respect of the vessels in catalytic crackers, using radioactive tracer gas such as krypton 85 in small diameter vessels. In large diameter vessels large bubbles of gas form in deep beds unless the gas velocity is kept around that for incipient fluidisation and then there will be considerable bypassing of gas, though some downward flow also occurs within the bed.

Theory and practice have shown that gas distribution is improved with increase in gas velocity and therefore with incipient velocity u_f. Since the latter varies with $d_m^{1\cdot8}$, a relatively small increase in particle size will substantially improve internal gas distribution.

11.9.2 Heat transfer

In most cases the heat for drying the solid in the fluidised bed can be provided by the sensible heat of the incoming gas. If this is the case it has been shown that the gas leaving the bed will be at substantially the same temperature as the bed, provided it is more than a few centimetres deep.

In some cases it may be necessary to provide heat from internal coils or from heated sections of the bed wall. This may be the case if the material to be dried is thermally sensitive, i.e. becomes degraded or fused at high temperatures. It may not then be possible to provide sufficient heat in the inlet gas.

If internal coils are used in the bed, they may alter the uniformity of fluidisation. In addition the heat-transfer coefficient from a coil is not easily estimated. Van Heerden[8] shows that no published correlation for heat-transfer coefficient is satisfactory as different correlations will give results differing by a factor of 50.

In practice only experimental work can decide the best positioning and surface area of any heating pipes required. The use of a vertically-finned internal surface to the chamber may prove effective for most efficient heat transfer but the possibility of the fins upsetting good fluidisation or being clogged by the material should not be overlooked.

11.9.3 Mass transfer

Evidence suggests that in some cases single bed type dryers behave as co-current units, the denser moist material migrating to the bottom of the bed. If this occurs, the mass-transfer coefficient and hence the rate of drying, may be less than expected in the top portion of the bed due to the lower driving force. Hence not all the sensible heat of the gas can be utilised and the exit gas cannot approach saturation.

This consideration will influence the decision whether to use a single-stage unit, or a multi-stage dryer in which the material being dried cascades from one bed to the next (Fig. 11.2). In the latter case each bed can be supplied with fresh drying gas or with recycled gas from the preceding bed. If recycled gas is used, it should be expected that fines will be carried by the exit gas from one bed through the base disperser of the next bed.

11.9.4 Solids residence time

A fluidised bed behaves similarly to a stirred reactor and there will always be a considerable variation of solids residence times about the mean in any continuous dryer. Tailby[13] concluded from a review of the literature that reliable design can only be achieved from data derived from study of large diameter pilot or full-scale plants.

It is clear that cascade units must be used in continuous plant if all the material must be uniformly dry.

11.10 GAS VELOCITY AND PRESSURE FOR INCIPIENT FLUIDISATION

A widely used empirical formula for the gas quantity required for incipient fluidisation is that due to Leva.[1] This is

$$G = 1{\cdot}41 \times 10^{-2} \,.\, \frac{d_m^{1\cdot 82}}{\eta^{0\cdot 88}} \cdot [\sigma(\sigma - \rho)]^{0\cdot 94} \text{ g/s m}^2 \qquad (11.1)$$

As an example, for particles of 3 mm diameter, density 2 kg/m^3 fluidised with gas of density 1·2 kg/m^3 and viscosity 200 μp (0·02 mNs/m^2), G = 4 900 g/s m^2 equivalent to a velocity of 4·5 m/s air measured at OTB. The terminal velocity of these particles in air at 90°C and one bar abs. is 13 m/s.

The pressure required for fluidisation is equal to the weight of solids over unit area of bed. Hence

$$\Delta P = 9{\cdot}8 \times 10^{-5} t_m \rho_b \text{ bar} \tag{11.2}$$

To this pressure must, of course, be added the pressure drop through the distributor, which should be at least as great as ΔP.

In all cases the minimum fluidising gas velocity must be checked in a small-scale fluidised bed of at least 120 mm diameter. The velocity will be found to vary with the moisture content and surface properties of the particles, such as their tendency to agglomerate, as well as with their diameter and density. For particles smaller than 150 μm diameter the fluidising gas velocity is very low, about 2 cm/s. Considerations of the water vapour loading of the exit gas, due to the drying of the solids, may show that a considerably higher gas velocity is required. The bed would then cease to be gently fluidised and become a boiling or spouted bed. A high gas velocity of this order would also produce entrainment of fines in the exit gas.

Any appreciable carry-over will influence heat and mass balance calculations and the situation may become very complex. This can be overcome by increasing the cross-sectional area of the bed (and thus the hold-up in the bed) to a value greater than that indicated by drying rates or residence time; the use of a fluidised bed dryer may then become uneconomic.

11.11 A TENTATIVE THEORETICAL DESIGN METHOD

11.11.1 Introduction

The design of a fluidised bed dryer is a two-fold problem. In the first place, the surface area required to transfer heat at the rate necessary to maintain the specified evaporation rate must be calculated. Secondly, the volume of the fluidised bed must be determined so as to provide sufficient residence time for mass transfer.

The heat-transfer surface required may be calculated by conventional means in terms of a temperature driving force and an overall heat-transfer coefficient determined from small-scale experiments. It is possible to design the fluidised bed volume on the basis of equating the mean residence time with a drying time observed either in the laboratory or on a large scale. The disadvantage of this method is that it is not possible to allow for the effect on drying time of variations between the experimental conditions and those of the design. Nor is it possible to analyse the effect of variations in conditions and thereby optimise the design.

For these reasons it is considered that a better approach is provided by a method which treats the mass transfer in terms of a drying-rate coefficient and a humidity driving force, according to the theory given in Chapter 2. The theoretical basis of a design method of this type applied to the particular case of a continuous, isothermal, steady state fluidised bed dryer is outlined below. It is assumed throughout that the gas velocity through the bed has been previously fixed by the fluidisation characteristics. The method could then be used to determine the proportions of a pilot plant or for preliminary comparative costing against another method of drying. It cannot, with present knowledge, be used yet for safe design of a full-scale plant without appreciable risk.

11.11.2 Drying-rate coefficient

The rate of drying of an individual particle suspended in a fluidised bed of drying material is controlled by the balance between rates of heat and mass transfer in exactly the same way as in every other type of dryer. The generalised theory of drying put forward in Chapter 2 can therefore be applied (within the limits of the assumptions made in its deduction) to the drying of individual particles within the fluidised bed. Consequently the individual particle drying-rate coefficient defined by the equation

$$K_1 = \frac{\text{rate of drying of an individual particle}}{\text{humidity driving force}}$$

is dependent only on the moisture content of the particle.

The drying rate per unit mass of fluidised bed is a mean of the individual drying rates for the total number of particles in the unit mass. Hence the drying-rate coefficient per unit mass of fluidised bed is defined by the equation

$$K_H^* = \frac{\text{rate of drying per unit mass of fluidised bed}}{\text{humidity driving force}}$$

K_H^* will be a function, not only of the moisture content per unit mass of the bed, but also of the manner in which this moisture is distributed between individual particles in the bed. Thus consider two beds (a) and (b) of equal moisture content containing N particles of uniform mass some of which are dry and the rest as follows: case (a) containing n particles at the critical moisture content, case (b) containing $2n$ particles at half the critical moisture content. Fig. 2.8 shows that the drying coefficient at half the moisture content is less than half that at the critical moisture content. Consequently the overall drying coefficient K_H^* of the bed in case (b) will be less than that of the bed in case (a).

11.11.3 The recycle ratio

A fluidised bed dryer is normally operated continuously at a steady state on a feed of constant moisture content and a high degree of mixing of the particles in the bed can be assumed. It is assumed that under these conditions the moisture content per unit mass of the fluidised bed is substantially uniform throughout the bed (but see Section 11.9) and remains constant at a value depending only on the ratio of the mass of solids in the bed (W) to the mass rate (w_d) at which dry solids are fed to and removed from the bed.

It is convenient to define a recycle ratio by the equation

$$W_d(1+R_r) = 10^3 W \tag{11.3}$$

$$\text{whence } R_r = 10^3 \frac{(W-W_d)}{W_d} \tag{11.4}$$

With the assumptions given above, the distribution of moisture among the individual particles in the bed, at any given bed moisture content, will depend only on the value of R_r. Hence, under normal operating conditions both the moisture content per unit mass of fluidised bed and the manner in which it is distributed depend only on R_r. R_r is therefore both necessary and sufficient to specify completely the value of K_H^*. The relation between R_r and K_H^* may be determined from laboratory experiments carried out at the steady state. It is essential in these experiments that the rate at which wet material is fed to the bed is exactly matched to the rate of evaporation.

Although it is not necessary to the development of the above argument for the bed to be perfectly mixed, it is necessary to attain a steady state so that although the flow patterns may be in a continual state of flux, the mean pattern does not change either with time or with the overall size of the bed, i.e. the distribution is constant with respect to time and scale.

It is reasonable to expect the distribution to remain constant with time but the effect of scale is doubtful. This difficulty is fundamental in the design of all equipment involving fluidised beds and in order to minimise the risk of introducing serious errors in scaling-up, care must be taken during the small-scale experiments to ensure that fluidisation is as uniform as possible and that no radical departures from geometrical similarity are made.

11.11.4 Humidity driving force

The humidity driving force at any point in the bed is the difference between the saturation humidity H_{ss} at the temperature of the fluidised solids T_s and the humidity H_g of the fluidising gas at that point. T_s and hence H_s, may therefore be taken as constant and equal to the mean bed temperature (observed from thermocouples inserted in the bed). Carrying out a water balance on an element of bed depth dx then gives the following equation:

$$A'G\,\mathrm{d}H = \frac{10^3 K_H^* W\,\mathrm{d}x(H_{ss}-H_g)}{t_m} \tag{11.5}$$

When $x = O$, $H_g = H_1$ = inlet humidity of drying gas.
$x = t_m$, $H_g = H_2$ = humidity of gas leaving the bed.

Integrating gives

$$\frac{K_H^* W}{A'G} = 10^{-3} \ln \frac{H_{ss}-H_1}{H_{ss}-H_2} \tag{11.6}$$

But by definition $A'G(H_2-H_1) = 10^3WK_H^*\Delta H_{lm}$ where ΔH_{lm} = log mean humidity driving force and W is taken as unity.

Hence
$$\Delta H_{lm} = \frac{H_2-H_1}{\ln\dfrac{H_{ss}-H_1}{H_{ss}-H_2}} = \frac{(H_{ss}-H_1)-(H_{ss}-H_2)}{\ln\dfrac{H_{ss}-H_1}{H_{ss}-H_2}} \tag{11.7}$$

If the fluidising gas were perfectly mixed throughout the bed, the mean humidity driving force to be used would be $(H_{ss}-H_2)$. It is believed that there is some backflow of gas within the bed and hence the use of the logarithmic mean is more correct. In practice it will make little difference provided either one or the other is used consistently.

11.11.5 Method of design calculation

The value for the mass gas rate G is first determined to give a reasonable compromise between uniformity of fluidisation and carry-over of dust, together with heat balance requirements. This value of G is then the value to be used in the final design and must be used throughout the experimental work. Then using the arguments put forward and assuming that the dried material and gases leaving the fluidised bed are at the bed temperature T_s, it is possible to write down the relations giverning the design of a fluidised bed dryer for any one substance as follows:

Water balance

$$W_d(M_2-M_1) = A'G(H_2-H_1) \tag{11.8}$$

Mass transfer

$$W_d(M_2-M_1) = 10^3WK_H^*\Delta H_{lm} \tag{11.9}$$

Heat transfer

$$W_d(c_s+M_1)T_s - W_d(c_s+M_2)T_f + A'G(S_2T_s-S_1T_1) + W_d(M_2-M_1)\lambda + \text{losses} = UA\lambda\Delta T_{lm} \tag{11.10}$$

Theoretical thermal efficiency (11.11)

$$\eta_t = \frac{W_d(M_2-M_1)\lambda}{A'G(S_2T_s-S_1T_1)+W_d(M_2-M_1)\lambda+W_d(c_s+M_1)T_s-W_d(c_s+M_2)T_f}$$

By definition

$$W_d(1+R_r) = 10^3W$$

Also

$$K_H^* = f_1(R_r G) \tag{11.12}$$
$$U = f_2(R_r G) \tag{11.13}$$
$$\frac{W}{A't_m} = f_3(G) \tag{11.14}$$
$$S = c_p+0{\cdot}48H \tag{11.15}$$
$$H_s = f_4(T_s) \tag{11.16}$$
$$\lambda = f_5(T_s) \tag{11.17}$$

where f refers to a function.

Equation 11.10 is only of interest if heating coils are to be used in the bed. The procedure for evaluating the unknowns will vary from case to case, but the following example is given, as it is believed that the conditions it meets are likely to be useful.

11.11.6 Example

Flowsheet quantities

Output as dry solids W_d
Specific heat of dry solids c_s
Moisture content of feed M_2
Moisture content of product M_1
Inlet temperature of feed T_f

Specified variables

Fluidising velocity G	*Fixed by fluidisation trials*
Bed temperature T_s (hence H_{ss} and λ).	This should normally be fixed as high as possible consistent with material temperature limitations.
Fluidising gas inlet temperature T_1 and humidity H_1.	Assumed to be at atmospheric temperature and humidity.
Log mean temperature differential from heating coils to bed ΔT_{lm}.	This applies when heating coils are required. ΔT_{lm} from T_{h} and T_s.

1. Choose a value for the outlet humidity H_2 which gives a reasonable value for the cross-sectional area of the bed, A', in equation 11.8 but such that the dewpoint of the exit gas is at least 10°C below T_s.
2. Find ΔH_{lm} from H_s, H_1 and H_2, and use in equation 11.9 to find the required value of WK_H^*.
3. In a small-scale experiment find the relationship between R_r and K_H^* for the given inlet and outlet moistures and gas rate and temperature. (This will mean varying the bed depth of the small-scale bed in order to vary R_r.)
4. From equations 11.3 and 11.12 make a plot of WK_H^* against R_r and find the value of R_r required.
5. Find the value of W in equation 11.3 and use this to fix the bed height. Check that this bed height is reasonable compared with the cross-sectional area A'.
6. Repeat the calculation using different values of H_2, if desired, and correlate the bed size against the power required for the fans, in order to find the economic size of the bed.
7. If heating coils are required, U must be found by small-scale experiments.
8. Estimate the heat losses from the bed.
9. Find the heat-transfer surface necessary in equation 11.10 and check that this can be fitted conveniently into the size of bed.

Comment

It has been assumed that a reasonable thermal efficiency can be obtained without raising H_2 above the danger limit for condensation. This may not always be possible and in such cases consideration should be given to either reducing the bed temperature T_s, which will mean that a larger bed must be used, or recovering some of the heat in the exit gases.

Alternatively it may not be possible to design for a reasonable thermal efficiency because the cross-sectional area A' may be reduced to ridiculous limits to give the desired value of H_2. In this case recycling of the fluidised gas should be considered. Any of these courses will lead to an increase in capital cost which must be weighed against the saving of operating costs.

CHAPTER TWELVE

Pneumatic conveying dryers

12.1 PRINCIPLES

In the pneumatic conveying dryer the solid is dried while being transported in a stream of warm gas, usually air. The pneumatic conveying dryer differs from the fluidised bed dryer, in which the solid is also suspended in the air stream, in that the density of the suspension is much lower.

The simplest form of pneumatic dryer is illustrated in Fig. 12.1. Drying takes place in the vertical tube up which warm air is flowing with a velocity greater than the free-falling speed of the largest particles in the wet feed. The wet feed is introduced at the bottom of the tube by means of a suitable feeder. It is common practice to employ a paddle wheel 'slinger' which projects far enough into the drying tube to throw the feed vertically on the centre line of the drying tube. Large agglomerates are broken up while smaller ones are thrown up the tube where they rapidly disintegrate as they dry. The handling characteristics of the feed can often be improved by recirculating a proportion of the dried product.

At the top of the tube the dry product is collected by means of a cyclone, and bag filters or electrostatic precipitators are used to prevent the emission of dust in the effluent air. On the larger dryers, two fans are employed so that the pressure differential is negligible at the feed point, thus avoiding either blowout or infiltration, both of which are equally undesirable.

In a simple dryer of this type the residence time in the drying tube is in the region of 0·5–3·0 s so that only easily dried materials can be handled. Most commercial dryers make some provision for recirculation of the feed in order to increase the average residence time up to about one minute. The methods adopted to effect recirculation are indicated later in the descriptions of individual dryers (Section 12.4). It must be appreciated that when recirculation is employed, there is no guarantee that each particle will go round the same number of times, and that there is a spread of residence times about the nominal value. The higher the ratio of recirculation to through-put the greater will be the spread. The recirculating pneumatic dryer is therefore unsuitable in cases where it is important that every particle should be given the same drying time.

Some published performance data[1–6] for pneumatic conveying dryers are given in Table 12.1.

12.2 TYPES OF MATERIALS HANDLED

(a) The wet feed must be reasonably free flowing and non-sticky so that it can be easily dispersed in the air stream.

(b) Because of the short contact time only materials which have high drying coefficients* can be handled. For example, sand has drying coefficients ranging from 0·11 to $2{\cdot}8\,s^{-1}\,bar^{-1}$ compared with $2{\cdot}8 \times 10^{-5}$ to $2{\cdot}8 \times 10^{-2}\,s^{-1}\,bar^{-1}$ for typical granular fertilisers for which the pneumatic conveying dryer is known to be unsuitable.

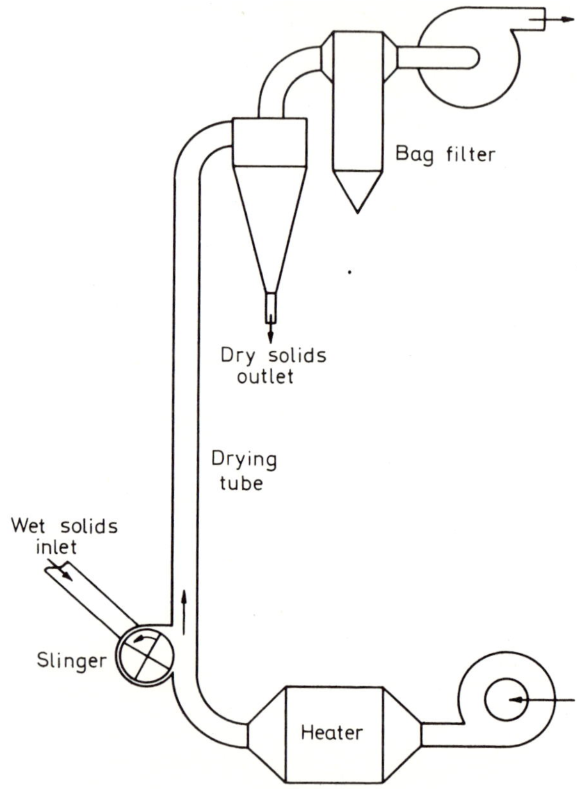

Fig. 12.1. Diagram of typical pneumatic conveying dryer

(c) In order to avoid high air velocities and consequently high pressure drops the maximum particle size should not be too great. Depending on the density, the maximum particle diameter is probably about 1–2 mm corresponding to a maximum air velocity of about 25 m/s.

(d) In general, the pneumatic conveying dryer is suitable for non-porous, crystalline (sand-like) materials, with a maximum particle size in the

*The units used for the drying coefficient (KA_u) are: (kg water evaporated)/(kg dry solid) (second) (bar pressure driving force) = 1/(second) (bar)

Table 12.1 Some published performance data on pneumatic dryers

Material	*Location*	*Tube dia* (cm)	*Tube ht* (m)	*Gas rate* (m^3/h) (NTP)	*Gas temp* (°C)		*Solid rate* (kg/h)	*Solid temp* (°C)		*Moisture* (%)		*Air/solid ratio*		*Water evaporated* kg/h
					In	Out		*In*	*Out*	*In*	*Out*	m^3/kg (NTP)	kg/kg	
Ammonium sulphate	Japan	18	1	1 100	215	76	950	38·5	63	2·75	0·28	1·2	1·5	23·5
Sewage sludge filter cake	U.S.A.	—	—	1 200	700	121	2 270	15	71	80	10	5·3	7·2	1 590
Coal 6 mm	U.S.A.	—	—	50 000	371	80	51 000	15	57	9	3	1·0	1·3	4 350
Hexamethylene tetramine	Germany	30	38*	3 600	93	50	2 500	—	48	6–10	0·08–0·15	1·4	1·9	18·1

*23 m vertical, 15 m horizontal

region of 1–2 mm diameter, and which do not tend to agglomerate to any great extent.

12.3 ADVANTAGE AND DISADVANTAGES

The advantages of a pneumatic dryer are:

(a) The very short contact time and co-current flow often permit satisfactory drying of heat-sensitive materials.
(b) For the same reasons, high air inlet temperatures and low air rates can often be used, resulting in high thermal efficiency.
(c) The dryer transports the solid so that its use may avoid the need for separate conveying equipment.
(d) The dryer requires little floor space and may be installed outdoors. The final dust collector should, however, be housed in a warm enclosure to prevent dewing and dust deposition in the flues and choking of bag filters if installed.
(e) The dryer is readily controlled as the low solids hold-up gives a quick response to changing conditions.
(f) There are few moving parts so that the maintenance requirements are low.
(g) The capital cost is usually low compared with other types of dryer.

The disadvantages include:

(a) Efficient dust collection is essential; difficulties with this may make the dryer unsuitable for some materials.
(b) Because of dust emission, the dryer is unsuitable for toxic materials; the use of superheated steam as the drying medium may overcome this difficulty in a few cases.
(c) Operating difficulties will result if the feed contains many large agglomerates which are difficult to disperse.
(d) In general, individual particles do not receive exactly the same treatment, especially in a dryer with recirculation, and this may be a disadvantage in some cases.

12.4 DESCRIPTION OF SOME COMMERCIALLY AVAILABLE PNEUMATIC CONVEYING DRYERS

Below are given diagrams and brief descriptions by code number of some pneumatic dryers manufactured in the U.K. Each has its merits in terms of the combination of performance and cost. Before considering any design, laboratory tests should be made on the suitability of the material for this type of dryer using samples covering the range of qualities which are to be expected under works conditions. One form of apparatus for the purpose is described in Appendix 12.1. Thereafter a trial on a pilot unit at the manufacturer's works should be made with freshly prepared samples of material—not with material that has been dried and then wetted.

Code P.C.2 (Fig. 12.2). This is essentially a simple straight-through dryer with a vertical dryer tube. Provision is made externally for recycling a proportion of the product if desired. The recycle stream is not fluidised and must pass through a rotary valve, so that the product must be dry enough to be free-flowing after one pass through the dryer. A paddle wheel 'slinger' is

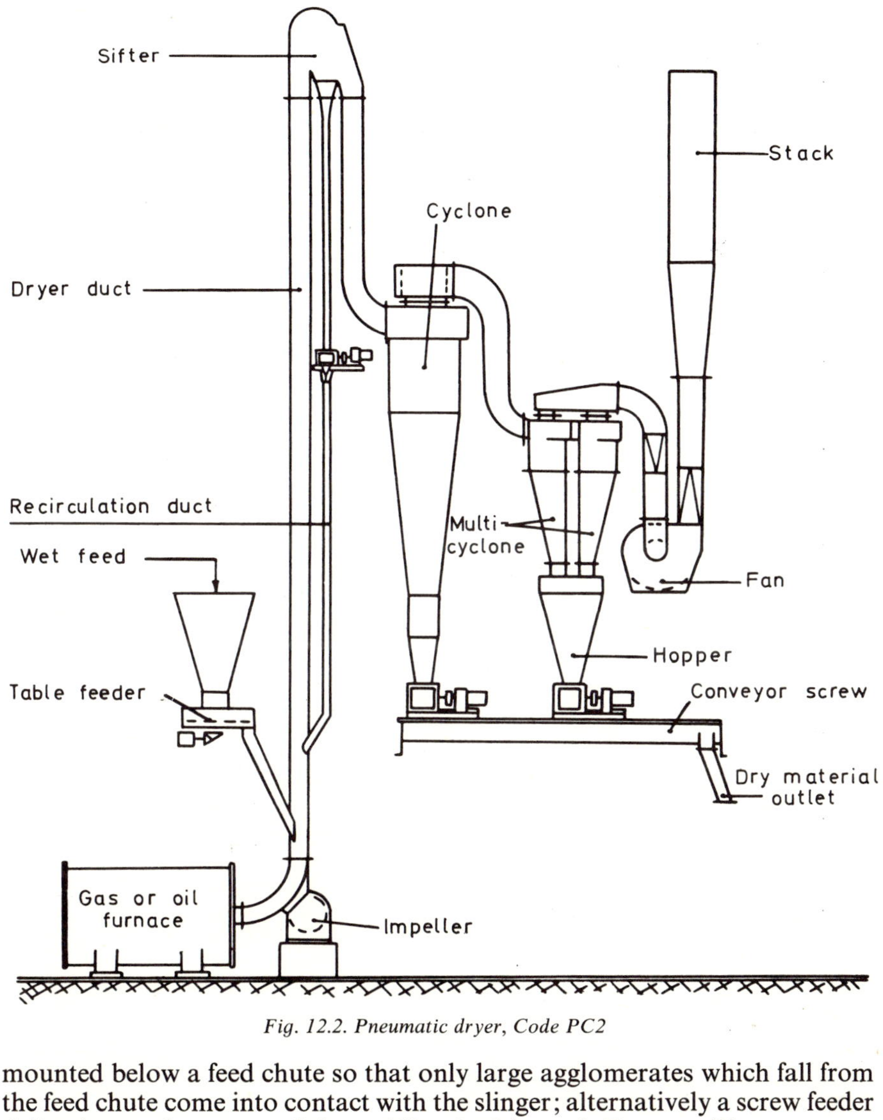

Fig. 12.2. Pneumatic dryer, Code PC2

mounted below a feed chute so that only large agglomerates which fall from the feed chute come into contact with the slinger; alternatively a screw feeder delivering to the centre of the paddle wheel is used. Dryers with outputs up to 150 t/h dried product are in operation.

Code P.C.3 (Fig. 12.3). This dryer is similar to an early design of P.C.2. It is a once-through dryer divided into two stages to reduce the height. The

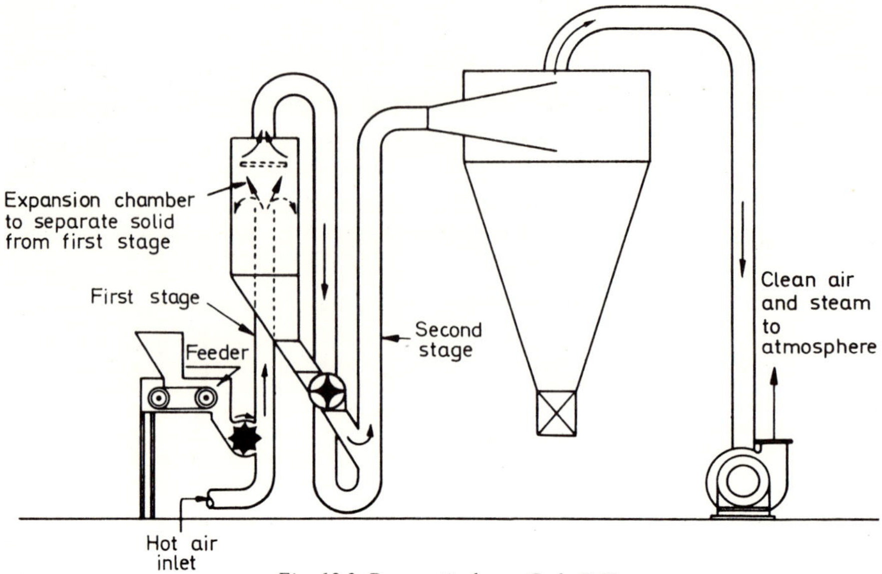

Fig. 12.3. Pneumatic dryer, Code PC3

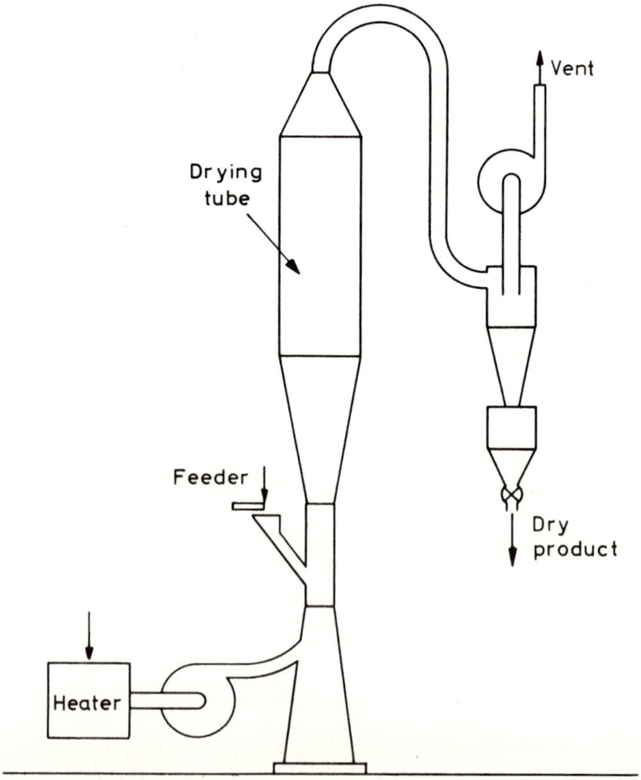

Fig. 12.4. Pneumatic dryer, Code PC4

drying tube is trombone shaped but only the risers are used for drying. The solid is separated from the air stream in an expansion chamber at the top of the first stage and transferred to the second stage by gravity through a star feeder. This arrangement avoids chokes at the bottom of the second stage.

The wet feed is introduced into the first stage by means of a band feeder and a 'slinger'. There are seven standard sizes available with dry solid capacities from $\frac{1}{2}$ to 20 t/h and water evaporation capacities from 100 to 2 000 kg/h.

*Code P.C.*4 (Fig. 12.4). This is a vertical tube dryer in which the tube has

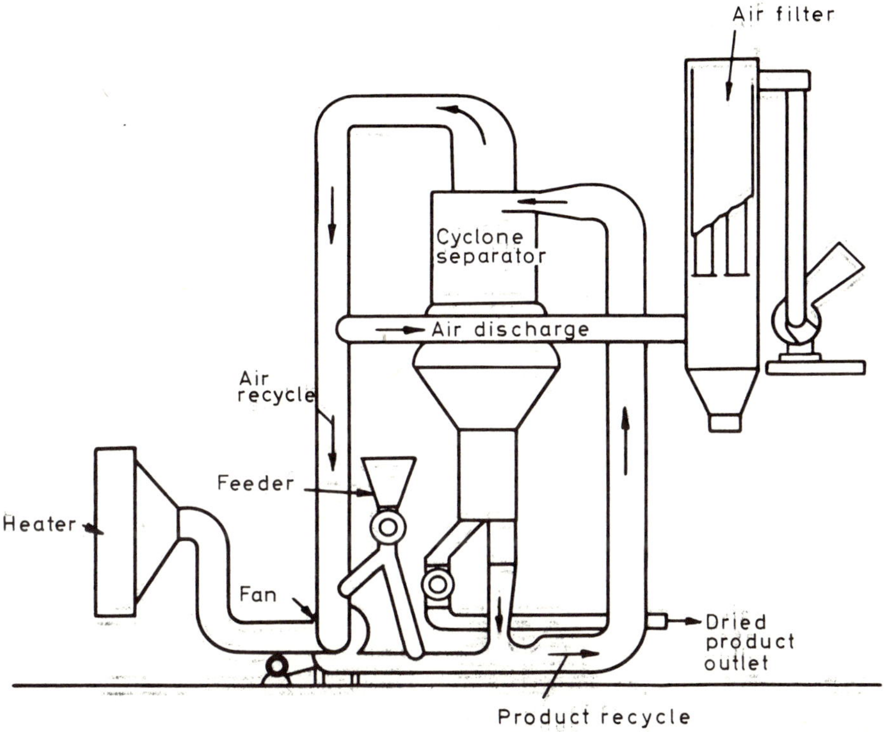

Fig. 12.5. Pneumatic dryer, Code PC5

a marked venturi shape; the air velocity at the feed point is about 30 m/s while in the upper part of the tube it is about 6 m/s.

In the upper part of the tube the mean velocity may be less than the free falling speed of the larger particles. As a result a circulation develops, downwards at the walls and upwards at the centre. Conditions in the parallel section of the dryer may be better described as a 'dilute phase fluidised bed'. Any particle may pass through without recirculation, but the conditions are such that large particles which need a longer residence time are more likely to recirculate.

This type of dryer cannot readily be designed by the principles to be outlined in Section 12.5 because the air velocity distribution and solid recirculation ratio are not known with sufficient accuracy. It appears that in general the height of this design of dryer may be considerably less than that of a straight through dryer designed for the same air rate and temperatures. Dryers are in operation for outputs of up to 20 t/h dry solids, maximum evaporation capacity of 5 000 kg/h.

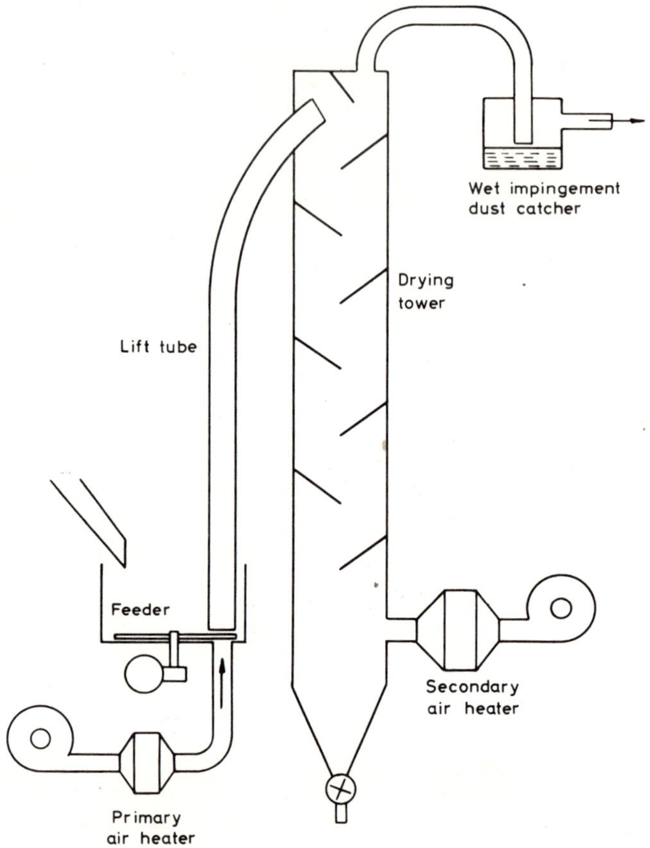

Fig. 12.6. Pneumatic dryer, Code PC6

*Code P.C.*5 (Fig. 12.5). This dryer is mainly a recirculating dryer. The wet feed is introduced into the air stream by means of a star feeder. The main part of the drying tube is vertical and discharges into a cyclone. Part of the solid material is returned to the drying tube and the product is taken off through a second star feeder. The main air stream is recycled; part is bled off and made up by fresh heated air. Six standard sizes are available, with capacities from 50 to 7 500 kg/h of dry solid.

*Code P.C.*6 (Fig. 12.6). This is only in part a pneumatic dryer. Drying is carried to the stage where the product is free flowing in a single pass conveying tube. This tube is nearly vertical, but is slightly curved to enter the top of the secondary drying tower. In this tower the solid cascades down over a series of deflector plates which also direct an ascending stream of warm air through the curtains of solid. A good counter-current action is thus obtained. By introducing cold air, the secondary tower could also be used as a cooler. A simple scraper feeder is used on this dryer.

This dryer has been used for drying by-product ammonium sulphate. It is available in only one size, with a capacity of 3 000 kg/h of dry salt. A double unit having two pneumatic conveying tubes feeding a single secondary tower is also made.

*Code P.C.*7 (Fig. 12.7). In this dryer both air and solid circulate round a

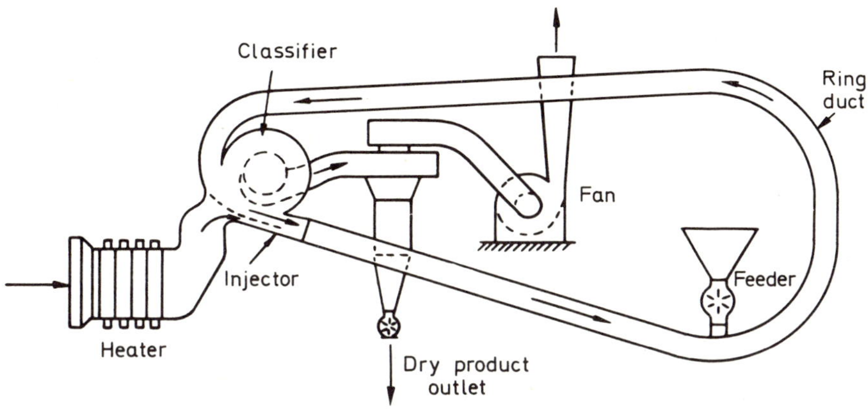

Fig. 12.7. Pneumatic dryer, Code PC7

closed circuit, which can be made almost any shape to provide the most convenient layout of the plant. A proportion of the air and solid are drawn off at the centre of the 'classifier'. A make-up of heated air is introduced through an injector nozzle and maintains the circulation round the 'ring'. Wet feed is introduced at a later point in the circuit by a simple star feeder. If the feed is likely to agglomerate, a mechanical disintegrator can be included in the circuit after the feed point.

The closed circuit makes it possible to provide a relatively long drying tube in a limited space. Circulating velocities and hence residence times can be estimated by the methods applicable to pneumatic conveying problems.

This dryer is usually run with a high solids recirculation ratio but the air recirculation is small. The classifier is said to have a selective action which tends to increase the recirculation ratio of the larger particles. A high re-circulation ratio may result in excessive attrition.

Only approximate design calculations are possible.

12.5 CAPITAL COSTS

Approximate prices (corrected to 1968) of a few dryers available from three British Manufacturers are given in Table 12.2 as a guide. These prices exclude installation and connection to services, which may add 50–100% to the total cost.

The dimensions of the dryers and their ancillary equipment are primarily

Table 12.2 Costs of pneumatic dryers
(Costs are adjusted to 1968 prices and exclude installation)

	Type	*Throughput* (kg/h)	*Evaporation* (kg/h)	*Air rate* (m^3/h) (NTR)	*Material of construction*	*Cost* (£)
1	P.C.4.	1 000	20	1 700	M.S.	3 200
2	P.C.4.	22 000	550	11 000	S.S.	22 000
3	P.C.4.	50 000	1 800	16 000	M.S.	24 000
4	P.C.4.	33 000	1 200	10 500	M.S.	17 500
5	P.C.2.	45 000	1 575	17 500	S.S.	35 000
6	P.C.7.	140	95	6 300	S.S.	11 000
7	P.C.7.	620	400	19 000	A.A.	24 000

M.S. = Mild Steel S.S. = Stainless Steel A.A. = Aluminium Alloy

determined by the air rate so that a close relationship between the cost and the air rate is to be expected. Thus for each material of construction the logarithm of the plant cost varies approximately directly as the logarithm of the air rate.

12.6 INSTRUMENTS AND CONTROL

Because the residence time in the pneumatic dryer is small, a change in the evaporation load quickly results in a change in the exit air temperature. This results in a change in the mean-temperature driving-force which in turn affects the product moisture content. The exact relationship between evaporation, exit air temperature and moisture content of the product is extremely complex. Obviously, for a constant heat input, an increase in the evaporation load results in a decrease in the air exit temperature and an increase in the moisture content of the product. Since there is little delay in the response of the air exit temperature on a change in load, this temperature forms a convenient variable on which to base the control of the dryer.

There are three ways in which the control system may react to a change of load:

(a) the inlet air temperature may be changed, with the air flow remaining constant,
(b) the feed rate may be varied in order to keep the evaporation load as nearly as possible constant,
(c) the air rate may be varied, with the air inlet temperature remaining constant.

The first method is that most commonly used. It usually lends itself to a simple form of control system and has the advantage that the major variations in temperature are confined to the wet end of the dryer. At the dry end, where the solid is drying in the falling-rate period and is therefore more likely to be affected by the drying conditions, there is only a small change in temperature. Because of the increased inlet temperature and almost constant exit temperature, the thermal efficiency increases with increasing load.

The second method can only be used where it is possible to vary the throughput of the dryer independently of any other process. Given this condition it may well be an attractive method in some circumstances. It would be especially useful if the dryer were operated from a source of waste heat which was subject to uncontrolled variations.

The third method is less flexible than the others in that it can only deal with variations below the normal load, and only over a limited range. If the load on the dryer increases, not only the heat input but the mean temperature difference must be increased. If the inlet temperature is fixed then the exit temperature must be increased. If the exit temperature is used as the control variable, however, it can only be restored to its original value, and consequently the necessary increase in driving force does not take place. Hence this system of control under-corrects. This may lead to trouble when an increase in load occurred, but apart from using more heat than is strictly necessary, no harm is likely to follow a decrease in load. The lower limit is fixed by the need to maintain a sufficient velocity for conveying. This method of control is the least useful but may be of value in some circumstances.

Whichever method of control is used and whether it is automatic or manual, it is important to ensure the steadiest possible feed rate to the dryer, and to try to avoid sudden changes in load. The response time of the control system is likely to be much greater than the residence time in the dryer. A sudden increase in load may therefore result in a substantial quantity of wet material entering the dry sections of the dryer and causing operating difficulties. Hunting of the control system may occur and could have a similar result.

It is important to avoid a large heat capacity in the heating system as this will increase the response time. If a source of heat having a large capacity must be used, then it should be run at a constant temperature and the required variation of air temperature obtained by the use of an automatically controlled by-pass on the heater.

In addition, the following measurements are of value in assessing the performance of the plant and in trouble shooting.

(a) Air inlet temperature.
(b) Air flow-rate.
(c) Pressure at the feed point (to avoid excessive air leakage through the feeder).
(d) Pressure differentials across the main components of the dryer.
(e) A fuel flowmeter in direct fired dryers.

12.7 INFORMATION REQUIRED FOR DESIGN

As with most types of dryer the design of pneumatic conveying dryers ultimately rests on practical experience, though theoretical principles can give guidance in some aspects of the design. Drying involves simultaneous heat and matter transfer, and in principle, a dryer can always be designed if appropriate values of the heat- and matter-transfer coefficients are available. A laboratory experimental method for determining these coefficients is described in Appendix 12.1 and the application of the results to the design of dryers is discussed in this section. A pilot plant experiment to confirm the results of the calculations is always desirable; this will normally be carried out by the dryer manufacturer.

The simplest form of dryer is the once-through vertical tube type. The questions to be answered in designing such a dryer are:

(a) What is the minimum air velocity to ensure that all the solid is carried up the tube?
(b) What is the inlet temperature of the air, and also the maximum inlet and outlet temperatures that can be used?
(c) What air/solid ratio is required to satisfy the requirements of drying and conveying the solid?
(d) What residence time is required to achieve the required exit moisture content of the solid?
(e) What heat input will be required?

The designer will require the following information:

(a) The dry solid throughput.
(b) The initial moisture content and temperature of the solid.
(c) The required exit moisture content.
(d) A fairly detailed size analysis of the solid (say in 5 size-ranges).
(e) The particle density of the solid material.
(f) The specific heat of the solid.
(g) The maximum temperature to which the solid can be exposed in safety for a short time and the maximum permissible temperature of the solid at the dryer exit.
(h) Heat- and matter-transfer coefficients from laboratory tests.

With this information it is possible to arrive at answers to the questions above and it is then possible to fix all the dryer dimensions.

The exit air temperature of the dryer will not be much greater than the exit solid temperature; experience suggests that in most cases a difference of 10–30°C may be expected. Given a maximum exit temperature of the solid, the exit air temperature can be assumed to be 20°C higher. The air/solid ratio must be high enough to ensure that the air can hold, at its exit temperature, appreciably more than all the moisture evaporated from the solid. The air must not be near saturation at exit because the short residence time in the dryer demands a high driving force.

This air/solid ratio must now be checked for its suitability for conveying; it is usually found, however, that the air required for drying is adequate for conveying.[3-7]

12.8 A DESIGN PROCEDURE FOR SIMPLE CASES

12.8.1 Outline of design procedure

A design procedure for a simple non-elaborated pneumatic dryer is described in Sections 12.8.2–8. It will be found valuable for checking first whether to arrange for pilot plant trials and then for checking against the pilot plant data. The outline of the procedure is as follows:

(a) From the given information on permissible temperatures and the quantity of water to be evaporated, choose the air rate and its inlet temperature to the dryer.
(b) Decide on the air velocity required for conveying and so fix the drying tube diameter.
(c) Determine the heat- and matter-transfer coefficients for a sample of the material to be dried in a suitable laboratory dryer (see Appendix 12.1).
(d) Calculate the drying time required.
(e) Estimate the height of the drying tube and decide whether it is desirable to recycle the feed.
(f) Estimate the heat requirements.
(g) Estimate power requirements.
(h) Decide on a method of feeding.
(i) Provide suitable instruments and controls.

12.8.2 The air inlet temperature and air rate

The heat load in the dryer is made up of the following components:

(a) Sensible heat to raise the solid from its inlet temperature to its exit temperature.
(b) Sensible heat to raise the water temperature from the feed temperature to the drying temperature.
(c) The latent heat of evaporation of the water vapour at the drying temperature.
(d) Sensible heat to raise the water vapour from the drying temperature to the exit air temperature.

External heat loss can be ignored at this stage.

It will usually be found most convenient to choose the air inlet temperature and calculate the air rate by heat balance. Although all solids can be expected to show a falling-rate period with rising solid temperature, for the purpose of this calculation it can be assumed that the drying temperature is the wet-bulb temperature corresponding to the state of the air at the beginning of the drying stage.

The inlet temperature must now be compared with the maximum permissible temperature to which the solid can safely be exposed. Although in principle the temperature of the solid should not exceed the exit temperature, it is possible for a small proportion to be heated to appreciably high temperatures. Some may touch the hot walls of the lower part of the tube, while some particles dry much faster than larger ones. The smallest particles quickly approach air temperature when they are dry and this will be quite high if the bulk of the moisture in the feed is yet to be evaporated. On the other hand, the particles will be exposed to the high temperatures for only a very short time so that usually there will be considerable latitude in the choice of air temperature.

The partial pressure of water vapour in the exit air must be compared with the vapour pressure over the exit solid. Provided the former is substantially less than the latter (say less than 50 %) the air rate is adequate for drying.

12.8.3 Air velocity for conveying

No precise rules can be given for choosing the air velocity because fine particles tend to form agglomerates of uncertain size. The minimum upward velocity in any part of a vertical tube in the dryer should be 2·5–3 m/s greater than the free-falling speed in still air of the largest particles. The calculation of this velocity (see Appendix 12.2) should be made for the coolest part of the air stream in a vertical tube. Under the turbulent conditions within the tube the actual settling velocity of a particle will be much less than its free-falling speed in still air. In practice most pneumatic dryers are operated with an air velocity of 10–30 m/s in the vertical tube.

12.8.4 The drying time

Estimating the drying time required is a problem in simultaneous heat and matter transfer and involves a knowledge of the appropriate transfer coefficients. This heat-transfer coefficient depends mainly on the properties of the particles and the drying medium and it can be estimated by means of a published correlation.[8] The mass-transfer (or drying) coefficient is related to the heat-transfer coefficient during the constant-rate drying period, but unfortunately the greater part of the drying time is taken up by the falling-rate drying period and experimental determination of the drying coefficient is necessary during this latter stage. An apparatus which allows simultaneous measurement of the heat-transfer and drying coefficients at successive stages in the drying process, under conditions similar to those in a pneumatic conveying dryer, is described in Appendix 12.1. The results of these measurements take the form of an average value of the heat-transfer coefficient (which should be largely independent of the moisture content) and a curve of drying coefficient versus moisture content.

The drying range required must now be divided arbitrarily into a number of sections and this can conveniently be done by assuming that the moisture content is halved in each section. The inlet conditions are fixed; the first step is to assume a temperature of the solid at the exit from the section. Then the temperature and humidity of the air at the exit from the section are found by a heat and matter balance, and hence the mean temperature difference and the mean partial pressure driving force can be calculated. If the solid is appreciably soluble in water, the heat of crystallisation should be taken into account (see Appendix 12.1). The water vapour pressure over the solid is that of its saturated solution at the same temperature.

Two estimates are made of drying time (= residence time in the hot air stream), namely the drying time, θ_h, based on heat transfer and the drying time, θ_m, based on mass transfer. The calculations are made by successive approximations for each of the above mentioned sections (in which the moisture content is halved) until an exit section temperature is found which gives $\theta_h = \theta_m$.

This is then the drying time for the first section. The calculated exit conditions then form the inlet conditions for the next section and the process is repeated. The total drying time is then the sum of the times for the sections. The procedure is illustrated in the worked example given in Appendix 12.3.

The conditions at the dryer exit should correspond with the overall heat and matter balance previously carried out, which provides a check on the arithmetical accuracy of the calculation. The calculation also shows the temperature distribution along the dryer for both the air and the solid, which is helpful in estimating heat losses and deciding whether conditions are satisfactory for a heat sensitive material.

Drying time based on heat transfer

In order to avoid evaluation of the surface area A_u per unit mass of solid, the heat transfer coefficient is expressed by the equation

$$(hA_u) = \frac{Q'_g + Q'_c + Q'_\lambda}{\theta_h \Delta T_{lm}} = \frac{\Sigma Q'}{\theta_h \Delta T_{lm}} \tag{12.1}$$

The sensible heat term is defined by

$$Q'_g = c_s(T_2 - T_1) + (M_2 T_2 - M_1 T_1) \tag{12.2}$$

The heat of crystallisation term is defined by

$$Q'_c = \frac{M_2 s_2 Q'_{c2}}{100 - s_2} - \frac{M_1 s_1 Q'_{c1}}{100 - s_1} \tag{12.3}$$

The heat of crystallisation is often negligible but must be considered for very soluble materials. When heat is evolved during crystallisation, as is usually the case, Q'_c is negative since the evolution of heat on the particles reduces the net heat transfer from the gas.

The term Q'_λ may be called the latent heat of evaporation because this is the dominant component; strictly it is the total heat of the vapour from the inlet temperature of the solid to the exit temperature of the gas. The value of the latent heat of evaporation of the water to be taken is that at the mean drying temperature of the solid.

Using the above equations, the drying time based on that transfer may be written as

$$\theta_h = \frac{\Sigma Q'}{\Delta T_{lm}(hA_u)} \tag{12.4}$$

Drying time based on mass transfer

Similarly the drying time based on the mass-transfer coefficient is expressed by an equation referred to unit mass of the solid

$$\theta_m = \frac{10^{-3}(M_1 - M_2)}{\Delta p_{lm}(KA_u)} \tag{12.5}$$

The vapour pressure of water over the solid is required for the calculation of the driving force in successive sections of the dryer. The vapour pressure of water is appropriate for insoluble solids but the vapour pressure of the saturated solution must be used for soluble solids.

During the constant-rate drying period the ratio $(KA_u)/(hA_u)$ is constant; it then decreases steadily during the falling-rate period. The results of the laboratory experiment are plotted as a curve of (KA_u) vs $\frac{1}{2}(M_1 + M_2)$ for each pass as in Fig. 12.10 in Appendix 12.1.

12.8.5 The height of the drying tube

The height of the drying tube is given by the product of the average velocity of the particle and the drying time. Under ideal conditions the particle velocity is the difference between the air velocity and the free falling speed of the particle. In practice the flow in a dryer is so highly turbulent that the effective particle velocity is close to that of the air stream, especially for the fairly small particles commonly dried in this type of dryer. Since the air temperature falls rapidly at first, then more slowly over the greater part of the drying tube, corresponding to the changing rate of drying, it appears that a reasonable, though conservative, estimate of the tube height will be given by the product of the drying time and the mean of the air velocities at the top and the bottom of the drying tube.

12.8.6 Recirculation of material in the dryer

If the drying tube in the simple designs shown in Figs 12.1 and 12.2 is found by the design calculations given above and in Appendix 12.3 to be inconveniently tall, it becomes necessary to consider recirculation of the material within the drying tube as shown in the diagrams of other proprietary designs.

Because the number of passes made by an individual particle is governed entirely by chance, the effective increase in residence time is less than the apparent increase suggested by the recycle ratio. A method of calculating the probability of a given number of recycle passes is given in Appendix 12.4.

Referring now to the general design method given above and illustrated in the worked example in Appendix 12.3, the stepwise method of calculating residence time yields a curve of product dryness against residence time. This can be used in conjunction with the probability Table 12.5 to estimate the product dryness for a given recycle ratio.

12.8.7 Heat requirements

The basic heat requirement is obtained from the inlet air temperature and the gas rate. To this must be added the heat losses calculated from the temperature profile of the tube and the thickness and quality of the lagging. A heat loss of 10% is a useful first approximation. Compensation for this loss can be made by increasing the air inlet temperature if permissible; alternatively by increasing the gas rate.

It should be appreciated that in this type of dryer the thermal efficiency decreases rapidly as the load increases. The air rate cannot be reduced because a sufficiently high velocity for conveying must be maintained. Consequently the heat losses in the stack gases are relatively large when the rate of input of material is reduced substantially. Thus in the dryer considered in the design example in Appendix 12.3, the fuel consumption per unit of product is 43% higher at half load than at full load. If there are prolonged periods of operation at partial load it may be more economic to instal two smaller dryers than one large dryer.

12.8.8 Power requirements

The total pressure drop over the drying tube is the sum of the dynamic pressure drop of the air stream and the static head due to the solid suspended in the drying tube. In practice it is usually found that the pressure drop across the drying tube is about 2·5 mb compared with a pressure drop of about 25 mb for the rest of the air system including the dust arrestor. This corresponds with an installed power of 2·0–2·5 kW per 1 000 m^3/h of air (NTP). To this must be added about 0·5 kW for ancillaries. To allow for transformer and transmitting losses, it is necessary to add about 12% to the plant kWh consumption of electricity when estimating the cost at the factory meters.

Small dryers may be operated with one fan but it is desirable to use a forced-draught and an induced draught fan on large dryers. These are operated to give a balanced pressure of one atmosphere at the feed point in order to avoid air in-leaks at this point.

APPENDIX 12.1 A LABORATORY MODEL OF A PNEUMATIC CONVEYING DRYER

A laboratory apparatus suitable for determining heat- and mass-transfer coefficients appropriate to the design of pneumatic conveying dryers is shown in Fig. 12.8. In this apparatus the sample is allowed to fall freely down a short column in a downward flowing stream of air. Transfer thus takes place under conditions very similar to those in a pneumatic conveying dryer but the difficulties associated with sampling and measuring the temperature of the dried solid which are encountered in a true pneumatic conveying dryer are largely avoided. The apparatus is operated at a relatively low air temperature, consequently a number of passes are necessary to complete the drying and obtain the figures to construct a curve of mass transfer coefficient versus moisture content.

The drying tube is approximately 1·25 m high and 8 cm in diameter. It is lagged and wound with an electric heater to minimise heat losses. Warm air is supplied from an electric heater and introduced into the top of the drying

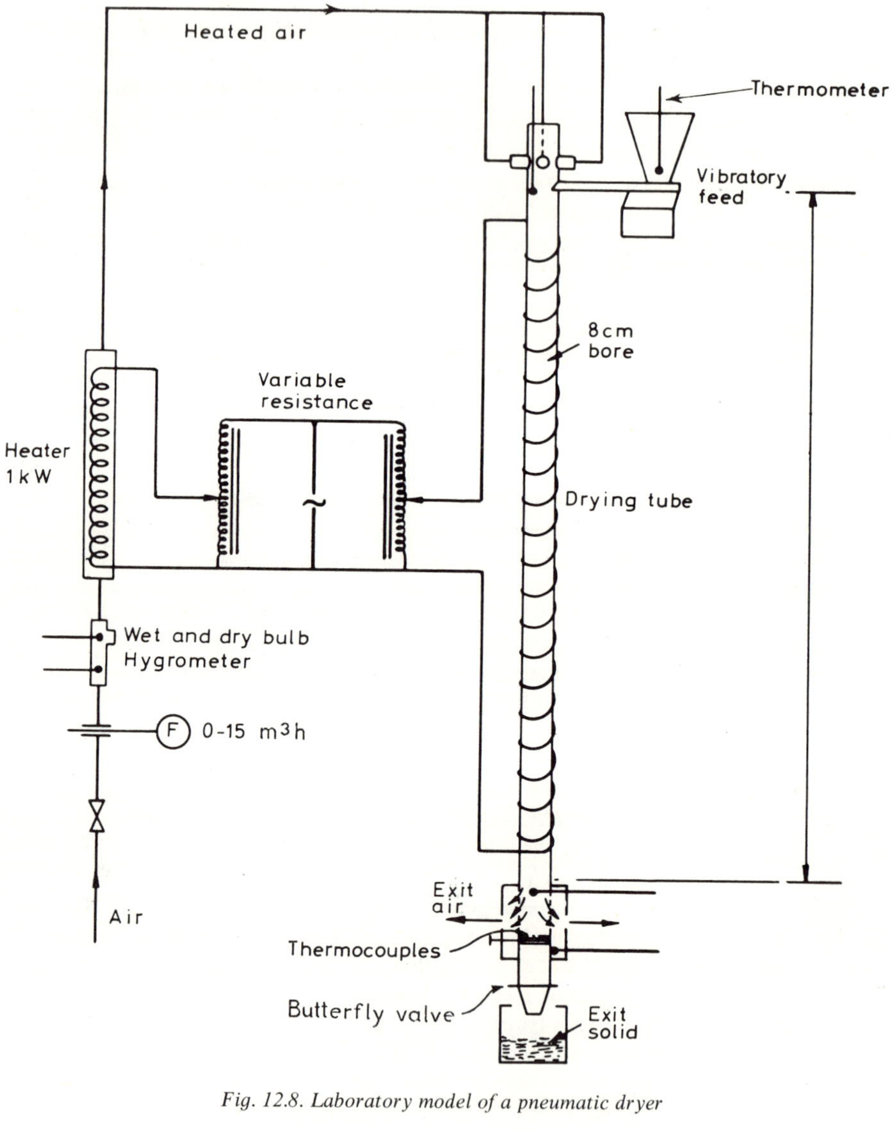

Fig. 12.8. Laboratory model of a pneumatic dryer

tube through four inlet tubes. The air leaves either from the bottom of the drying tube or from lateral holes just above the butterfly valve near the bottom of the tube; the purpose of this valve is explained below. The tube itself is a standard Pyrex pipe; the top and bottom units are of Perspex.

The solid feed is introduced into the top of the column by means of a small vibratory feeder. The feeder chute must have a low thermal conductivity as it was found that appreciable heating, and probably also drying, of the feed occurred in a metal chute used in the first model. The chute enters the drying tube through a soft rubber diaphragm shaped to minimise the free area available for air leakage.

The air temperature is measured by thermometers at the top and bottom of the drying tube. The concentration of solids is low at the bottom of the tower and does not have any significant effect on the measurement of the air temperature. The feed temperature is measured by a thermometer in the feeder hopper. This thermometer should be moved frequently until it shows a steady reading, as the heat capacity of the thermometer tends to cool the granules adjacent to the bulb, and heat transfer through a mass of loose granules is poor.

To measure the feed exit temperature three thermocouples are mounted about 6 mm above the butterfly valve and over one half of it only, so that the valve can be fully opened without interference. To measure the solid's temperature, the valve is closed; the thermocouples are then covered by the solid and soon show a steady reading. Thermocouples are used because of their low heat capacity compared with a thermometer, while the use of three connected in series minimises instrument errors. The thermocouples should project about $2\frac{1}{2}$ cm into the tube to minimise errors due to heat conduction along the wires. To avoid heat losses from the solid during temperature measurement, a warm air jacket is provided round the butterfly valve.

The initial moisture content of the feed is determined by taking samples from different parts of the feed hopper. Samples of the exit solids are collected directly in weighing bottles.

A.12.1.1 Test procedure

An initial sample of about 6 kg is required. If it is sticky and difficult to handle, it should be mixed with a suitable quantity of dry material. This can be taken as an indication that similar back mixing of dried product will be required on the full-scale dryer. The feed rate to the dryer should be about 100 g/min. The air rate should be about 4·2 l/s (15 m^3/h) at a temperature of about 70°C. This gives an air velocity of about 1·1 m/s in the drying tube. The air temperature may need to be varied so that the number of passes to achieve the desired dryness lies between 6 and 12. With fewer than 6 passes a satisfactory analysis of the results is not possible while with more than 12 the test becomes unduly prolonged and loss of sample may cause embarrassment. During a run, the feed rate, air rate and air temperature should be kept as steady as possible.

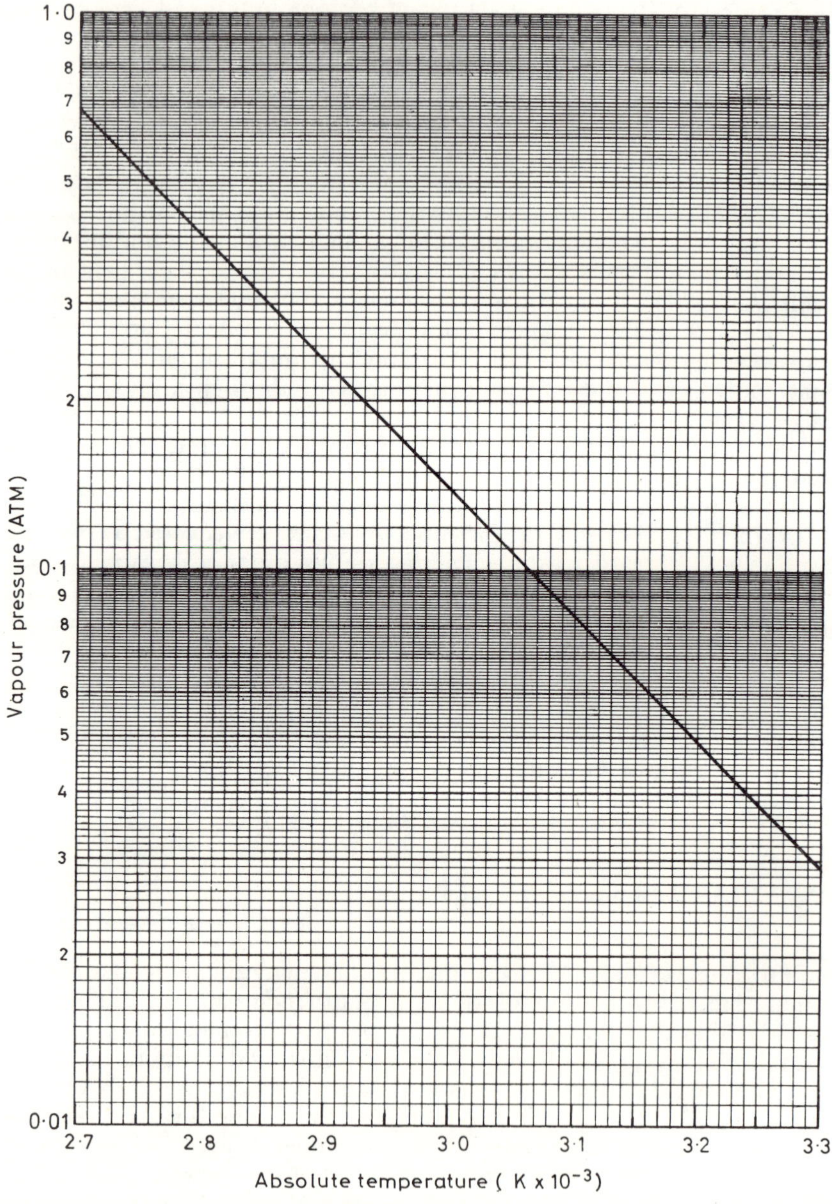

Fig. 12.9. Vapour pressure of water over the moist salt used in the laboratory experiments and for the example of the design of a full-scale pneumatic dryer

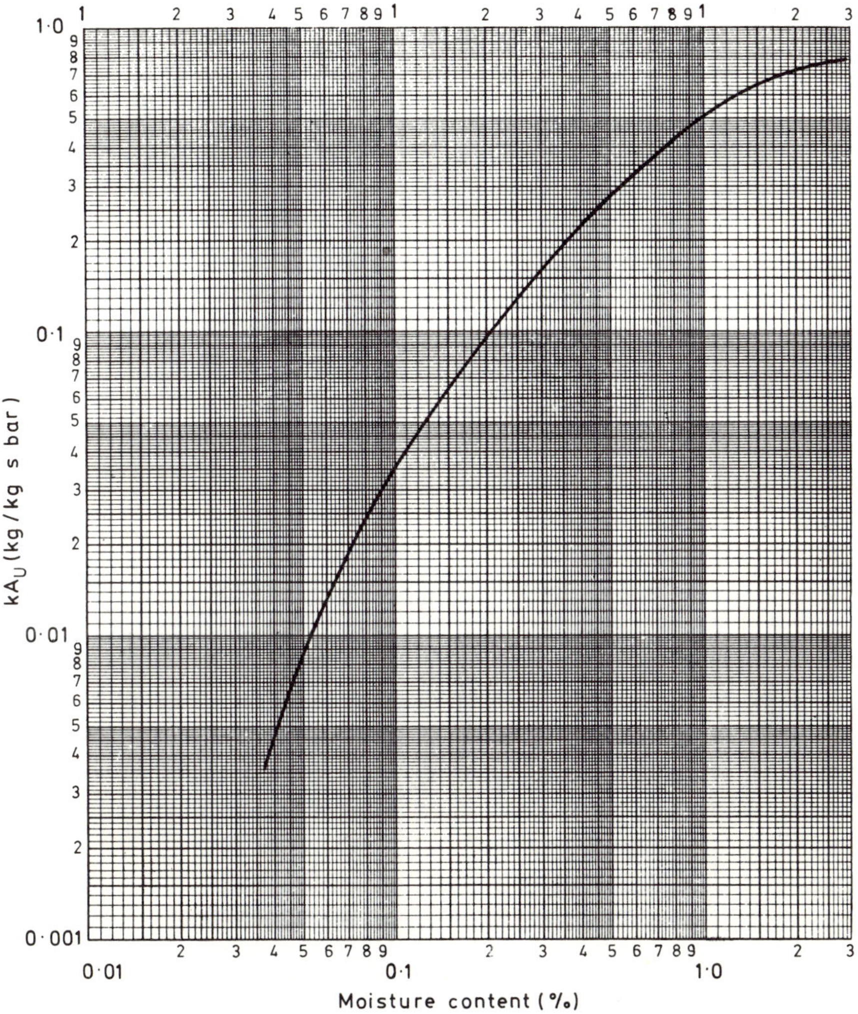

Fig. 12.10. Relation between the mass-transfer coefficient (KA_u) and the moisture content of the salt determined in the laboratory model of a pneumatic dryer. This curve is used in the design of a full-scale pneumatic dryer for the same salt as described in Appendix 12.3

A.12.1.2 Heat-transfer coefficient in laboratory dryer

The data provided by each pass through the dryer is used to calculate from equation 12.1 a heat-transfer coefficient (hA_u). Results for a trial are given in Table 12.3. These results are averaged to give a value for the heat-transfer coefficient (hA_u) to be used in design of a full-scale dryer.

Table 12.3 Drying tests on an inorganic salt in laboratory model conveying dryer

Pass number	*Symbol*	*Unit*	2	3	6	7
Initial moisture	M_1	% w/w	3·36	2·40	0·415	0·139
Final moisture	M_2	% w/w	2·53	1·92	0·158	0·069
Mean moisture		% w/w	2·94	2·16	0·290	0·104
Mass transferred		g/kg	8·3	4·8	0·70	0·70
Initial solid temperature	T_1'	°C	23	24	25	21
Final solid temperature	T_2'	°C	34	34	43	45
Inlet air temperature	T_1	°C	73	73	74	70
Exit air temperature	T_2	°C	68	68	69	66
Log mean temperature difference (air solid)	ΔT_{lm}	°C	41	41	36	33
Sensible heat to dry solid		kJ/kg	18·1	16·7	29·2	40·3
Sensible heat to residual water		kJ/kg	1·2	0·8	0·1	0·1
Latent heat of evaporation		kJ/kg	19·5	11·1	6·0	1·6
Total heat transfer		kJ/kg	38·9	28·6	35·3	42·2
Water vapour pressure over solid at temperature T_1'		mb	19	21	22	17
Water vapour pressure over solid at temperature T_2'		mb	36	36	60	66
Partial pressure of water vapour in the air		mb	5	5	4	4
Log mean v.p. driving force	p_{lm}	mb	21	23	34	31
Heat-transfer coefficient	(hA_u)	kJ/kg s °C	1·88	1·40	1·96	2·57
Mass-transfer coefficient	(KA_u)	kg/kg s bar	0·791	0·417	0·152	0·045

Residence time = 0·5 s
Specific heat of dry solid = 1·64 kJ/kg °C.
Latent heat of evaporation of water = 2 390 kJ/kg (= 570 cal./g)
Operating Pressure = 1 atmosphere = 1·013 bar
Initial moisture to Pass No. 1 = 4·75% by weight
Final moisture from Pass No. 8 = 0·046%.
Average value of (hA_u) for all 8 passes = 1·93

A.12.1.3 Mass-transfer coefficient in laboratory dryer

Similarly the mass-transfer coefficient is calculated for each pass from equation 12.5. The vapour pressure of water over the moist salt used in the example is given in Fig. 12.9 for the range of temperatures employed. This is, of course, the vapour pressure of water over the saturated solution.

The value of (KA_u) decreases as drying proceeds during the falling-rate period. The results of the experiment are plotted in Fig. 12.10 as a curve of (KA_u) against the mean moisture content $\frac{1}{2}(M_1+M_2)$ for each pass. This curve is used for the calculation of the residence times by mass transfer in the stages of design of a full-scale dryer—See Appendix 12.3.

A.12.1.4 Calculation of transfer coefficients from laboratory test results

Operating figures for 4 passes out of 8 in a test run on the laboratory apparatus with an inorganic salt are given in Table 12.3 together with their subsequent calculations of heat- and mass-transfer coefficients.

A.12.1.5 Residence time

The residence time θ depends on the acceleration of the particles under the combined action of gravity and air resistance and also on the velocity of the air stream. The recommended air flow gives a velocity of about 1·2 m/s. Particles smaller than 50 μm have a terminal velocity of less than 10% of this, so that for such small particles it will be sufficient to assume that their velocity is equal to that of the air stream. At the other extreme particles larger than about 500 μm will be virtually unaffected by air resistance over the short length of the drying tube, so that the residence time can be calculated on the assumption that the particles are falling with the acceleration due to gravity. For particles of intermediate size it is necessary to allow for the effect of air resistance.

It may be assumed that the particles fed into the drying tube are accelerated to the (downward) velocity of the air (1·2 m/s) in a distance of 75 mm. Thereafter the particles accelerate under the action of gravity until they reach their terminal velocity relative to the downward velocity of the drying air in the tube. The method of calculation for the residence time θ is given at the end of Appendix 12.2 for particles of sizes likely to be found in materials to be dried in pneumatic dryers. The effective length of the dryer to be taken in these calculations is the distance from the feeder to the butterfly valve.

APPENDIX 12.2 TERMINAL VELOCITIES OF PARTICLES

A useful summary of the empirical relationships governing the free-falling speeds of particles in air is given in Reference 9.

In S.I. units, the terminal velocity for small particles falling in still air is given by Stokes Law in the form of the equation.

$$v_s = 5{\cdot}45 \times 10^{-10}\, d_m^2 \frac{(\sigma - \rho)}{\eta} \text{ m/s} \qquad (12.6)$$

where d_m is in microns, μm.
For particles falling in air, ρ can be ignored.

Stokes Law does not apply to particles falling in air which are larger than 50 μm, density 2 000 kg/m^3. The terminal velocity, v_t then varies not as the square of the diameter but as some smaller power between 2 and $\frac{1}{2}$. Table 12.4 gives the terminal velocities v_t for a range of particle sizes in air within the range of temperatures covered in this chapter. This table is based on the equations derived by C. N. Davies.[10]

A.12.2.1 Non-spherical particles

The terminal velocities of particles which are not spherical can be estimated with reasonable accuracy, provided they are approximately equidimensional and not plate-like or acicular, by multiplying v_t by a shape factor F_p. This is defined as the ratio d_v/d_s where d_v is the diameter of a sphere having the same volume as the particle and d_s is the diameter of a sphere having the same surface area as the particle d_v can be taken as the diameter corresponding

Table 12.4a Terminal velocities v_t of spheres in air at temperatures from 15°C to 540°C (m/s)
(Abstracted from Reference 9 and converted to m/s and °C)

diameter μm	°C 15	90	205	305	425	540
50	0·074	0·062	0·052	0·045	0·041	0·037
100	0·25	0·22	0·19	0·17	0·16	0·14
250	0·91	0·87	0·83	0·81	0·78	0·74
500	1·98	2·01	1·99	2·00	1·99	1·95
600	2·39	2·43	2·46	2·45	2·46	2·44
700	2·79	2·87	2·91	2·93	2·92	2·93
800	3·14	3·23	3·35	3·42	3·42	3·42
900	3·48	3·66	3·79	3·81	3·90	3·90
1 000	3·84	4·00	4·21	4·27	4·33	4·42
1 250	4·73	4·91	5·12	5·27	5·44	5·56
1 500	5·37	5·76	5·98	6·25	6·50	6·68
2 000	6·71	7·18	7·71	8·09	8·45	8·65
2 500	7·71	8·45	9·20	9·73	10·32	10·55
3 000	8·75	9·56	10·5	11·3	12·0	12·3
5 000	11·6	13·1	14·6	15·9	17·1	18·0

Density difference $(\sigma - \rho)$ of 1 g/cm^3

Table 12.4b

diameter μm	°C 15	90	205	305	425	540
50	0·14	0·12	0·10	0·09	0·08	0·74
100	0·46	0·41	0·37	0·33	0·30	0·30
250	1·51	1·48	1·43	1·40	1·36	1·34
500	3·17	3·20	3·26	3·26	3·26	3·30
600	3·78	3·84	3·94	4·02	4·02	4·05
700	4·33	4·51	4·70	4·73	4·82	4·85
800	4·85	5·07	5·27	5·43	5·49	5·61
900	5·42	5·64	5·91	6·10	6·25	6·25
1 000	5·97	6·19	6·50	6·80	6·89	7·05
1 250	7·10	7·53	7·86	8·24	8·62	8·79
1 500	7·78	8·65	9·40	9·62	10·0	10·4
2 000	9·72	10·6	11·7	12·3	13·0	13·4
2 500	11·3	12·5	13·7	14·7	15·5	16·0
3 000	14·8	14·0	15·5	16·6	17·7	18·6
5 000	16·7	18·6	21·2	23·0	25·0	26·4

Density difference $(\sigma - \rho)$ of 2 g/cm^3

to the sieve size on wire mesh sieves. The value of F_p for ordinary grit and dust lies between 0·75 and 0·85.

When the Reynolds number $(Re) = 10^{-3}\, v d_m \rho / \eta$ is used in particle velocity calculations, v and d_m refer to the particle velocity and diameter and not to the gas velocity and diameter of the tube in which the particle is falling.

A.12.2.2 Velocity of particles after falling a given distance

A particle falling from rest under the influence of gravity falls an appreciable distance before it has accelerated to its terminal velocity v_t. The time taken for the velocity v_t to be reached is insignificant in full-scale apparatus. It is necessary, however, to calculate this time with the aid of graphs in order to arrive at the total residence time of a particle falling through the short distance of 1–2 m in laboratory co-current drying apparatus described in Appendix 12.1.

The size of particle considered should be the most frequently occurring

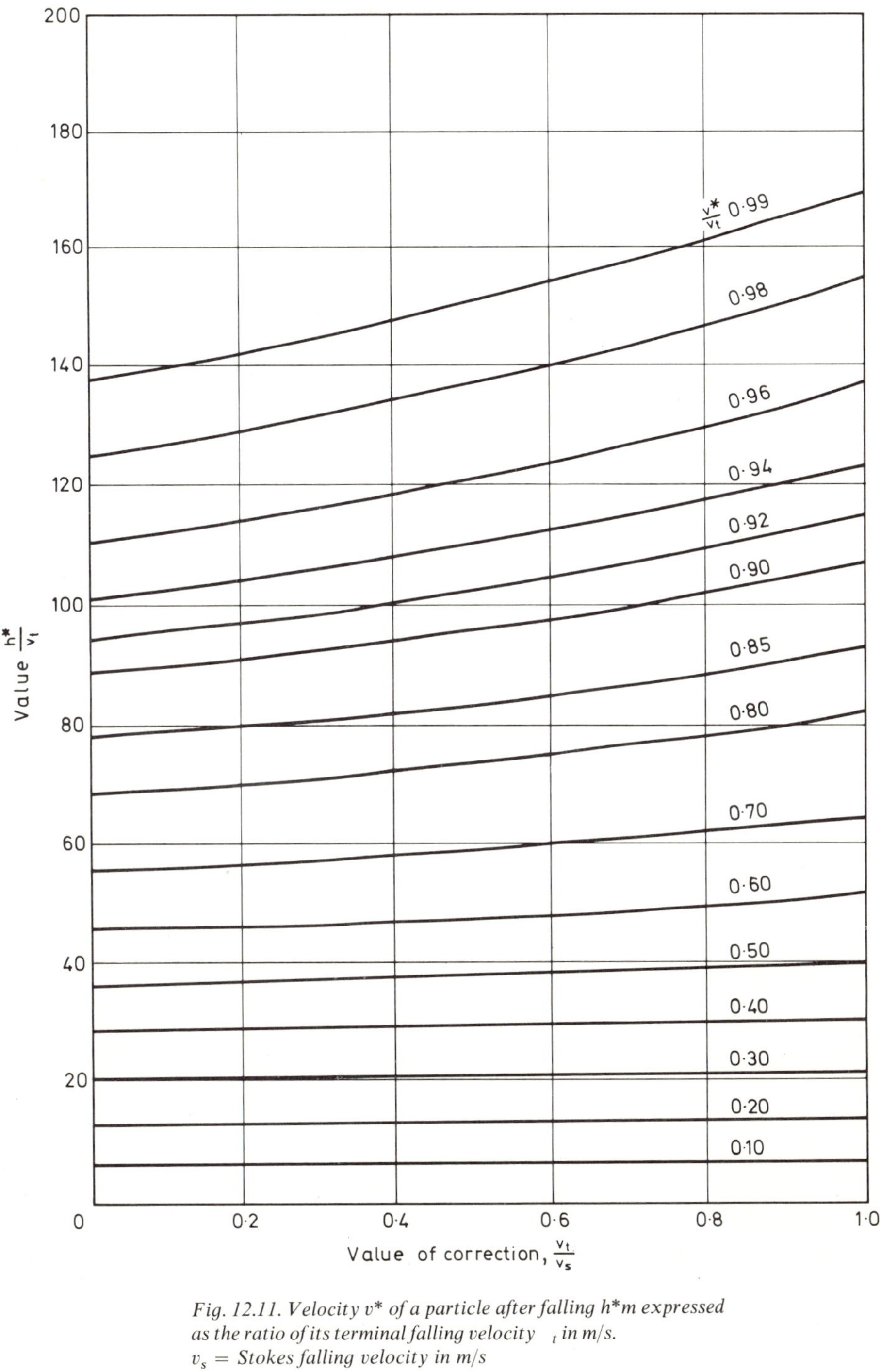

Fig. 12.11. Velocity v^* of a particle after falling h^*m expressed as the ratio of its terminal falling velocity $_t$ in m/s.
v_s = Stokes falling velocity in m/s

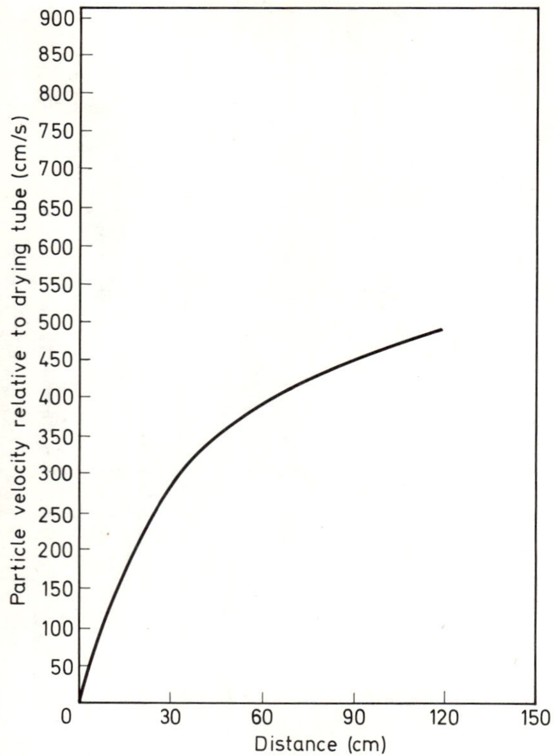

Fig. 12.12. Particle velocity vs distance fallen

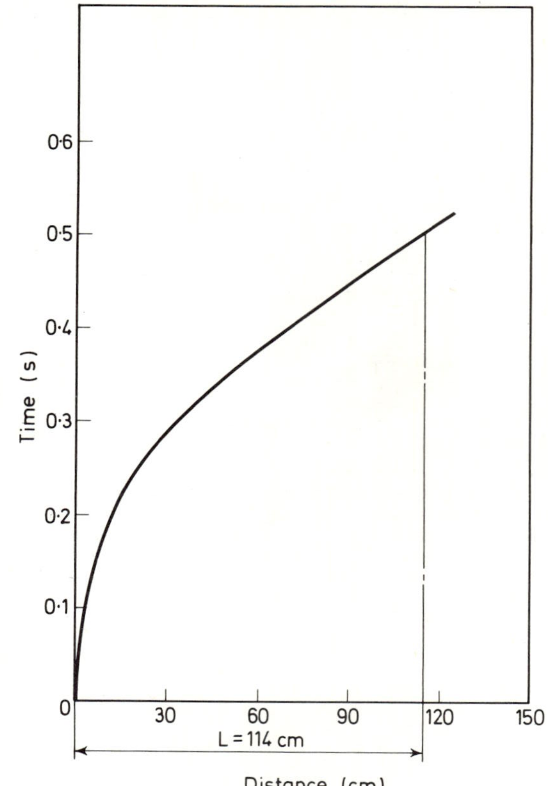

Fig. 12.13. Elapsed time vs distance fallen

size in the size-weight concentration curve or the largest size present in significant quantities. The ratio v_t/v_s is then calculated. v_t is corrected by shape factor if necessary. The graphical solution is based on the well-known equation for acceleration from rest, $v^* = \sqrt{2gh^*}$. Fig. 12.11 gives the ratio of v^*/v_t for various values of h^*/v_t and the ratio v_t/v_s. The value of v^* is then obtained from the ratio v^*/v_t.

The time taken for a particle to fall down through the length of the tube h^*, of the co-current laboratory apparatus (Appendix 12.1) is then obtained by stepwise calculation thus:

(a) Calculate the free falling particle velocity v^* relative to still air at intervals of distance h^* of 20 cm.

(b) Add the mean velocity of the air stream to each calculated velocity, thus obtaining the particle velocity relative to the tube.

(c) Construct a curve of velocity v_s distance, assuming that the particle velocity is equal to the air velocity at a distance of 8 cm below the feeder. An example is given in Fig. 12.12.

(d) Average the velocity of the particles over successive short distances. Then construct curve of elapsed time v_s distance fallen for distance slightly longer than the length of the laboratory tube. An example is given in Fig. 12.13. The residence time in the tube is then read from this curve.

APPENDIX 12.3 A WORKED EXAMPLE OF THE DESIGN OF A LARGE PNEUMATIC CONVEYING DRYER

Requirement: To dry the inorganic salt for which data obtained in a laboratory model of a pneumatic dryer has been recorded in Table 12.3 of Appendix 12.1.

A.12.3.1 Given conditions

Dry salt rate	35 t/h = 9 720 g/s
Inlet moisture	3·5 kg H_2O/100 kg dry salt
Moisture in dried salt	0·1 kg H_2O/100 kg dry salt
Inlet temperature of salt	60°C
Maximum permissible salt temperature	90°C
Specific heat of dry salt	1·63 kJ/kg
Density of dry salt	1·77 g/cm^3
Diameter of largest particles present	2 500 μm
Most frequent size range	855–1 205 μm (Mean size 1 030 μm)
Pressure in drying tube	1 atmosphere (1·013 bar)
Vapour pressure curve for saturated solution of salt	Fig. 12.9

Mass-transfer coefficient kg/s kg atm	Fig. 12.10
Heat-transfer coefficient (from Table 12.3)	1·97 kW/kg °C

A.12.3.2 Choice of air temperatures and air rate

In order to keep within permissible limits, let the exit temperature of the dry salt be 85°C and the air temperature be 450°C. With these relatively high temperatures, experience indicates that there is likely to be a substantial temperature between the air and the solids at the outlet of the pneumatic tube; an exit air temperature of 120°C is assumed in the calculations. Then:

	$W(=$ kJ/s)
Sensible heat to raise 9·72 kg/s dry salt from 60°C to 85°C	397
Sensible heat added to residual water	36
Latent heat of evaporation of 0·33 kg/s of water	790
Sensible heat of water vapour from 60°C to 120°C	39
Total	1 262

The heat of crystallisation has been neglected. The latent heat of vaporisation of water has been taken as 2 390 kJ/kg corresponding to the average vapour pressure at which evaporation occurs.

Taking specific heat of air as 1·28 kJ/m^3, the air rate necessary to provide the above heat is

$$\frac{1262}{1{\cdot}28(450\text{–}120)} = 2{\cdot}93 \text{ m}^3/\text{s (NTP)}$$

If the partial pressure of water vapour in the inlet air is 0·020 atm, the volume of water vapour entering with the air is 0·06 m^3/s. At the exit of the drying tube, the partial pressure of water vapour will be 0·15 atm. The vapour pressure of water over the saturated salt solution of 85°C is 0·410 atm. Hence the air will be far from saturated with water vapour.

The air/solid ratio by weight is about 40. This is adequate for conveying.

A.12.3.3 Diameter of drying tube

	m^3/s
Volume of inlet air	2·93 at 0°C
Volume of water evaporated (0·33 kg/s)	0·41 at 0°C
Total volume of exit temperature	4·80 at 120°C

The terminal falling velocity of the largest particles (2 500 μm) would be at most 13·5 m/s at 200°C if they were spherical (Appendix 12.2). It is more likely to be 10 m/s.

If the internal diameter of the drying tube is 60 cm, the velocity of the exit wet air will be 17 m/s. This should be satisfactory.

Table 12.5 Calculation of residence time in a full-scale pneumatic dryer for the inorganic salt for which the drying characteristics have been determined in Table 12.3

Item	*Stage*	*Symbol*	*Units*	*1*		*5*	
A	Initial moisture	M_1	% w/w	3·5		0·2	
B	Final moisture	M_2	% w/w	1·6		0·1	
C	Mean moisture		% w/w	2·55		0·15	
D	Evaporation		kg/kg	0·019		0·001	
E	Ratio (dry air)/dry solid		m^3/kg	0·3086		0·3086	
F	Initial water vapour in air		m^3/kg	0·0057		0·0502	
G = E+F	Inlet gas rate		m^3/kg	0·3143		0·3588	
H	Evaporation (volume)		m^3/kg	0·0254		0·0014	
I = G+H	Exit gas rate		m^3/kg	0·3397		0·3702	
J	Inlet solid temp.	T_1'	°C	60	60	81	81
K	Assumed exit solid temp.	T_2'	°C	70	80	85	90
L = 1·6 (K–J)	Sensible heat to dry solid		kJ/kg	16·3	32·6	6·52	14·7
M = 2390D	Latent heat of vaporisation	λ	kJ/kg	45·4	45·4	2·39	2·4
O	Vapour superheat		kJ/kg	9·2	7·5	0·08	0·1
P = L+M+O	Total heat given up by incoming air	ΣQ	kJ/kg	70·9	85·5	8·99	17·2
Q = P/1·21G	Temp. drop of incoming air		°C	185	223	21	39
R	Inlet air temperature	T_1	°C	450	450	134	134
S = R−G	Exit air temperature	T_2	°C	265	226	113	94
T = R−J	Inlet temp. difference		°C	390	390	53	53
U = S−K	Exit temp. difference		°C	195	146	28	4
V	Log mean temp. difference	ΔT_{lm}	°C	281	238	39	19
W	Heat-transfer coefficient	$(h\,A_u)$	kW/kg °C	1·97	1·97	1·97	1·97
X = P/V W	Residence time by heat transfer	θh	s	0·128	0·175	0·117	0·46
Y	V.P. water over solid at inlet		atm (Fig. 12.13)	0·140	0·140	0·345	0·345
Z	Partial press at inlet		atm	0·020	0·020	0·139	0·139
a = Y−Z	Inlet V.P. difference		atm	0·120	0·120	0·204	0·204
b	V.P. water over solid at exit		atm	0·210	0·330	0·405	0·480
c	Partial press at exit		atm	0·094	0·094	0·139	0·139
d = b−c			atm	0·116	0·236	0·266	0·341
e	Log mean p.p. difference	Δp_{lm}	atm	0·118	0·172	0·236	0·269
f	Mass-transfer coefficient for mean H_2O content	(KA_u)	$s^{-1}\,bar^{-1}$	3·23	3·23	2·77	2·77
g = D/ef	Residence time by mass transfer	θ_m	s	0·209	0·143	0·064	0·056
Plot X and g against K	Solution of simultaneous eqns. for θh, θm and solid exit temps., see K						
h	Residence time		s	0·161		0·065	
h*	Total ditto for all 5 stages					0·453	
i	Solid exit temp.	T_2'	°C	76		83	
j = 1·63(i–J)	Sensible heat to solid		kJ/kg	26·2		3·27	
k = 2390 D	Latent heat of evaporation	λ	kJ/kg	45·4		2·39	
l	Vapour superheat		kJ/kg	7·55		0·46	
m = j+k+l	Total heat transfer	Q	kJ/kg	79		5·70	
n = m/1·21 G	Temp. drop of inlet air		°C	206		13	
o = k−n	Exit air temp.	T_2	°C	244		121	

Operating pressure assumed = 1 atmosphere = 1·013 bar absolute

A.12.3.4 The drying time

The calculation of the residence time is shown in Table 12.5. The range of moisture content to be covered has been divided into five sections. The average value of the laboratory-determined heat-transfer coefficient $(hA_u) = 1{\cdot}97$ kW/kg °C, was used in each section.

The residence time is defined by the solution of a pair of simultaneous equations, one in terms of heat transfer and the other in terms of mass transfer, namely equations 12.4 and 12.5. The inlet conditions to the first section were known; two temperatures were assumed for the solid at exit from the section and the corresponding exit temperatures calculated for the air stream. Allowance has been made for the superheating of the vapour; since a high degree of precision was not required this was done by trial and error. The two equations are functions of the solid exit temperature. If they are assumed to be linear over a small range of temperature, then plotting residence time against solid exit temperature gives a value of the residence time and the solid exit temperature which satisfies both transfer equations. The assumption that the equations are linear appears to be approximately true over a small temperature range and the error is minimised by using log–log paper.

The air exit temperature corresponding to the solid exit temperature thus calculated is found by a heat balance over the section. The conditions at the exit of the first section are now fixed and form the inlet conditions to the next section. The process is then repeated through each section in turn. The solid and air temperatures at exit should agree with the values used in the calculation to find the air rate; and the differences in the complete set of calculations outlined in Table 12.4 are in fact −2 and +1°C respectively, indicating that there are no serious cumulative errors in the calculation.

A.12.3.5 The height of the drying tube

The sum of the residence times for the separate sections is 0·45 s. The diameter of the duct has been fixed at 60 cm. Consequently the inlet air velocity at 450°C is 31·3 m/s. Assuming that the average particle velocity is equal to the arithmetic mean of the inlet and exit air velocities, the height of the drying tube is

$$L_d = \frac{31{\cdot}3 + 17}{2} \times 0{\cdot}45 = 10{\cdot}9 \text{ m (say 11 m)}$$

A.12.3.6 Heat requirements

Assuming that approximately 10% of the heat given up by the air stream is lost, then the air temperature at the heater should be 490°C. If the ambient air temperature is 15°C, then the fuel consumption will be 17·3 MW.

A.12.3.7 Power requirements

The power cost at factory meter will be about 40 kW.

A.12.3.8 Approximate cost

Interpolating from Table 12.2 the approximate cost of the dryer in stainless steel, excluding installation charges, will be £20 000.

A.12.3.9 Summary of design

Dry salt throughput	35 t/h
Initial moisture	3·5 %
Final moisture	0·1 %
Initial temperature of wet salt	60°C
Final temperature of dry salt	63–85°C
Inlet air temperature	490°C
Exit air temperature	120°C
Air rate	$2{\cdot}93\ m^3/s = 10\,500\ m^3/h$ at 0°C
Effective height of drying tube	11 m
Diameter of drying tube	60 cm
Heat consumption	17·3 MW
Power consumption (total at factory meter)	40 kW
Approximate cost (not installed)	£20 000

APPENDIX 12.4 THE EFFECT OF RECIRCULATION ON RESIDENCE TIME

Denote the recycle ratio by R_r' defined as the ratio of recirculated solid to the product take off. Let x_s be the fraction of the total flow in the drying tube taken off as product; then

$$x_s = \frac{1}{1+R_r'}$$

Consider a parcel of Y particles entering the drying tube in the feed. Assuming these are uniformly dispersed in the drying tube then $x_s Y$ will leave in the product after only one pass and $(1-x_s)Y$ will be recirculated. Of these $x_s(1-x_s)Y$ will be taken off in the second pass and $(1-x_s)^2 Y$ will be recirculated; $x_s(1-x)^{n-1}Y$ will be taken off after the nth pass and $(1-x)^n Y$ will be recirculated. A 'recirculation table' is given in Table 12.6 for $R_r' = 1$, 4 and 9. Used in conjunction with a curve of moisture content *vs* drying time it is possible to make an estimate of the moisture content of the product from a dryer having a known single-pass residence time and recirculation ratio.

Table 12.6 Recirculation table: fraction leaving after n passes

R'_r *Recirculation ratio*	*1*	*4*	*9*
$x_s = \dfrac{1}{R'_r+1}$	0·5	0·2	0·1
No. of pass n			
1	0·500	0·200	0·100
2	0·250	0·160	0·090
3	0·125	0·128	0·081
4	0·063	0·102	0·073
5	0·031	0·082	0·066
6	0·016	0·065	0·059
7	0·008	0·052	0·053
8	0·004	0·042	0·048
9	0·002	0·033	0·043
10	0·001	0·026	0·039
11		0·021	0·035
12		0·017	0·031
13		0·013	0·028
14		0·010	0·025
15		0·008	0·023
16		0·006	0·021
17		0·005	0·019
18		0·004	0·017
19		0·003	0·015
20		0·002	0·014
20–∞		0·021	0·120
	1·000	1·000	1·000

CHAPTER THIRTEEN

Spray dryers

13.1 GENERAL DESCRIPTION

In the spray drying process, the feed material, which is in liquid or mobile form, is sprayed into a large chamber into which hot air (or other gas) is introduced. The liquid in the spray is rapidly evaporated leaving particles of solid which must then be separated from the air. The motion of the air and material is more or less co-current.

A spray dryer therefore consists of the following essential parts:

(a) Drying chamber.
(b) Means for spraying the feed material into the drying chamber.
(c) Means for introducing hot air into the drying chamber.
(d) Means for removing air from the drying chamber at a point normally remote from the entry of the feed and the hot air.
(e) Means for separating the product from the exit air.

Figs 13.1 and 13.2 show types of spray dryer installations. Conical shaped chambers as in Fig. 13.1 are commonly employed, but other shapes are also in use.[1]

Three methods are used for spraying or atomising the feed liquor. These are:

(a) Nozzles in which the feed liquor is forced under pressure through small holes.
(b) Nozzles in which spray is generated by the action of a secondary fluid such as compressed air.
(c) Spinning discs.

Various methods are used for introducing the hot air (or gases) into the chamber. In some designs it enters through the roof of the chamber and in others through special ports. For the conical type of chamber, the air is usually given a swirling motion by vanes in the inlet opening. This assists intimate mixing of the air with the spray and also promotes a cyclone action, centrifuging the product to the chamber walls from which it falls down to be collected at the bottom of the cone. It is necessary to have a separator in the exit air stream to collect the product carried over with the exit air.

In some designs all the product leaves the chamber with the exit air to be collected in separators. Cyclones or bag filter installations are the means usually employed for this.

Some spray dryers are fitted with vibrators, moving rakes or sweeper arms to prevent the product sticking to the chamber walls and to assist in collecting the product.

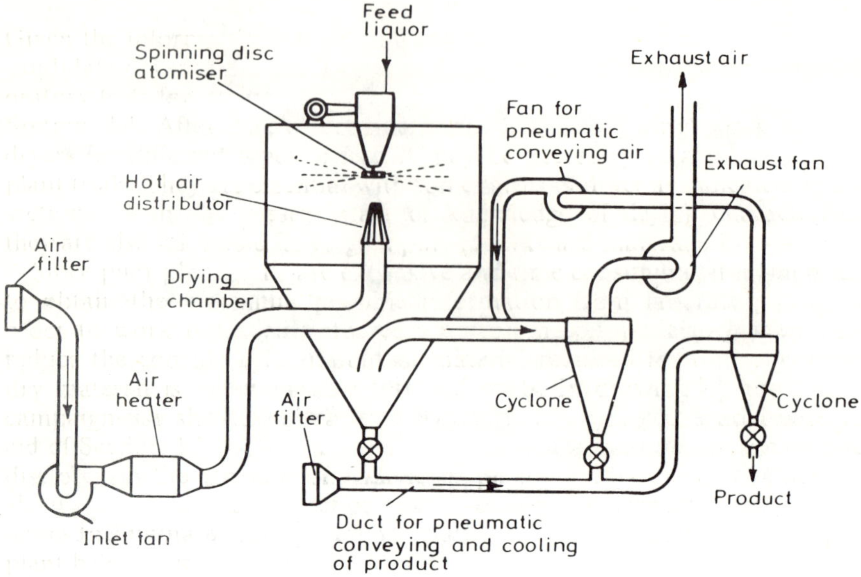

Fig. 13.1. Spray dryer: Type I

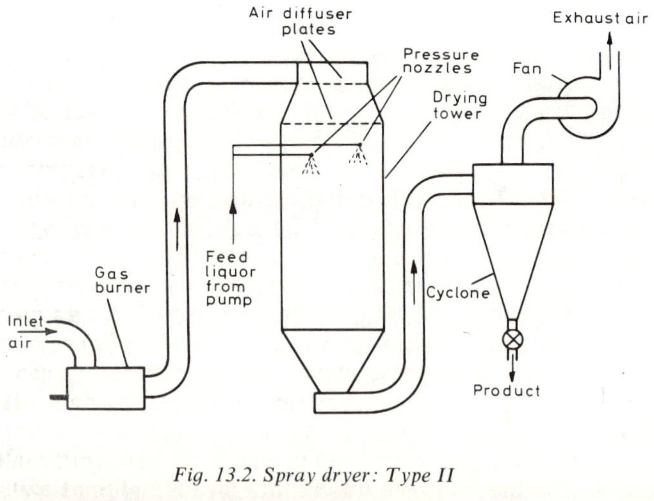

Fig. 13.2. Spray dryer: Type II

13.2 FORM OF MATERIALS HANDLED IN SPRAY DRYERS

In general, substances which can be handled as liquids in pipes can be fed into spray dryers and dried. Thus solutions, emulsions, non-settling suspensions and slurries can be spray dried. Gravity feed systems are often adequate, the head tank being stirred if necessary: otherwise the feed liquor may be pumped to the atomiser. In certain cases granular materials such as filter cakes may be successfully spray dried, provided that the cohesive forces between the particles are low. Powders subject to explosion hazards need special provision, for example closed-circuit inert gas flow or flame suppression devices. Where the final product is sticky it is possible to cover the particles with dusting agents to render them more free flowing.

The product is always a powder, but the form of particles depends upon the substance being dried. When the feed is in solution or emulsion form, the dried particles are usually spherical or nearly so, and are frequently hollow, the dried cenospheres being much larger than the primary particles in the latex. Cenospheres have a hole in the wall through which evaporated water has escaped.

The products of spray drying are usually in the size range 1–300μm. Where the product is in the form of hollow spheres, the bulk density is low. The properties of spray dried materials are further discussed in Reference 2.

13.3 ADVANTAGES AND DISADVANTAGES OF SPRAY DRYING

In cases where a liquid must be dried directly, the choice is usually between a drum dryer (Chapter 9) and a spray dryer. However, in special circumstances, spray drying may be considered as an alternative to other dewatering treatment such as filtration or crystallisation followed by drying in other types of equipment. Occasionally, the feed material may be specially prepared as a liquid if spray drying is particularly desired.

The main advantages of spray drying are as follows:

(a) Because of the quick drying times, many heat sensitive materials can be satisfactorily spray dried where other equipment would be unsuitable.

(b) In spray drying, the material does not contact the walls of the equipment until it is dry and, moreover, the walls are at only about the same temperature as the exit air. Problems of sticking and corrosion are thus reduced.

(c) The product is obtained as a finely divided, free-flowing powder, and in a form readily dissolved in an appropriate solvent.

(d) The particle size of some products is adjustable within limits by variations in the atomising conditions.

(e) The process is suitable for the continuous drying of reasonably high tonnages of material.

(f) The process may, in certain cases, eliminate the necessity of filtration or grinding, which would be necessary in an alternative process.

(g) In certain cases, where a low bulk density is desirable in the product, spray drying is advantageous.
(h) Cleaning to semi-sterile conditions is more easily achieved than on most equipment.

Some disadvantages of spray drying are:

(a) The heat requirement per unit weight of product is high because:
    (i) The moisture content of the feed must be high compared with most other types of dryers. (But see Section 13.2.)
    (ii) The thermal efficiency is often rather low because of restricted air inlet temperatures and rather high exit air temperatures.
(b) In certain cases, low bulk density of the product may be a disadvantage. (However, spray drying does not necessarily produce a low bulk density product, and it should not be assumed in every case that the bulk density will be low.)
(c) The capital cost of the equipment is high per unit annual weight of product, particularly for low capacities.
(d) The equipment requires a large amount of space.
(e) The recovery of dusty product from the exit gases may be a troublesome problem or may involve the use of expensive equipment.
(f) The spray dryer is not readily usable for toxic products, unless special precautions are taken.
(g) All impurities in the liquor remain in the product.

13.4 THEORY

13.4.1 Limitations

While normal heat and matter transfer and resistance concepts apply to spray drying, they are of limited application due to the steady-state assumptions which must be made. A successful theory of spray drying requires the analysis of a number of interconnected problems, namely:

(a) atomisation of the feed liquor into droplets by a nozzle or rotating disc;
(b) time to evaporate a droplet of decreasing surface area and decreasing velocity relative to a medium of increasing humidity and decreasing temperature during the constant and falling-rate periods;
(c) trajectory of a droplet of decreasing mass.

No satisfactory theoretical treatment has yet been deduced and the development of a spray dryer for a particular purpose is therefore based on experiment and past experience. Many empirical relationships have been proposed but are not given in this section because of their limited theoretical basis. The literature also contains numerous attempts to give a satisfactory theory and a few of these are discussed below since they may provide a means of extrapolating from empirical data to cover other conditions of spray dryer operation. (For a fuller survey see Marshall's monograph.[3])

13.4.2 Atomisation

In the case of disc atomisers, Bär[4] derived an expression for the maximum diameter of droplets formed at the periphery of discs by the interaction of centrifugal and surface tension forces ('direct drop' formation):

$$d_d \propto \frac{1}{N_r} \cdot \frac{(\gamma)^{\frac{1}{2}}}{\rho D_s} \tag{13.1}$$

where ρ is the density of the liquid.

In addition he suggested that large drops, once formed from sheets or ligaments of liquid issuing from the periphery, are further shattered by their velocity relative to the air ('velocity spray' formation):

$$c_\rho u^2 \rho_g \propto \gamma / d_d \tag{13.2}$$

Although it is the mean drop size which is most useful for heat-transfer calculations, the maximum size determines the diameter of chamber to contain the spray.

Bär's equations predict smaller maximum drop sizes than those observed in practice and the theory does not therefore fully explain the mechanism of drop formation.

13.4.3 Heat transfer and evaporation of drops and small particles

Marshall and Seltzer[1] obtained theoretical expressions for time required to evaporate pure liquid drops in static air or when moving at their terminal velocity. Ranz and Marshall[5] studied the evaporation of drops (including those containing dissolved or suspended solid matter) and obtained theoretical relations between dimensionless numbers for the determination of heat- and matter-transfer coefficients; the equations are backed by experimental results. From these they derive an expression for the drying time of drops containing solid matter: this expression requires a knowledge of the stage at which the drying particle forms a solid crust which determines its final size. Kirschbaum[6] arrived theoretically at an expression for determining drying times in a spray dryer fed with particles of constant diameter. This can only apply where the dried product is in the form of cenospheres and crust formation at the droplet surface is almost instantaneous.

From a theoretical analysis, Sjenitzer[7] has produced graphs for determining the braking distance of a droplet and extra evaporation caused by this braking. He pointed out that the braking distance in the case of cenosphere-forming solutions or emulsions will be shorter than those derived from the graphs which assume droplets of water only.

13.5 GENERAL CONSIDERATIONS ON DESIGN AND OPERATION OF EQUIPMENT

13.5.1 Size and shape of chamber

The size of the drying chamber is determined mainly by the requirement that particles must be sensibly dry before they contact the chamber walls.

For the same reason the shape of the chamber should roughly conform to the shape of the spray. Thus a spinning disc atomiser giving an umbrella shaped spray requires a chamber of large diameter, but not particularly deep, while pressure nozzles normally require a high chamber of only moderate diameter.

For these reasons the product from a very small spray dryer must be different from that normally produced in a full-size dryer, since in a small dryer, in order that the particles shall be dry before they contact the walls, the spray must be very fine. There is at present no reliable information to enable results obtained on a small laboratory-scale dryer to be correlated to the design of full-size equipment. It is also doubtful whether manufacturers of proprietary spray dryers have any sound method of scaling up equipment by such a large ratio as would be required in order to design a full-size plant from laboratory-scale spray drying trials. Thus, although experiments on laboratory spray dryers may be useful to indicate the feasibility of spray drying a particular substance, it is always recommended that, before deciding on any particular spray drying installation, trials should be carried out on a dryer as near to full size as possible.

13.5.2 Types of atomiser

The pressure nozzle type of atomiser is generally cheaper than the spinning disc type and for substances with which there is little danger of blocking the injection holes, they should give satisfactory service, though some erosion may be expected. The nozzle atomiser is relatively inflexible in operation and is not advisable where a range of products is to be dried or when the dryer is to be controlled by varying the feed rate. In some cases atomising nozzles can produce dry material of larger particle size and hence greater bulk density than is possible with a spinning disc, and may be preferred for this reason.

Various types of spinning atomiser are in use. One type consists of an inverted spinning bowl with liquid fed to the inside surface of the bowl and escaping round the periphery. In another type the material passes through radial (or nearly radial) passages in a wheel-shaped atomiser. Rotational speeds are quite high, 15 000 rev/min (3 m/s peripheral speed) and over being common.

The condition of the product can be influenced to some extent by the atomising conditions, higher atomiser speeds giving smaller particles, quicker drying due to the larger surface area, and usually higher bulk density. A word of warning is, however, necessary; above a certain speed with some designs of disc some suspensions begin to coagulate. There are then large irregular particles in the spray.

13.5.3 Air inlet temperature and control

From considerations of thermal efficiency, it is desirable to operate with a relatively high air inlet temperature – a round figure of 250°C can be used for preliminary estimates. However, in certain cases it may be necessary to limit the inlet temperature because of explosion hazards or thermal instability, or to avoid contamination due to burning of stray particles. To prevent burning of this sort, certain areas of the chamber adjacent to the air inlet louvres are sometimes cooled, either by means of cold air jets or by water jacket cooling.

The atomiser may also be cooled by a cold air jet. Thermally labile materials can usually be handled by restricting the air exit temperature rather than the air inlet temperature since it is the exit temperature which the product will attain. However, in cases where an exceptionally low exit air temperature is essential it may be necessary to restrict the air inlet temperature, since otherwise, the quantity of water evaporated would tend to saturate the exit air and result in condensation and loss of evaporative driving force during the final stages of drying.

Any normal method of air heating may be used. Heating by direct mixing with products of combustion is quite frequently practised. For special cases where the product would be harmed by contact with hot oxygen, it is possible to use an inert gas such as nitrogen as the drying agent instead of air.

In a spray dryer working at constant air rate and constant inlet air temperature, the air-exit temperature can be varied by changing the liquid feed rate, and this generally results in changing the dryness of the product. Too low an air-exit temperature will result in the product being too wet, but a high air-exit temperature is not desirable because heat and dryer capacity are wasted and because product degradation may occur with heat sensitive materials. In order to maintain a uniform product, the feed rate may be automatically controlled to maintain the air-exit temperature at a constant value; to achieve good control a high grade instrument with proportional and integral terms and a good rate of response is required. Variation of feed rate is not usually possible with atomising nozzles and a spinning disc system will be required.

Alternatively, in certain cases it may be preferable to keep the feed rate constant and vary the air-inlet temperature. For steam inlet air heaters, the slow rate of response makes this procedure difficult unless a damper-controlled air bypass round the heater is used to vary the resulting inlet temperature.

13.5.4 Air volume rate

In a given spray dryer, a high air volume rate results in a high evaporative capacity. However, high air rates also result in shorter dwell-times in the dryer and in order to obtain satisfactory drying the air rate must be limited to some extent. Normal retention times in spray dryers, based on the chamber volume and the air flow, are up to about one minute.

13.5.5 Pre-evaporation and preheating of feed liquor

To increase the thermal efficiency of drying, it may be desirable in certain cases to thicken the original liquor by evaporation, which is a more thermally efficient and cheaper process than spray drying. Also in certain cases it may be advantageous to preheat the feed liquor before spraying it into the dryer, e.g. to give better atomisation; preheating will also increase slightly the capacity of a given size of dryer.

13.5.6 Blending of substances in spray dryers

Simultaneous drying and blending of substances can sometimes be conveniently done in a spray dryer. This may be possible even if it is undesirable to mix the feed liquors, by bringing the liquors to the atomiser in separate pipes, or by the use of more than one atomiser in the dryer – one for each separate feed liquor. Intimate mixing then takes place in the drying chamber.

13.5.7 Cleaning of spray dryers

Spray dryers must normally be cleaned at fairly frequent intervals which must be found by experience; at least 10% downtime should be allowed in production estimates. The parts needing most regular attention are the atomiser and the air inlet ports of louvres. A bridge or light scaffolding may be provided for gaining access to these parts, or they may be made so that they can be readily removed for cleaning. The walls of a spray dryer do not usually need so much cleaning, and blowing with compressed air may be generally adequate. Some makers provide a built-in 'air broom' system for this purpose. However, if the product is to be changed it may be necessary to hose the chamber down. If it is necessary to scrape the chamber walls, an elaborate scaffolding may be required but some dryers have a rail round the top corner from which a ladder, small cage or boatswain's chair can be suspended.

13.5.8 Operating variables

In the operation of a spray dryer, certain conditions may be varied to ensure that the best production efficiency and product physical properties are achieved. In many cases a primary variable is found to affect a factor which in turn alters other factors, e.g. an increase in feed solids content increases the feed rate which increases the particle size of the dried product.

Table 13.1 shows the effect of the main variables assuming that all other variables in the table are kept constant. It also assumes that the feed rate is the main controlled variable.

13.6 RUNNING COSTS

The cost of spray drying varies so much with the feed material that it is desirable to obtain estimates of services and labour from the manufacturers after trial on a pilot plant.

Table 13.1 Effect of variables in spray dryer operation

Variable increased	*Factors increased*	*Factors decreased*
Chamber inlet temperature	*Feed rate* and thus: Product rate Particle size (b) Product moisture content Chamber wall build-up (a)	Bulk density (b)
Chamber outlet temperature	Product thermal degradation (a)	*Feed rate* and thus: Product rate Particle size (b) Product moisture content Chamber wall build-up
Gas volume rate	*Feed rate* and thus: Product rate Particle size (b) Product moisture content Chamber wall build-up (a)	Residence time
Feed concentration	Product rate Bulk density (b) Particle size (b)	
Atomiser speed Atomiser disc diameter *For stable lattices*	Bulk density	*Particle size* and thus: Product moisture content Chamber wall build-up
For unstable lattices	Coagulation (a) and thus: Particle size Product moisture content Chamber wall build-up	
Atomiser vane depth Atomiser vane number	Bulk density (b)	Particle size (b) and thus: Product moisture content Chamber wall build-up
Atomiser vane radial length		*For unstable !atices* Particle size Chamber wall build-up
Feed surface tension	Bulk density (b)	Particle size (b)
Chamber inlet gas humidity	Product moisture content Chamber wall build-up (a)	

(a) This factor will only occur if a critical value of the variable is exceeded
(b) Not for suspensions

13.6.1 Air heating

Costs will depend on the service chosen to heat the air, i.e. steam, gas, electricity, fuel oil, etc., either singly or in combination. From the required water evaporation rate and the permissible chamber inlet and outlet temperatures, the heat required can be calculated. For most cases an extra 10°C of air temperature drop is allowed for radiation and other thermal losses from the dryer chamber.

13.6.2 Power

As a first approximation a power consumption of 5 kW per 1 000 m^3/h of main inlet air may be assumed. This does not include power for outlet conveyors.

13.6.3 Other services

Demineralised water or clean condensate is usually required when starting or stopping a dryer and for use in cases of feed blockage, etc. The supply should be sufficient to provide water up to the maximum possible evaporation rate; the duration of use is, however, short and water costs are usually negligible in the total services cost.

Small amounts of compressed air may be required for instruments and for air lances for chamber cleaning.

13.6.4 Labour and supervision

The chief jobs of a spray dryer operator are:

(a) Starting up and shutting down the dryer while water is fed to the atomiser.
(b) Changing over from water to feed and back again.
(c) Checking the atomiser, temperature control and other instruments.
(d) Maintaining the feed and the discharge of dried material, and
(e) Carrying out periodic cleaning.

Where reasonable automatic control of dryer temperature is possible, operation of a dryer should not normally occupy more than half an operator's time where direct packing of product is not required. For larger dryers two men are normally required for cleaning the chamber and atomiser parts so as to complete the work within a reasonable time. The second man should not be required for cleaning work for more than 5% of his time; this may be exceeded if cleaning is required between batches of less than, say, 36 h or if the material dried is difficult to remove from the dryer walls.

Supervision will depend on the number and location of the dryers and upon preceding and following processes. One eighth to one quarter of a junior supervisor's time should be sufficient for one spray dryer.

13.6.5 Maintenance

Fitters are required to check the atomiser lubrication frequently (because of its high speed of rotation and high operating temperature) and for removing atomiser and air disperser parts to allow normal process cleaning work. Routine maintenance to the atomiser, air fans and powder collection system (normally cyclones or bag filters) forms the greater part of the balance of mechanical work. Bag filter installations tend to require costly maintenance if losses of expensive dried products are to be prevented.

Checks on the instruments must be frequent where close temperature control is required.

The cost of maintenance is about 10% of the capital cost of the dryer.

13.7 CAPITAL COSTS

For initial estimating purpose, the capital cost of a spray dryer may be taken as £90 per kg/h of rated water evaporation for a minimum of 1 000 kg/h evaporation (1968 basis). This cost excludes foundations, buildings and ancillary equipment not directly connected with the drying process which together amount to about an equal amount.

Depreciation should be calculated on the basis of a life not exceeding 10 years.

CHAPTER FOURTEEN

Miscellaneous continuous dryers

14.1 INTRODUCTION

A drying process can be viewed as a heat- and matter-transfer operation or as a material transport problem. The dryers described in this chapter are distinguished by their mode of material transport; for each type the method of heat transfer can usually be varied. The limited field of useful application of these dryers does not demand a detailed treatment, and hence they are only briefly described and the field of application indicated.

Often the dryers described here have been traditionally applied in a certain industry rather than as a choice after a rational investigation. The technically and economically correct choice may involve a major investigation and a knowledge of the answers to many complex questions.

14.2 SCREW CONVEYOR DRYERS

14.2.1 General description

Essentially the screw conveyor dryer consists of a jacketed conveyor in which material is simultaneously heated and dried as it is conveyed. The dryer may be left open to the atmosphere, but more often the top of the dryer is closed and is air-swept to carry off the evaporated moisture. A unit of this type can be designed to operate batchwise under vacuum. The heating medium is usually hot water, steam, or a high temperature heat-transfer medium such as pot oil, fused salts or Dowtherm. Electrical heating elements have also been used, embedded in the conveyor casing.

The drying process may be carried out in a series of conveyors. In the multi-pass arrangement the conveyors are usually mounted horizontally one above the other with the material dropping from one unit to the next and travelling in a direction opposite to the pass above (Fig. 14.1).

In one type the heat-transfer area is increased by heating the screw thus giving the maximum heat exchange area with minimum space requirements. The helical flights are hollow and the heat-transfer agent is circulated as the flights move the material to be dried.

The usual materials of construction are mild steel, cast iron or stainless steel. The castings which form the ends of the trough generally carry stuffing boxes and often the shaft bearings are mounted on brackets also carried by the end castings.

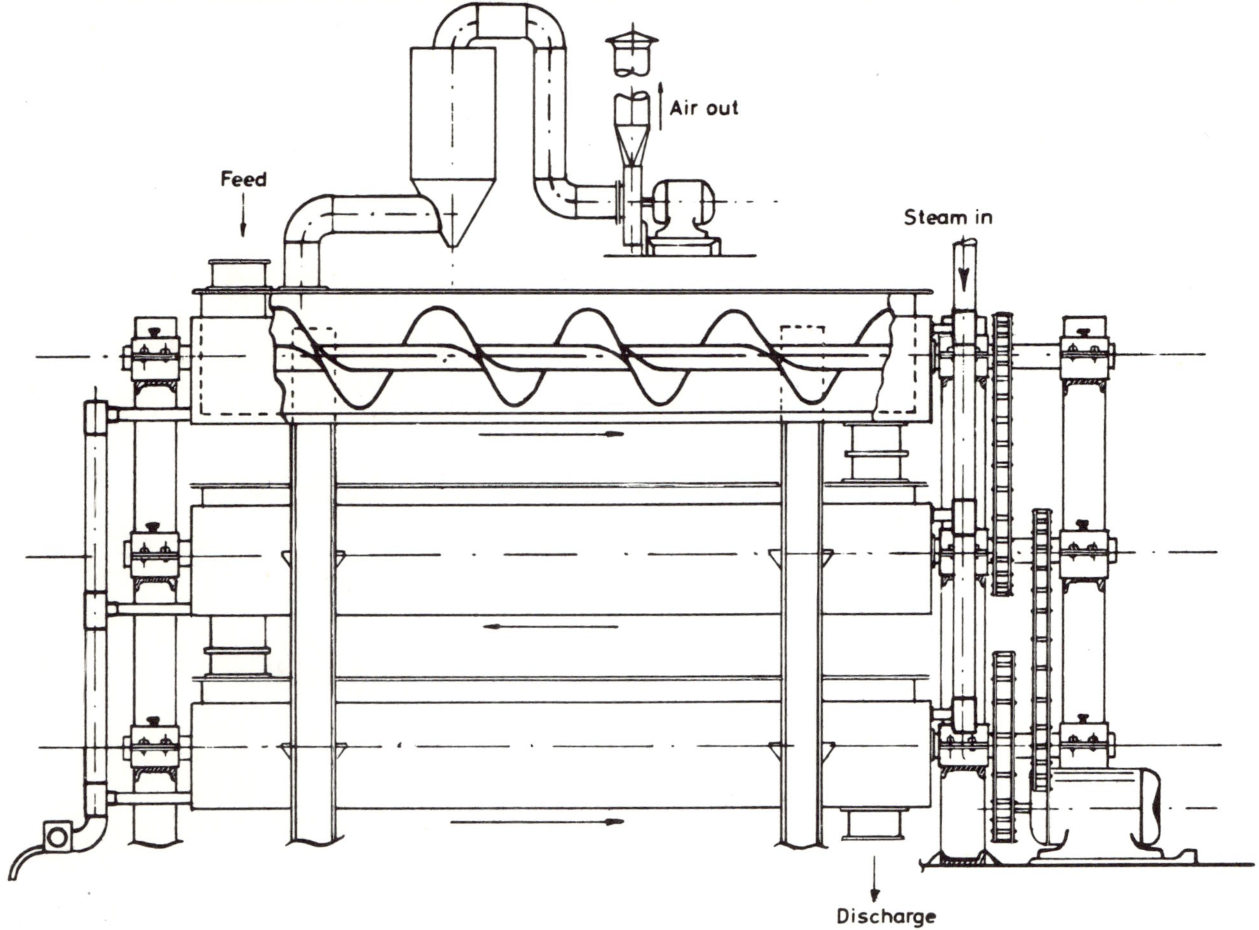

Fig. 14.1. Screw conveyor dryer

Ancillary equipment may simply consist of mechanisms for the feeding and discharge of material. Special feeding and discharge arrangements are required for vacuum operation. If the dryer is air swept, it is generally necessary to provide means for the separation of finely divided material entrained in the air stream. Heated bag filters are commonly employed.

14.2.2 General application

Screw conveyor dryers are not widely used, probably because they have a limited efficiency as dryers. They can be usefully applied to the drying of granular free-flowing materials or to materials which become friable at some stage during the drying process; they should not be used if the material has a tendency to foul the heat-transfer surface and so reduce the rate of heat transfer (though a recycle of dried material will often minimise or even eliminate this problem).

They can be used with advantage when material must be conveyed some distance to the next stage in the process or when the drying must be carried out in an inert atmosphere.

In any drying problem for which the screw conveyor dryer might find application, tests should be done to examine the behaviour of the material as it is simultaneously dried and conveyed. Such questions as fouling of the heat-transfer surface and particle size degradation may be of great importance. In addition, drying tests should provide design information and secure data on the effect of operating variables on product quality.

14.2.3 Mechanical performance

Design procedures for screw conveyors are so well established by plant manufacturers that few difficulties are associated with their operation when handling non-abrasive, non-corrosive materials. For some sticky materials a ribbon helix should be used to prevent build up between the helix and shaft. Thrust bearings should if possible be located at the discharge end of the conveyor so that the shaft is in tension.

14.2.4 Heat transfer

The heat-transfer coefficient from the hot surface to the material handled is mainly a function of the moisture content of the material and of the speed of the conveyor. For the usual range of conveyor speeds the heat-transfer coefficient lies in the range 6 to 70 $W/m^2 {}^\circ C$. At high conveyor speeds of up to 200 rev/min somewhat greater heat-transfer coefficients are obtained.

14.2.5 Performance of a typical screw conveyor dryer

The following figures are available for the drying of a fine crystalline solid which is toxic and requires a safe handling procedure. The material is insoluble in water and has little tendency to foul the heat-transfer surface.

The conveyor trough is heated electrically by elements in the form of pads secured to the outside of the casing. The screw is not heated.

Plant

Length of screw (one intermediate bearing)	7 m
Diameter of screw	480 mm
Depth of casing from cover to centre line of screw	400 mm
Speed of screw	16 rev/min
Motor shaft power	9·5 kW
Material of construction of plant	Aluminium

Performance

Throughput expressed as dry solids	250 kg/h
Initial moisture	9 %
Final moisture	0·1 %
Evaporation	2·25 kg/h
Inlet temperature of solid	20°C
Outlet temperature of solid	100°C
Temperature of heated wall	400°C
Power consumption per ton of product	410 kWh
Thermal efficiency (based on sensible and latent heat)	15·9 %

14.2.6 Capital cost

The installed cost (1968) of a stainless steel screw conveyor dryer, jacketed in mild steel for low pressures, varies from £300 to £500 per square metre of heat-transfer surface.

14.3 VIBRATING CONVEYOR DRYERS

14.3.1 General description

In this type of dryer, the heat is applied in one of several ways to a moving bed of material in a vibrating conveyor. In the type shown in Fig. 14.2, heat is supplied by convection from a stream of hot gas which passes through and over the bed of material. Frequently heat is transferred by conduction through the bottom of the conveyor. This indirect heat can be supplied by condensing steam or any other suitable vapour in a jacket fitted under the conveyor. Alternatively, the sensible heat in a hot gas stream can be utilised, but this particular method has several basic disadvantages. The low gas film co-efficient limits the rate of heat transfer and the low heat capacity of the gas stream reduces the driving force. A gas flame which impinges directly on the bottom of the conveyor has, however, been used as a heat source. The sensible heat in a liquid heat-transfer medium has not found application for this duty; presumably the additional weight added to the vibrating deck

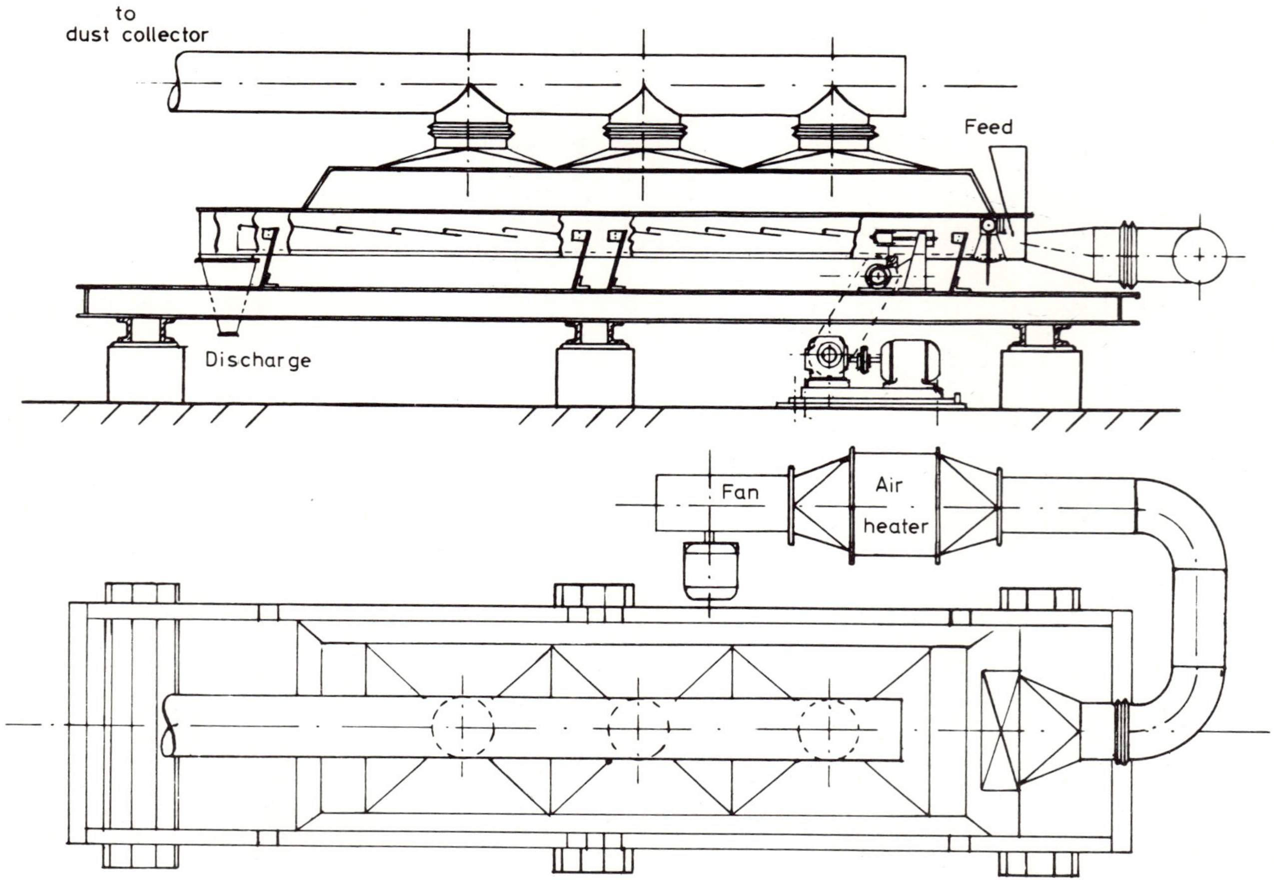

Fig. 14.2. A vibrating conveyor dryer employing a hot air stream

almost completely precludes its use. To supplement or even replace either the indirect or direct heating methods, a source of infra-red radiant heat is sometimes mounted above the moving bed of material.

The vibrating conveyor can be excited by (a) a rotating unbalanced mass, (b) two synchronous contra-rotating unbalanced masses which give a resultant horizontal reciprocating thrust, (c) electromagnetically induced vibrations from one or more solenoids, (d) vibrations induced by a crankshaft, eccentric or cam mechanism, (e) vibrations induced by a pneumatic or hydraulic mechanism.

The methods of vibration are limited to those which will allow control of the throughput rate. For example, in a vibrating conveyor which is excited by out of balance weights, no control of the residence time in the dryer is possible, unless the weights have adjustable eccentricity, and this is mechanically complicated and so rarely employed. Alternatively, the frequency of the conveyor could be varied, but this precludes the possibility of operation at the natural frequency. For small dryers the excitation can be conveniently supplied by an electro-magnet. These high frequency units operate at a low amplitude on a.c. mains frequency or a sub-multiple of it. If the solenoid is energised from the a.c. mains in series with a half wave rectifier, an operating frequency of 50 Hz is obtained rather than 100 Hz which would result if the rectifier were omitted. The amplitude is used to control the residence time of the solids in the dryer. This can be regulated by providing a variable resistance in series with the solenoid and rectifier. The dryer is often mechanically tuned to the resonance frequency.

Ancillary equipment may simply consist of chutes for feeding and discharging material, although on some dryers a distributor mechanism is necessary at the feed end. If a direct heating method is used, gas ducts are required and it is usual to fit either a cyclone separator or heated bag filter on the gas exit.

14.3.2 General application

The field of application of this type of dryer is somewhat limited as it is primarily a conveyor and secondarily a dryer. It can only be used for granular or free flowing materials which can be conveyed without sticking, unless a high recycle is used. Many finely divided materials tend to agglomerate even at fairly low moisture contents. On the other hand, the conveyor causes very little particle degradation, and it is almost free from dust nuisance when indirect heating is applied. A factor which severely limits the use of this dryer is the excessively large equipment and high power requirements for slow drying materials. However, if a natural frequency vibrating conveyor is used, the power consumption may not preclude its use. For certain crystalline materials the vibrating conveyor can be usefully employed, since the moisture is located on the surface of the material and high diffusional resistances are absent. It has a particular field of usefulness when high temperatures can be used.

If a vibrating conveyor dryer is being considered for a drying application, a conveying test is advisable to examine questions of handleability. When the conveying behaviour is being examined, heat should simultaneously be applied to raise the temperature of the material. Some wet materials which have a high temperature coefficient of solubility can be satisfactorily handled when cold, but as the temperature is increased, the quantity of liquid phase present increases and agglomeration may take place.

14.3.3 Mechanical performance

The vibrating conveyor dryer usually needs little maintenance; this is a reflection of its relatively simple design. Nevertheless, the suspension springs require to be replaced at regular intervals as they are susceptible to fatigue. In addition, some exciters of the unbalanced type have a particularly destructive effect on bearings.

Some designs of vibrating conveyors need to be placed on fairly substantial foundations, though balanced units are available which have only a very small reaction on the foundation.

14.3.4 Power requirements

It is not possible to estimate even roughly the power consumption of a conveyor which does not operate at its natural frequency, since the power consumption is a function of the method of excitation and of the operating frequency in relation to the natural frequency.

For a natural frequency conveyor the approximate power requirement is given by the relationship:

$$\text{Power required at shaft kW} = 8{\cdot}6 \times 10^{-2} W_d L_d \qquad (14.1)$$

To determine the natural frequency of a conveyor excited by an unbalanced motor a variable speed drive is fitted, and the frequency of oscillation is varied until the lowest ammeter reading is obtained. Except for the first few moments when the conveyor is brought up to its natural frequency, the current taken by the motor is not greater than is necessary to run the motor unloaded (that is, with the driving belt unattached). This is illustrated in Figs 14.3 and 14.4.

14.3.5 Heat transfer

The mean heat-transfer coefficient for materials which can be handled in an indirectly heated vibrating conveyor dryer falls within the limits (10–100 W/m^2°C). The heat-transfer coefficient from the surface of the tray to the solid is roughly proportional to the initial moisture content. In general the heat-transfer coefficient is somewhat greater for a vibrating conveyor than for a screw conveyor.

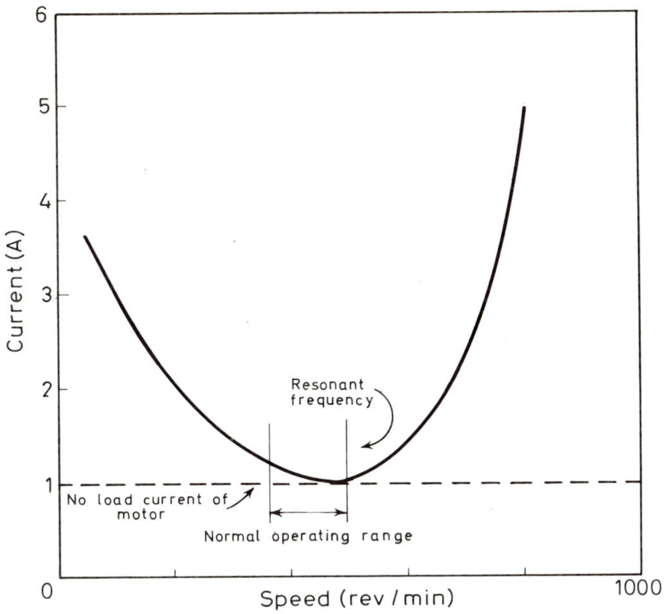

Fig. 14.3. Typical graph of power consumption versus frequency for vibrating conveyors

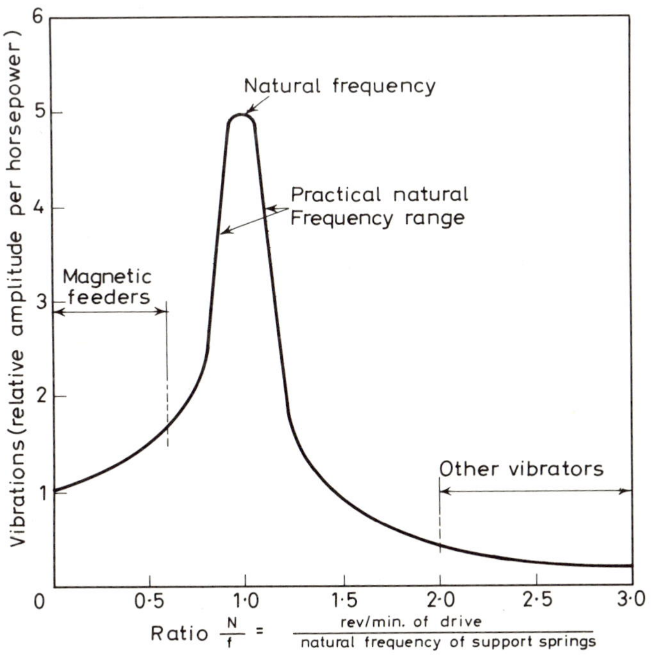

Fig. 14.4. Graph of magnitude of vibration versus excitation frequency ratio

14.4 SIMULTANEOUS DRYING AND GRINDING

14.4.1 General description

This simultaneous drying and grinding of coal in pulverised coal installations is a long established practice, and it is also used in the chemical and process industries.

Certain mills lend themselves to combined drying and grinding. Hammer

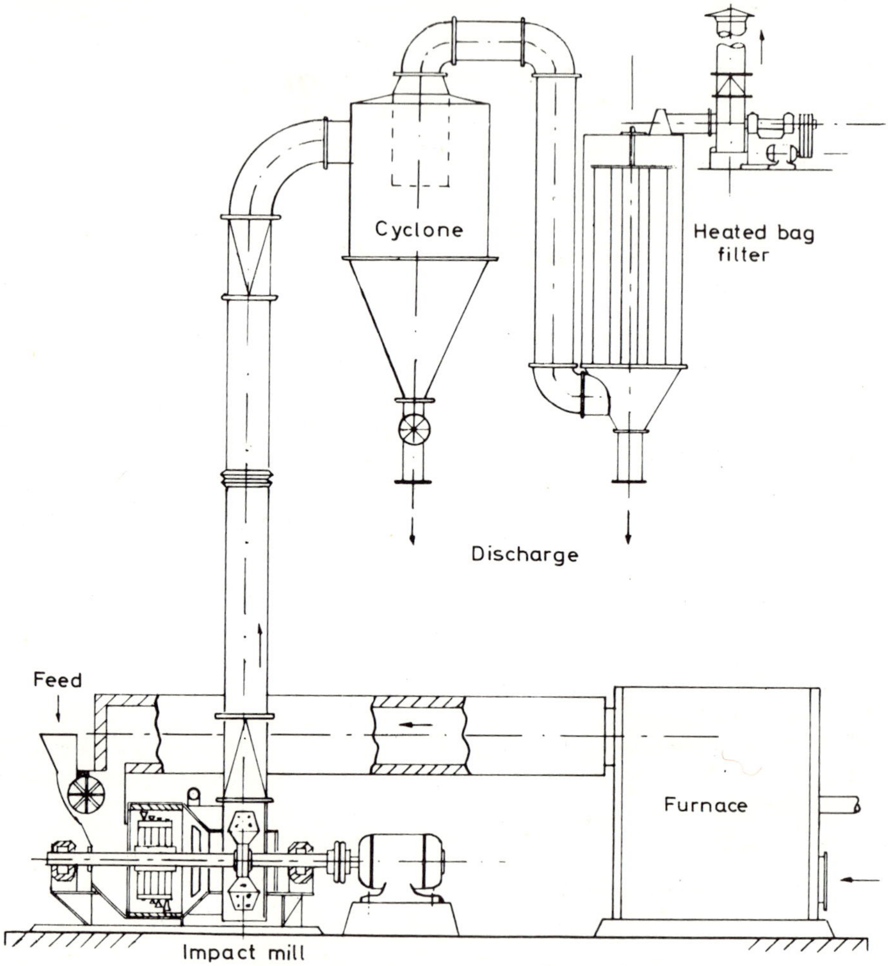

Fig. 14.5. Combined dryer and pulveriser

mills, attrition mills, ball mills and ring roll mills have all been used for this duty. The drying is achieved by the introduction of hot air or flue gases into the mill casing, and in some mills the casing is heated externally. Fig. 14.5 shows a typical installation which can be used with an air temperature up to 450°C.

Frequently the grinding is done in closed circuit, but simple air sweeping of the inside of the mill is also practised. The air sweeping usually effects some classification.

The source of hot gas for drying may simply be a coke-fired stove, or flue gases from an oil or gas-fired combustion chamber. For low temperature drying applications air may be preheated by heat transfer from condensing steam.

14.4.2 General application

Simultaneous drying and grinding is practised for convenience or to facilitate grinding, as for example, when water of crystallisation is released during the grinding operation. The combined operations of drying and grinding can often be integrated with conveying and classification in a unit of this type. These multiple functions often justify the use of this equipment, as savings in labour and mechanical handling accrue.

14.4.3 Mechanical performance

If a ball mill is used, mill liners are usually provided. These are expendable and protect the shell of the mill. In a ball mill the charge of steel balls or flint pebbles has to be renewed from time to time. High speed mills, such as that shown in Fig. 14.5, require renewal of pins, hammers and liners.

If the material handled is abrasive in nature, the conveyor tube must be carefully designed to avoid erosion at bends. The designed gas velocity should not be too high and bends should be of large radius.

14.4.4 Power requirements

The power requirements for simultaneous drying and grinding are the sum of two quantities: firstly, the energy required for the milling operation and secondly for the pneumatic conveying stage. Total power is within the range 20–50 kWh/t.

14.4.5 Explosion risk

Care must be taken when drying organic materials with a current of hot air in grinding machines that the concentration of suspended dust in the oxygen present does not fall into the explosive range. During normal running the dust concentration is, of course, well above the explosive limit in the mill and product collecting systems. During start-up, however, the dust concentration will be within the explosive limit unless the mill and collecting systems are filled with inert gas and dry ground material fed to the system for about a minute before introducing the wet feed and hot air.

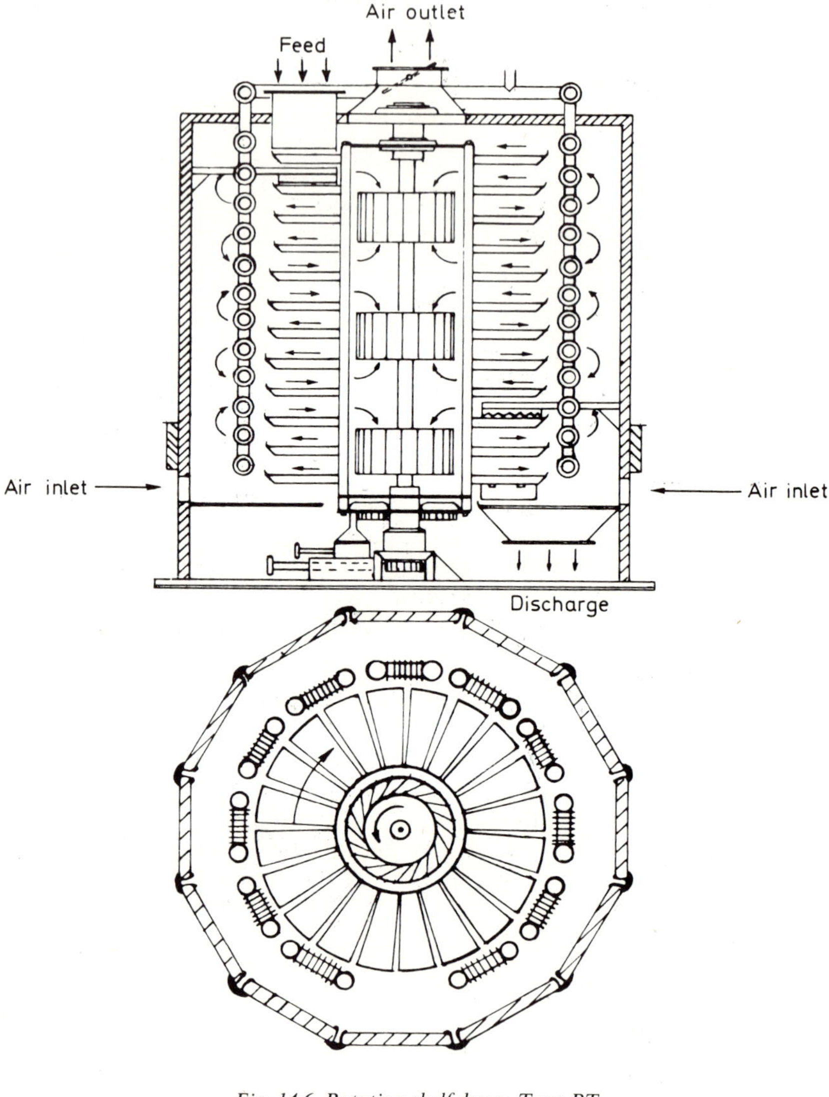

Fig. 14.6. Rotating shelf dryer, Type BT

Explosion relief discs are essential in the product collecting system.

It is most important that layers of fine dust should not be allowed to accumulate anywhere within the building, for example on ledges out of sight near the roof, since a slight harmless 'pop' in the drying system is capable of raising this dust into a cloud which, if ignited, would cause a catastrophic explosion.

14.5 ROTATING SHELF DRYERS

14.5.1 General description

Two dryers which are typical of this class are described; the common feature is the rotating shelf assembly.

In Type BT the dryer consists of a stack of annular shelves, each of which is made up of a number of segmental plates with slots between as shown in Fig. 14.6. The shelves, which rotate, pass under stationary transfer arms, which sweep the material off the segments and through the slots. The slots

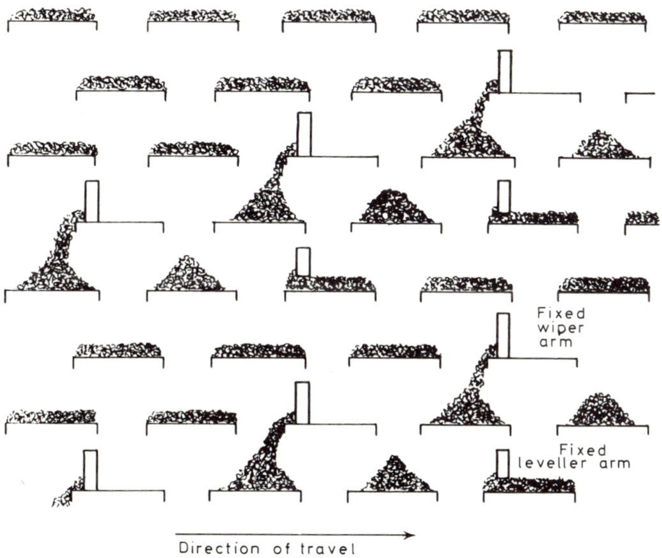

Fig. 14.7. Diagrammatic view of transfer of material in Type BT rotating shelf dryer

are staggered, so that transfer takes place regularly from shelf to shelf as shown in Fig. 14.7. This transfer process is repeated until the dried product is discharged from the bottom of the dryer. A bank of steam-heated tubes surrounds the shelves, while a vertical shaft on the axis of the dryer carries several radially bladed fans to cause hot air to flow over the shelves. After circulation, the air is discharged to atmosphere. The dryer is mounted in an insulated body with removable panels for access.

A modification of this type is made up of a spiral assembly of hinged trays, which are loaded and unloaded outside of the dryer body. The air is circulated in the same way as its more common counterpart.

Type J dryers consists essentially of a tower fitted internally with a series of truncated cones and an axially mounted rotating shaft carrying a series of discs which alternate with the stationary cones. The material to be dried is fed on to the top plate and is thrown off by centrifugal force on to the walls of the first cone and from there it falls to the second plate where the process is repeated. At the bottom of the tower the dried material is discharged. To effect the drying, hot air is blown up the tower, and any dust entrained is removed in a cyclone or other type of separator. The cones in the upper part of the tower, where the material is wet, are cleaned by scrapers which are carried on the central shaft.

Dust separation equipment is always required for the Type J dryer and is frequently necessary for the Type BT dryer.

14.5.2 General application

Type BT is ideally suited to handle granular or crystalline materials. It can also dry pastes and filter cakes provided they become friable during drying, though pre-forming is sometimes necessary.

Type J is ideally applied to the drying of crystalline materials such as salt and sugar. As the material has only a short residence time in this unit, its use is confined more or less to drying in the constant-rate period. It causes little damage to crystalline materials through particle breakdown.

14.5.3 Mechanical performance

Although rotating shelf dryers are relatively complex, they give rise to few mechanical problems when correctly applied. Problems do arise when the dryers are required to handle sticky solids. Otherwise, if a robust design is obtained, few troubles are encountered.

14.5.4 Power requirements

The power consumed by Type BT is surprisingly low.

The power required for material handling is that needed for wiping it off the tray and for levelling the piles to form a layer of uniform thickness at each transfer. For example, in a unit in which the conveyor belt is spaced 200 mm vertically and moves in a spiral 5 m in diameter, a 450 m length of conveyor only requires about 1 kW.

The power required for the fans is also low since the path of the air flow is relatively unrestricted.

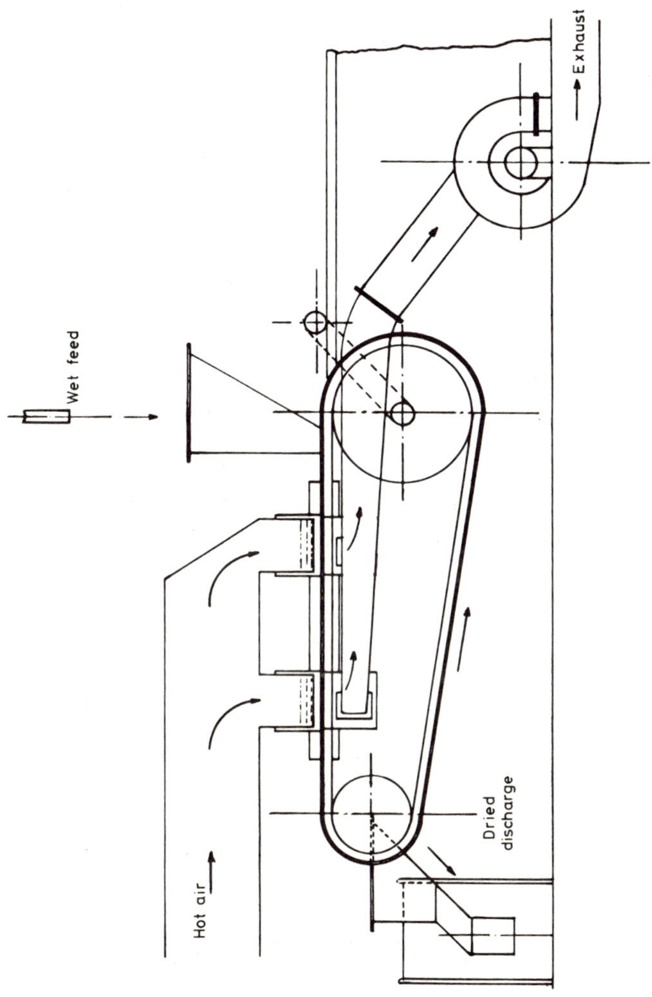

Fig. 14.8. Horizontal screen dryer

14.6 SPECIAL VARIANTS OF THE CONTINUOUS THROUGH-CIRCULATION DRYER

14.6.1 General description

The dryers described in this section are of two types:

(a) the continuous rotary top feed filter dryer,[1]

(b) a special form of the conveying screen dryer.

The continuous rotary top feed filter is totally enclosed in a housing into which a fan blows heated air. This air is drawn from the filter chamber through the bed of material into the centre of the drum. This equipment is used for dewatering and washing in addition to drying. The crystal magma is fed on to the top of the filter drum and as much as possible of the perimeter of the filter is utilised for drying. Dry crystals are taken off near the bottom of the drum or from a point on the ascending side. The special construction in this design has two functions:

(a) it removes liquid by displacement from the crystal surfaces by passing large volumes of air through the material,

(b) the hot air in the filter housing evaporates the residual moisture from the crystal surfaces producing dry material.

The conveying screen dryer is a special form of a horizontal through circulation dryer in which hot air or other gases are drawn through a bed of material as shown in Fig. 14.8. The design is such that material can be mechanically dewatered by the gases as well as dried. A series of separate wire covered horizontal screens or pallets are used, which are supported on rollers as they move in a straight line through the drying zone. The wet material is fed on to a screen which is moved slowly under the gas ducts. The hot gases are blown on top of the material with a velocity of up to about 1·5 m/s, so that the water is driven out and the sensible heat in the gas dries the material. Moisture is carried away in a duct under the bed, and the dried material is discharged at the end of the belt.

14.6.2 General application

Both types require to have a crystalline or granular feed, which falls within well defined limits of particle size. This is important as it is necessary to provide a permeable bed without an excessive pressure drop.

The filter dryer is only applicable to those materials which do not require a long drying time. It is ideally suited for coarse crystalline materials where the moisture to be removed is located on the surface. It is not applicable in general to precipitated materials.

The horizontal bed dryer is, in general, used for relatively coarse materials which can be supported on a wire mesh screen. For this reason it is used for coal drying and in the metallurgical industry but it has few applications in the chemical industry.

14.6.3 Mechanical performance

Little information has been recorded on the mechanical performance of these units. The rotary filter dryer requires skilled maintenance and in general has maintenance needs which are somewhat in excess of those in a rotary vacuum filter.

The maintenance costs of the horizontal bed type are 5–10% of the installed cost.

14.6.4 Power requirements

In dryers of this type power is required for the fan and for auxiliaries. The power requirement for the fan varies, of course, with the permeability of the head of material being dried, so that it is not practicable to give figures of general utility.

It is suggested that the total power required be taken as 2–20 kW per square metre.

14.6.5 Performance data

The following performance figures are available:

(a) Rotary drum filter dryer.

For a unit with 7 m^2 of drum area, 4 t/h of salt crystals were reduced to 0·1% moisture. The cycle time was 2 min and the cake thickness 4 cm. Under these conditions 5·7 m^3/s of air was used with a vacuum of 0·1–0·13 mb absolute pressure.[1]

(b) Conveying screen dryer.

For a unit treating material of 3 mm, the evaporating capacity was 6–13 g/s m^2. This was with an air velocity of 0·5–1·3 m/s, a bed thickness of 12–80 mm and a pressure drop of 13–75 mb. (See Chapter 8, Section 8.5.1.)

CHAPTER FIFTEEN

Specialised techniques

15.1 INTRODUCTION

This chapter describes three highly specialised drying techniques: freeze-drying; drying by infra-red radiation; drying by dielectric heating. Their applications are limited in the chemical industry but within their limitations they may provide the best, and in some cases the only, solution to a particular drying problem.

15.2 FREEZE-DRYING

15.2.1 Definition

Freeze-drying is desiccation by sublimation of vapour from the solid state in a vessel evacuated to an absolute pressure of around one millibar. It is relatively slow and expensive but in certain cases is the only means of obtaining a satisfactory product. It has been used for drying pharmaceutical products and some fine grained dyestuffs. Major advances in the technique have been made since 1960 in the freeze-drying of foodstuffs.

15.2.2 Reasons for Freeze-drying

(a) It is sometimes possible to take advantage of changes of relative volatility with temperature, and to freeze-dry a material without loss of another component which would have evaporated with the water at a higher temperature.
(b) At low temperatures it is possible to avoid undesirable changes of the material being dried such as skin formation, bacterial activity, organic growth or chemical change.
(c) Surface tension effects are suppressed during freeze-drying so that the product tends to have an open porous structure which facilitates quick re-absorption of water when required.

The following list shows the range of products which are freeze-dried regularly.

Foodstuffs	Meat	Fruit
	Certain vegetables	Fruit juices
		Milk

Pharmaceuticals	Virus solutions	Antibiotics
	Serums	Bacterial cultures
Animal tissues and extracts	Bones	Tumours
	Skin	Arteries
	Blood	Hormones

15.2.3 The freeze-drying process

Freeze-drying is usually a batch process. The initial freezing may be done outside the freeze-dryer or it may be accomplished by evaporation under vacuum. The container holding the substance is placed inside a vacuum chest or coupled to a vacuum manifold (used for large numbers of small containers). Normally inerts are removed by pumping, and water vapour by condensation. Heat must be supplied to the substance to provide the latent heat of vaporisation but at such a rate that the temperature of the material being dried is not raised above a set temperature below 0°C.

Conduction within the solid. In most substances left to dry undisturbed there are two stages of drying. In the first stage the main ice surface retreats from the solid boundary leaving behind a spongy product containing isolated ice crystals. In the second stage these crystals sublime at a very slow rate because of the resistance to heat flow to them.

This slow second stage can be speeded up by compression of the spongy product as drying proceeds. This is done by means of centrifugal force (applicable to fluids) or by compression between the tray and cover (applicable to solids).

With some substances the temperature can be raised above freezing after the first stage of drying without damage to the product. In this case the drying rate is greatly increased.

15.2.4 Initial freezing

Quick freezing normally results in:

(a) Smaller ice crystals being formed which lessens damage to the walls of organic cells, permits a higher sublimation rate, and reduces forces due to expansion on the container.

(b) Reduced risk of producing highly concentrated solutions before freezing is completed. (Such solutions might result in osmotic drainage of cells or chemical reaction.)

Slow freezing, however, may give a high survival rate for bacteria and cells.

Forces due to expansion on freezing may be reduced by the addition of certain salts, e.g. potassium or sodium chloride, or by weakening the bond with the container wall by treatment with an organo-silicone.

Freezing is done by conventional refrigeration techniques using solid CO_2, freon, etc, known as pre-freezing; or by evaporation in the vacuum freeze dryer, known as self-freezing. The second method is slower but may result in a simpler cycle. About one sixth of the water has to be evaporated to freeze the substance, more if the proportion of water is low. This initial drying before a low temperature is reached may result in skin formation, undesirable in the product and acting as a barrier to diffusion of moisture from the centre during drying, hence increasing the length of the drying cycle. Adding water initially may avoid this.

Spill. On application of the vacuum in the self-freezing procedure, air may be released from the liquid and produce froth. 'De-gassing' without producing excessive froth is accomplished by slow application of the vacuum or by rotation of the container at a speed which gives a high centrifugal force that effectively holds down the froth.

When using deep containers with a narrow outlet, the velocity of the departing vapour may be high enough to carry out particles of product. This is known as 'spill'. It can be reduced by rotating the container or fitting some form of shield over the opening.

Stoppering. Where it is undesirable for the dried contents of bottles to come in contact with the air, automatic stoppering devices are available.

Ampoules can be sealed before removal from the vacuum manifold.

15.2.5 Supply of heat for sublimation of vapour

For maximum drying rate at a given temperature, the substance should have a small thickness and a large free surface area. As the drying rate also varies with the saturation vapour pressure over the substance, its temperature should not be lower than necessary. The rate at which heat is added will be controlled by this temperature.

Methods of supplying heat to various types of material are:

(a) *Granules* are spread in a thin layer on a tray and heat is supplied by radiation from above, or by conduction through the tray. A ribbed blackened tray has been found most satisfactory.

(b) *Liquids* are contained in a shallow tray or rotated in a bowl during freezing in order to spread the freezing fluid round all or part of the container walls.

These two methods give a large surface area of sublimation but if the containers are small (i.e. ampoules), it may not be necessary to rotate them during freezing.

Heat may be supplied through the walls and base of the container from electric heating elements or water pipes. In the case of a manifold dryer sufficient heat may be obtained from the atmosphere; alternatively the air round the containers may be heated separately or they can be immersed in a heated bath.

(c) *Solids* are heated by conduction through the tray and sometimes through a cover as well. Better contact is achieved if a compressive force is applied between the tray and the cover. Good results have been obtained using expanded metal sheets as these leave a greater surface area for sublimation (see also Section 15.4).

15.2.6 High-frequency dielectric heating

This method of heating can only be used with special precautions, otherwise arcing and corona discharge occur in the vacuum. With large solids there is a risk that the product will be scorched internally. Nevertheless it may be permissible in the latter stages of drying to lower the vacuum and raise the solid temperature above freezing under which circumstances h.f. heating can be used. The theoretical advantage of h.f. heating is that heat tends to be generated within the material leading to quicker drying out of interstitial moisture.

15.2.7 Maintenance of the vacuum

Inerts are removed initially by ordinary vacuum pumps and/or steam ejectors.

On a small scale, vapour may be removed either by chemical means or by pumping, but the latter introduces the complication of separating water from the oil in the pump.

Normally vapour is removed from the system by condensation. For quick drying the vapour pressure in the system should be as low as possible. The following points ensure this:

(a) The condensing surface must be as cold as possible.

(b) It must be close to all surfaces where sublimation is occurring.

(c) Enough condenser surface must be available so that ice does not build up so thickly that it causes a resistance to heat transfer. Ice normally builds up on those parts of a condenser facing the substance being dried. In the drying of many substances a thick layer of ice on the condenser may not be a hindrance in the latter stages because the limiting factor is the rate at which molecules leave the solid. Room must be left for the build-up of ice between the substance and the condenser; the specific gravity of this ice may be as low as 0·7.

It is sometimes more convenient to have a separate condenser preceding a vacuum chamber.

15.2.8 Defrosting of condenser

The ice may be removed by passing hot water through the condenser. The condenser should be designed so that when the bond between the condenser and the ice is melted, the ice falls clear, thus shortening the time the dryer is out of use. The ice may also be removed by some form of mechanical scraper.

15.2.9 Types of freeze-dryer

To cater for the various factors discussed above there are three main types of dryer.

(a) Rotary spin freeze-dryers – used only for liquids.

(b) Manifold freeze-dryers – these are convenient for drying the contents of large numbers of small capacity containers. They suffer from the disadvantage that the breaking of one container destroys the vacuum over all the others.
(c) Cabinet shelf dryers—these are the most common and are used for a large variety of substances.

15.2.10 Quantitative data on freeze-dryers

There are so many factors involved in the design of freeze-dryers that it is normal to determine the time of a cycle and the optimum temperatures during it by a small-scale laboratory run.

The following quantities will serve as a rough guide. The vacuum is usually in the range 0·1–5 μb (0·1–5 N/m^2). Shelves are cooled to −40°C and the condenser coils to −70°C. The typical length of cycle for bacterial pellets is 2–20 h; foodstuffs take 5–10 h if the temperature is raised in the latter stages; biological fluids take 20–50 h.

Table 15.1 relates the size of shelf dryers with heat and power input. The

Table 15.1 Approximate relation between size and energy input to freeze dryers

Model	*Weight* (t)	*Capacity* (litre)	*Shaft power* (kW)	*Heating* (kW)	*Floor space* (m^2)
Small laboratory	$\frac{1}{4}$	0·1	0·2	0·2	1
Small production	$2\frac{1}{2}$	20	4	5	13
Large production	40	500	75	50	150

values for shaftpower and electric heating are maximum rates which may only be reached for short periods in the cycle. The power will vary with the kind of refrigerant used.

15.3 DRYING BY INFRA-RED RADIATION

15.3.1 Definition

The term 'infra-red drying' is applied to a drying process in which the heat input to the drying solid is effected by means of infra-red radiation. I.R. radiation is defined as electromagnetic radiation in the wavelength band 0·76 μm–400 μm. Radiation in this frequency range is excited by the thermal vibrations of molecules and conversely the absorption of i.r. radiation gives rise to thermal vibrations. I.R. radiation thus provides a means of transmitting heat between two solids and this heat may then be used for drying. 'I.R. radiation' and 'radiant heat' are the same thing. The diffusion of the vapour from the solid follows the same laws as in other forms of drying.

15.3.2 Theory

The theory of drying by radiant heat has been discussed in Chapter 2.

Although radiation from gases is important in furnaces, for drying the source of radiation is invariably a heated solid. The radiation is distributed over a wide band of frequencies with a well-defined maximum energy level at a wavelength which varies inversely as the absolute temperature according to Wien's Displacement Law:

$$\lambda^*_{\text{max}} = \frac{2\,885}{T_r^*} \tag{15.1}$$

The intensity of radiation emission is proportional to the fourth power of the absolute temperature according to the Stefan-Boltzmann Law.

$$Q_r^* = (\sigma)T_r^{*4} \tag{15.2}$$

A body which follows the Stefan-Boltzmann Law exactly is said to be a 'black body'. Actual surfaces emit less radiation than a black body and the ratio of the actual emission to that of a black body at the same temperature is termed the emissivity of the surface. Most dull and rough surfaces have emissivities of about 0·6–0·8, but bright metal surfaces have low emissivities in the range 0·01–0·1.

Some examples of maximum energy wavelengths and black body emissions are given in Table 15.2.

When i.r. radiation falls on a surface any combination of three things may happen.

(a) The radiation is absorbed and its energy is converted into heat.
(b) The radiation is reflected from the surface.
(c) The radiation is transmitted through the material.

For the purpose of drying the fraction of the incident radiation which is absorbed is the most important; this fraction is termed the absorptivity. The absorptivity of a surface is approximately equal to its emissivity at the same temperature as the source of the incident radiation.

Table 15.2 Relation between energy emission and maximum energy wavelength of radiation from sources at various temperatures

Temperature (°C)	(K)	*Maximum energy wavelength* (μm)	*Total energy emission* (kW/m^2)
100	373	7·73	1·1
500	773	3·73	2·1
1 000	1 273	2·27	150
1 500	1 773	1·63	570
2 000	2 273	1·27	1 520

The absorptivity of most surfaces varies considerably with the wavelength of the incident radiation. Dull, rough surfaces generally have a high absorptivity for long wavelengths independent of their colour, but for short wavelengths colour is important; dark colours have higher absorptivities than light colours. Extreme cases are rough black surfaces which have a high absorptivity, and highly polished metals which have a very low absorptivity, for all wavelenths.

When a material is partially transparent to the incident radiation, the absorptivity will also depend on the thickness of the material. Water is especially important in this respect because when a very moist material is exposed to i.r. radiation, a high proportion of the radiation may be absorbed in the water film. Water is almost completely transparent to radiation of wavelengths shorter than 2·5 μm; but it has strong absorption maxima at 3·1 μm and 6·1 μm and at these wavelengths a film of water 1 mm thick absorbs about 80% of the incident radiation. The source temperatures which give peak emission at these wavelengths are:

2·5 μm	881°C
3·1 μm	658°C
6·1 μm	192°C

As the material dries and the moisture film breaks up, the absorptivity of the solid material becomes more important and in the later stages most of the heat will be transferred to the water from the solid. In principle it should be possible to dry a relatively deep bed of granular material if the solid is largely transparent to a band of wavelengths which is strongly absorbed by water, but in practice this is an unusual situation and i.r. radiation can be regarded as essentially a surface heating effect.

15.3.3 Applications of infra-red drying

(a) Drying sheet material e.g. textiles, paper, films.
(b) Drying surface finishes e.g. paints, stove enamels.
(c) Surface drying of bulky non-porous articles.
(d) Laboratory testing of drying characteristics of any chemical and small-scale drying of industrial chemicals where other methods fail.

The majority of chemical installations fall into category (d).

15.3.4 Advantages and disadvantages of infra-red dryers

Advantages

(a) The apparatus is simple to operate and easily adapted to changing conditions. It occupies little space as there is no bulky ancillary equipment.
(b) Effluent can be kept to a minimum.
(c) In some cases infra-red dryers are faster than convective dryers because the heat can be concentrated on the work instead of being wasted in heating the surroundings.
(d) The directional characteristics of some radiators can be used to dry limited areas of large objects.

Disadvantages

(a) Running costs are high.
(b) Radiant dryers are essentially surface dryers, which limits the range of materials for which they are suitable.

15.3.5 Design of infra-red dryers

I.R. dryers may be operated batchwise but are usually continuous conveyor dryers. In the latter case they may be integrated with the assembly line conveyors. Radiation sources are supplied as units, and a continuous i.r. dryer consists of an appropriate number of radiators arranged around a suitable conveyor so that the maximum amount of radiation is intercepted by the work. A relatively open structure is normally used so that natural convection currents can remove the evolved vapour. A fume extraction hood is mounted above the dryer and, in the case of gas heated radiators it removes the products of combustion as well as the evolved vapour. If solvent recovery must be practised then a closed structure with controlled ventilation must be employed.

When sheets of material are to be dried it is easy to arrange directional radiators so that a high proportion of the radiation is intercepted by the material. Where miscellaneous small articles are to be dried the effective cross-section may be limited and in this case the radiators should be used to form as nearly as possible a complete enclosure. Non-radiating surfaces should be well insulated and if practicable lined with bright metal; then most of the incident radiation will be reflected or re-radiated. Openings should be kept to a minimum as all the radiation passing through an opening is lost. By these means the work surface can be made a large proportion of the total absorbing surface and a corresponding large proportion of the net radiant transfer will be usefully absorbed.

15.3.6 Types of radiator

Electric lamp radiators

These resemble conventional electric light bulbs, but a glass having a high i.r. transmission is used and the bulb is usually parabolic in shape and gold-plated to give a concentrated beam of radiation. A tungsten filament working at 2 200°C is used. The range of wavelengths is 0·6 μm–3 μm with maximum energy at 1·1 μm. The lamps are available in a single size rated at 250 W.

Electric sheathed filament radiators

In these radiators the radiating element is a long filament in a refractory sheath mounted in a parabolic mirror. A wide range of sizes is available.

The element operates at 750°C, which gives a range of wavelength from 1·2 μm–8 μm with maximum energy at 2·8 μm. These elements can, of course, be run at lower temperatures if desired.

High temperature gas-heated radiators

In these radiators a metal or refractory surface is heated by gas flames to about 800°C. In one design an air/gas mixture is fed to the back of a perforated refractory block and burns in a thin sheet of flame on the front surface. Metal tubular radiant heaters are also available. In these the products of combustion are circulated round a closed loop to give a uniform surface temperature.

Low temperature gas-heated radiators

In one commercial design the radiating surface is a metal sheet having a maximum temperature of 340°C giving maximum energy at a wavelength of 4·7 μm. The radiating surface is heated at the back by non-aerated burners. The unit is available in two forms; a unit for vertical or inclined mounting, having a radiating area of about 1 m × 0·5 m and a gas consumption of 26·4 MJ/h = 7·33 kW, and a horizontal unit with an effective area of 1 m × 0·6 m and a gas consumption of 46 MJ/h = 12·8 kW.

15.3.7 Choice of radiators

The choice of radiator may be determined by the absorption characteristics of the material to be heated. If strong absorption by the water film is required, then a radiator running at about 700°C is required as the radiation from this coincides with a strong absorption band of liquid water. Water is almost completely transparent to radiation from high temperature lamps.

When a solution is being dried a skin may tend to form which prevents the complete evaporation of the solvent. If a radiation band is chosen to which the solution is largely transparent then the film is heated indirectly from the surface below and this difficulty can often be overcome. This technique, employing high temperature lamps, is usual for drying many varnishes and cellulose finishes and gives a superior finish.

In cases where the wavelength is unimportant there is not much to choose between the different types of radiator and the final choice will largely depend on relative running costs and convenience.

15.3.8 Sizing infra-red dryers

Infra-red dryers differ from other forms of dryer only in the mode of heat transfer. The principles of radiant heat transfer are discussed in detail in most books on heat transfer and given suitable data it would be possible to design an infra-red dryer using the principles outlined in Chapter 2. In practice there are rarely sufficient data available and as with other types of dryers, laboratory test data are usually necessary for designing an infra-red dryer. The principal data required are the intensity of radiation and residence time necessary to achieve the desired result. All the manufacturers of radiant heaters have experimental equipment available, or a simple test rig based on a few radiators can be set up in the user's own laboratory. As the radiators used on the plant may not be the same as those in the test equipment, it is also necessary to know the efficiency of conversion of the energy input to i.r. radiation and the effective radiating area. No reliable data

Table 15.3 Energy conversion of infra-red radiators

Type of radiator	*Radiating area* (m^2/kW *of energy input*)	*Conversion efficiency* %
Electric lamp	$9{\cdot}2 \times 10^{-3}$	95
Sheathed filament projector unit	0·37–0·73	70
High temperature gas radiator	0·41	34
Low temperature gas radiator	0·55–0·64	34

are available for energy conversion but the estimated conversion efficiencies given in Table 15.3 are probably accurate enough for this purpose. The effective radiating area is expressed in terms of square metres per kilowatt of energy input.

The factors quoted in the standard text books on heat transfer may be used to estimate the effect of a different geometry but it is much more reliable to make the laboratory model as nearly as possible similar to the final unit, so that scaling-up is mainly a matter of increasing the length to give the required residence time at a given throughput.

15.3.9 Instrumentation

For the type of duty to which infra-red heating is normally applied it is not possible to obtain a control signal from the work. Control is a matter of finding suitable running conditions by trial and error and then maintaining them constant. Electric radiators are conveniently controlled by energy regulators; although these are on/off controls, a large number are employed and are unlikely to keep in step so that the average level of radiation remains steady. Low temperature gas heaters should have automatic control of the panel surface temperature.

15.3.10 Costs

Capital cost

Capital costs are given in Table 15.4. These prices refer to the units themselves in 1968 and the cost of the framework, conveyor and ventilating system must be added. These items will vary considerably but in general the radiators will be the main cost of an infra-red dryer.

Table 15.4 Capital costs (1968) of infra-red radiators

Type of unit	*£ per* kW *of radiant energy output*
Electric lamp	5·4
Electric sheathed filament	11·0
	5·5
High temperature gas	10·2
Low temperature gas	10·2
	9·0

Table 15.5 Costs of radiant energy (1968)

Type of unit	*£ per* MWh *radiant energy*
Electric lamp	6·3
Electric sheathed filament	8·6
High temperature gas	7·5
Low temperature gas	7·5

Energy costs

Table 15.5 gives the costs of energy for infra-red radiators on the basis of electricity at £6/MWh and gas at £2·5/MWh (about £0·07/New Therm).

It should be noted that when gas heaters are used in an enclosed type of oven there is a significant convective contribution from the products of combustion, so that running costs may prove to be appreciably lower than this.

15.3.11 Maintenance costs

Lamp radiators have a life of 2 000–3 000 h and sheathed filament elements a life of 5 000–10 000 h; the replacement of the elements is the major maintenance item. The cost will vary greatly according to the type of unit and the conditions under which it operates, but it appears to be around £2 for lamps and £0·8 for sheathed filament radiators, per MWh of energy input. There is little to go wrong in the gas units and the corresponding figure should be low.

15.4 DIELECTRIC DRYING

15.4.1 Definition

The term 'dielectric drying' refers to the method of supplying heat to a substance to be dried. Vapour and moisture transfer depends on the same laws as in other methods of drying.

When certain substances (usually insulators or poor conductors of heat or electricity) are placed in an electrostatic field their molecules are deformed and energy is stored within. If an alternating field is applied energy is continually absorbed and released. However a certain amount of electrical energy is transformed into thermal energy which results in the temperature of the substance rising.

The thermal energy released in a dielectric heater per unit volume =

$$8{\cdot}6 f^* V^{*2} \varepsilon^* \cos\phi \times 10^{-8}\ \mathrm{W/m^3} \qquad (15.3)$$

The dielectric constant of water (about 80) is considerably higher than that of most other substances so heat tends to be produced directly in the water to be evaporated.

ε^* and ϕ tend to vary with frequency and moisture content so exact theoretical analysis is impossible. Therefore tests must be run to find a suitable frequency and cycle for drying. It should be noted that techniques are always improving so that although in the past a certain substance may not have been dried successfully the possibility of successful dielectric drying should not now be ruled out. Generally the heat will be mainly liberated in the water, but there will be little difference in temperature between the water and the solid; this temperature in general will be the boiling point. This technique is therefore unlikely to be suitable for temperature sensitive materials.

15.4.2 Advantages and disadvantages

Advantages

(a) Dielectric dryers can dry bulk materials of low thermal conductivity without excessive temperatures. Their real potential application is for the complete removal of tightly bound water from materials for which the falling-rate period of drying is excessively long.
(b) The variation of dielectric constant with moisture content results in automatic reduction of heat input to a low level as the product becomes dry. As a result measurement of the water content of the material is inherent in the process which is therefore especially suitable for automatic control.
(c) Simple start-up and shut-down.
(d) The space occupied is fairly small and flexible.
(e) Minimum effluent.
(f) Either batch or continuous operation can be used.
(g) Non-contaminating.

Disadvantages

(a) Cannot be used for conducting materials.
(b) High capital cost.

(c) High running cost.
(d) Seldom suitable for vacuum drying because of arcing and corona discharge and the high cost of suitable equipment.
(e) Has seldom been used in the chemical manufacturing industry and every proposed application is regarded as experimental and liable to failure.

15.4.3 Electrical requirements

Owing to the high voltage all parts must be carefully screened and guarded from personnel. The electrodes do not touch the material so that high voltages may be used with safety.

The frequency used is in the radio range, therefore electrical screening must be installed to avoid interference with radio and television.

The electrical equipment performs the following functions:
(a) Converts mains power to the required voltage and frequency.
(b) Incorporates automatic control to maintain resonance between the oscillator and the product.
(c) Controls product dryness. For batch operation the cycle can be pre-set or the machine can sense product dryness by the change in the circuit impedance. For continuous operation power input can be fixed or controlled by circuit impedance as for batch operation or by temperature of the material at exit.

For a given power a lower voltage is required when a really high frequency is employed. As there may be limitations on voltage due to sparking and corona discharge, it may pay to use the highest possible frequency. Until recently, however, the difficulties of generation limited the practical upper frequency to 100 MHz or so. The advent of relatively high power magnetrons has now made use of micro-wave generators up to 10 000 MHz (10 GHz) practicable. Typical operating conditions are:

Frequency range—conventional valve generator 2–100 MHz
—magnetron generator 1–100 GHz
Voltage—1–30 kV
Power—up to 200 kW

15.4.6 Costs

Capital costs depend so much on the application that no useful figures can be given. Some advice on plant manufacturers with suitable experience can be obtained from the Central Electricity Research Laboratories, Leatherhead, Surrey.

Running costs depend on design of dryer, material being dried and electrical conversion efficiency. Overall figures are 0·5–1 kg water evaporated per kWh input.

Conversion efficiency to r.f. power is 50%–60%. Maintenance costs tend to be high mainly because of valve replacement costs.

References to all Chapters

The first numbers in the following list refer to the Chapter number. Not all the references for Chapters 9 and 11 are referred to in the text.

1.1. MOSS, A. A. H., 'The drying of solids: the gaps in our knowledge', *Chemistry and Industry,* 40, 21 June (1965)

1.2. NONHEBEL, G., (Editor), *Gas purification processes,* Geo. Newnes Ltd., London (1964)

1.3. Ninety-eighth annual report on alkali, etc., works for 1961. H.M.S.O. (1962)

1.4. JARMAN, R. T. and DE TURVILLE, C. M., 'The visibility and length of chimney plumes', *Atmospheric Environment,* 3, 257 (1969)

1.5. 'Problems of noise in chemical plant'—a series of papers published in *The Chemical Engineer,* page CE105 and 223 (1968)

GENERAL

1.6. KRISCHER, O., *Die Wissenschaftlichen Grundlagen der Trocknungtechnik, Berlin, 1956,* Springer Verlag, Berlin (1956)

1.7. KNEULE, F., 'Das Trocknen', *Verlag Sauerländer,* Aarau Switzerland (1969)

1.8. JACKSON, J., *Cooling Towers,* Butterworths (1951)

2.1 CREMER, H. W. and WATKINS, S. B., *Chemical Engineering Practice, Vol. 7, Heat Transfer.* This volume contains chapters on drying

COULSON, J. M. and RICHARDSON, J. F., *Chemical Engineering, Vols 1 and 2,* Pergamon Press, Oxford, second edition (1968)

2.2. PERRY, J. H., *Chemical Engineers' Handbook,* McGraw-Hill, New York, fourth edition (1963)

2.3. MORRIS, G. A. and JACKSON, J., *Absorption Towers,* Butterworths, London (1953) (This book is out of print: an abbreviated version of the theory of absorption, drying is the reverse process, has been given by Jackson, J., in Chapter 5 of *Gas Purification Process,* edited by Nonhebel, G., Geo Newnes, London (1964) new edition in preparation)

2.4. SHERWOOD and PIGFORD, R. L., *Absorption and Extraction,* McGraw-Hill, New York, second edition (1952)

2.5. MCADAMS, W. H., *Heat Transmission,* McGraw-Hill, New York, third edition (1954)

PSYCHROMETRIC TABLES AND CHARTS

Those published by Zimmerman, O. T., and Lavine, I., Industrial Research Service, Dover, New Hampshire, U.S.A. (1945) have the widest range–they are in degrees Fahrenheit. *Perry's Handbook* (Reference 2) contains useful graphs.

5.1. SHEPHERD, C. B., HADLOCK, C. and BREWER, R. C., 'Drying materials in trays', *Ind. Chem. Eng.,* **30**, 388 (1938)

5.2. SCHAFFNER, R. M. and KOEHLER, W. A., 'Performance data in atmospheric tray dryers', *Trans Am. Inst. Chem. Eng.,* **35**, 303 (1939)

5.3. GAMSON, B. W., THODAS, G. and HOUGEN, O. A., 'Heat, mass and momentum transfer in flow of gases through solids', *Trans. Am. Inst. Chem. Eng.*, **39**, 1 (1943)
5.4. PERRY, J. H., *Chemical Engineers' Handbook, fourth edition*, Chapter 15, McGraw-Hill Book Co., New York and London (1963)
5.5. WHITE, A. S., 'Drying of chemical products by atmospheric tray dryers', *Fuel and the Future, Vol. 1*, 255, Ministry of Fuel and Power, London (1948)

8.1. PERRY, J. H., *Chemical Engineers Handbook*, fourth edition, Chapter 15, McGraw-Hill Book Co., New York and London (1963)
8.2. GLOVER, H. C. and MOSS, A. A. H., 'A method of design of continuous through-circulation dryers', *Trans. Instn. Chem. Engrs.*, **35**, 208 (1957)
8.3. MARSHALL, W. R. and HOUGEN, O. A., 'Performance of through-circulation dryers', *Trans. Amer. Inst. Chem. Engrs.*, **38**, 91 (1942)
8.4. GAMSON, B. W., THODAS, G. and HOUGEN, O. A., 'Theory of through-circulation dryers', *Trans. Amer. Inst. Chem. Engrs.*, **39**, 1 (1943)
8.5. WILKE, C. R. and HOUGEN, O. A., 'The constant rate in through-circulation dryers', **41**, 445 (1945)
8.6. FILONENKE, G. K. and KORCHAGIN, V. D., 'Kinetics of through-circulation drying', *Inz. Fiz. Z. Akad. Nauk Belorus, SSR*, **1**, 39 (1959)
8.7. ARTHUR, J. R., LINNETT, J. W. and RAYNOR, E. J., 'The flow of air through a bed of granules', *Trans. Farad. Soc.*, **46**, 270 (1950)
8.8. AEROV, M. E., 'Heat and mass transfer in a granular bed', *Zh. Tekh. Fiz.*, **26**, 1233 (1950)
8.9. COBERLY, C. A. and MARSHALL, W. R., 'Temperature gradient in gas through a fixed granular bed', *Chem. Engng. Prog.*, **47**, 141 (1951)
8.10. WAGSTAFF, J. B., 'Air flow in beds of granular solids', *Ind. Eng. Chem.* **47**, 1129 (1955)
8.11. PLAUTY, D. A. and JOHNSTONE, H. F., 'Heat and mass transfer in packed beds', *J. Amer. Inst. Chem. Engrs.*, **1**, 193, (1955)
8.12. GUSEV, I. V., NIKITINA and AEROV, M. E., 'Heat and mass transfer in a granular bed', *Zh. Takh. Fiz.* **26**, 2005 (1956)
8.13. DITSWORTH, R. L. and GERGES, 'Through-circulation drying of frozen-cut sweet corn', *Agric. Engng.*, **34**, 251 (1953)
8.14. SIMMONDS, W. H. C., WARD, G. T. and MCEWEN, E., 'Through-circulation drying of wheat grain', *Trans. Inst. Chem. Engrs.*, **31**, 265, 279 (1953)
8.15. MITCHELL, T. J. and POTTS, C. S., 'Through-circulation drying of brewers spent grain and sugar beet', *J. Sci. Food Agric.*, **9**, 2929 (1958)
8.16. TOWEI, R., 'Design of tunnel dryers', *Trans. Soc. Chem. Engrs., Japan*, **8**, 85 (1950)
8.17. FRANKLIN, C. E. L. and SHARP, M. Q., 'Dryers for earthenware', *Brit. Ceram. Res. Assocn.*, Paper No. 123 (1951)
8.18. CLEGG, R. R., 'Conveyor dryer for ceramics', *Ceramics*, **3**, 464 (1951)
8.19. BROUGHTON, D. B. and MICKLEY, H. S., 'Design method for continuous tunnels', *Chem. Eng. Prog.*, **49**, 319 (1953)
8.20. KAMEI, S. and TOWEI, R., 'Performance of tunnel and band dryers', *Mem. Engng. Fac. Kyoto Univ.*, **16**, 14 (1954)
8.21. KAMEI, S. and TOWEI, R., 'Performance data on tunnel and tray dryers', *Chem. Ing. Tech.*, **26**, 1 (1954)
8.22. KILPATRICK, P. W., LOWE, E. and VAN ARDSEL, W. B., 'Review of tunnel dryers', *Advances in Food Research*, **6**, 313 (1955)
8.23. NEWTON, W. H., 'Controlled feeding device for granular materials', *Brit. Chem. Engng.*, **4**, 615 (1959)

9.1. LAVETT, C. O. and VAN MARLE, D. J., 'Performance data on drum dryers', *Ind. Eng. Chem.*, **13**, 600 (1921)
9.2. REAVELL, J. A., 'Review of drum drying', *Trans. Inst. Chem. Eng.*, **6**, 115 (1928)
9.3. HARCOURT, G. N., 'Performance data on drum dryers', *Am. Soc. Mech. Eng.* (Sept. 17, 1936)

9.4. HARTE, C. R., 'Performance data on drum dryers', *Ind. Eng. Chem.*, **28**, 7 (1936)

9.5. VAN MARLE, D. J., 'Performance data on drum dryers', *Ind. Eng. Chem.*, **30**, 1006 (1938)

9.6. HARCOURT, G. N., 'Review various types of dryer', *Chem. Met. Eng.*, **45**, 179 (1938)

9.7. MOREY, F. C., 'Equation for pick-up on drums', *J. Res. Nat. Bur. Stand.*, **25**, 385 (1940)

9.8. ANON, 'General review drum drying', *Chem. Age*, **45**, 169 (1941)

9.9. ANON, 'Vacuum drum dryers', *Chem. Age*, **48**, 557 (1943)

9.10. KIRSCHBAUM, E., 'Theory of evaporation and heat transfer on drum dryers', *Z. Ver. Deut. Ing.* **No. 3**, 73 (1943)

9.11. HAAS, T. K., 'Scale up of drum dryers', *Chem. Weekblad*, **45**, 330 (1949)

9.12. STRINGER, W. E., 'Performance double and twin drum dryers on whey', *Food. Inds.*, **21**, 892 (1949)

9.13. FOX, H. H., 'Automatic level control on drum dryers', *Chem. Eng.*, **57**, No. 9, 122 (1950)

9.14. REAVELL, B. M., 'Review of drum drying', *Chem. Trade J.*, **128**, 669 (May 23, 1951)

9.15. BAUMANN, R., 'Performance single and double drum dryers on food products', *Chem. Ing. Tech.*, **25**, 607 (1953)

9.16. MONTGOMERY, A. E., 'Paper on cylinder dryers', *T.A.P.P.I.*, **37**, No. 1, 1 (1954)

9.17. NISSON, A. H. and KAYE, W. G., 'Paper on cylinder dryers', *T.A.P.P.I.*, **38**, No. 7, 385, 395 (1955)

9.18. NISSON, A. H., 'Paper on cylinder dryers', *Chem. and Ind.*, 198 (April 17, 1956)

9.19. KIRSCHBAUM, E. and BAUMANN, R. F., 'Theory evaporation and heat transfer on drum dryers', *Chem. Ing. Tech.*, **28**, 520 (1956)

9.20. ANON, 'Fabric on cylinder dryers', *B.C.I.R.A. Shirley Inst. Bull.*, **30**, 42 (1957)

9.21. NISSON, A. H. and HANSEN, D., 'Evaporation and heat transfer on drum dryers', *Am. Inst. Chem. Eng. J.*, **6**, 606 (1961)

9.22. ANON, 'Drum dryer for powder material', *Chem. Process*, **7**, (10), 7 (1961)

10.1. STAIRMAND, C. J., *Gas Purification Processes*, Chapter 12, (ed. G. Nonhebel), George Newnes Ltd., London (1964)

10.2. SAEMAN, W. C. and MITCHELL, T. R., Jr., 'Analysis of rotary dryer and cooler performance', *Chem. Eng. Prog.*, **50**, 467 (1954)

10.3. SAEMAN, W. C., 'Air-solids interaction in rotary dryers and coolers', *Chem. Eng. Prog.*, **58** (June 1962)

10.4. FRIEDMAN, S. J. and MARSHALL, W. R., 'Studies in rotary drying', *Chem. Eng. Prog.*, **45**, 482–493 (1949)
See also *Perry's Chem. Eng. Handbook*, fourth edition, Section 20, McGraw-Hill Book Co. Ltd., London and New York (1963)

10.5. JACKSON, P. J. *et al.*, 'Automatic continuous measurement of sulphur trioxide in flue gases', *J. Inst. Fuel*, **43**, 10 (1970)

11.1. LEVA, M., *Fluidization*, McGraw-Hill Co., New York (1959)

11.2. ZENZ, F. A. and OTHMER, D. F., *Fluidization and Fluid Particle Systems*, Reinhold Publishing Co., New York (1960)

11.3. PERRY, J. H., *Chemical Engineers Handbook*, fourth edition, McGraw-Hill, London and New York (1963)

11.4. Symposium on Fluidization, Proc. Chem. Eng. Group of Soc., *Chem. Ind.*, **42** (1963)

11.5. VANECEK, V., MARKVART, M. and DRBOHLAV, R., *Fluidized Bed Drying*, Leonard Hill Ltd., London (1966)

11.6. HARRISON, D., DAVIDSON, J. F. and DE KOCK, J. W., 'On the nature of aggregate and particulate fluidization', *Trans. Inst. Chem. Engs.*, **39**, 202 (discussion 233) (1961)

11.7. STAIRMAND, C. J., 'Removal of dust from gases', *Gas Purification Processes*, Chapter 12 (ed. G. Nonhebel), Geo. Newnes Ltd. (1964). For more detail see also *J. Inst Fuel*, **29**, 58 (1956), and *The Chemical Engineer*, CE310 (Dec. 1965)

11.8. VAN HEERDEN, C., *J. of Applied Chemistry, Suppl. Issue 1*, **2**, 57 (1952)

11.9. CHERRETT, N., 'A fluidized bed dryer/cooler for polymer chip', *British Chemical Engineering* (Dec. 1964)

11.10. WATERHOUSE, D., 'Fluidized bed techniques', *J. British Steel Castings Research Association,* No. 71 (Feb. 1963)

11.11. RHYS-JONES, D. C., 'Fluidization in the coal industry', paper to Fluidization Symposium, *Proc. Chem. Enging Group, Society of Chemical Industry,* **42**, 73 (1963)

11.12. STEMERDING, S., DE GROOT, J. H. and KUYPERS, N. G. M., 'Engineering aspects of fluidization', paper to Fluidization Symposium, *Proc. Chem. Enging Group, Society of Chemical Industry,* **43**, 35 (1963)

11.13. TAILBY, S. R., 'The field for fluidization research', introductory paper to Fluidization Symposium, *Proc. Chem. Enging Group, Society of Chemical Industry,* **42** (1963)

GENERAL REFERENCES

11.14. STRATTON, R. A., 'Batch drying in industry', *Chemistry and Industry,* **76** (January 9, 1965)

11.15. Proceedings of the International Symposium on Fluidization–Eindhoven. Netherlands University Press. Amsterdam (1967) (Reviewed in *The Chemical Engineer,* CE 359 (Oct. 1968)

12.1. NAGANO, E. and AKIZAWA, I., 'Pneumatic conveying dryers for ammonium sulphate', *Coal Tar (Japan) C,* 394–6 (1954)

12.2. GORDON, C. W., 'Flash drying', *Chem. Eng. Prog.,* **45**, 477 (1949)

12.3. KROLL, K., 'Gas stream dryers', Z. V. Deut. Ing. **94**, 360 (1952)

12.4. HITCHCOCK, J. A. and JONES, C., 'The pneumatic conveying of spheres through straight pipes', *Brit. J. App. Phys.,* **9**, 218 (1958)

12.5. WOOD, S. A., *Proc. Inst. Mech. Eng.,* 141 (1939)

12.6. OTHMER, D. F., *Fluidization,* Rheinhold Publ. Corpn., New York (1956)

12.7. ZENZ, F. A., *Ind. Eng. Chem.,* **41**, 2801 (1949)

12.8. JOHNSTONE, H. F. *et al.,* 'Heat transfer to clouds of falling particles', *Trans. Amer. Inst. Chem. Eng.,* **37**, 95 (1941)

12.9. HAWKESLEY, P. G. W., BADZIOCH, S. and BLACKETT, J. H., *Measurement of Solids in Flue Gases,* Chapter 8 and Appendix 9, British Coal Utilisation Research Association, Leatherhead (1961)

12.10. DAVIES, C. N., 'Definitive equations for the resistance of spheres', *Proc. Phy. Soc.,* **57**, 259 (1945)

13.1. MARSHALL, W. R. and SELTZER, E., 'Principles of spray drying, Parts I and II', *Chem. Eng. Prog.,* **46**, 10, 501 and 575 (1950)

13.2. DUFFIE, J. A. and MARSHALL, W. R., 'Factors influencing properties of spray-dried material', *Chem. Eng. Prog.,* **49**, 417 (1953)

13.3. MARSHALL, W. R., 'Atomization and spray drying', *Chem. Eng. Prog. Monograph Series,* **50**, 2 (1954)

13.4. BÄR, P., 'Über die physikalischen Grundlagen der Zerstaübungstrocknung', *Doctorate thesis,* Karlsruhe University (1935)

13.5. RANZ, W. E. and MARSHALL, W. R., 'Evaporation from drops', *Chem. Eng. Prog.,* **48**, 141, 173 (1952)

13.6. KIRSCHBAUM, E., 'Fundamental principles and recent developments in spray drying', *Chem. Ing. Technik,* **24**, 3 (1952)

13.7. SJENITZER, F., 'Spray drying: theoretical considerations on the movement and evaporation of liquid droplets', *Chem. Eng. Sci.,* **1**, 101 (1952)

14.1. IRVIN, D. F., 'Filter drying', *Ind. Eng. Chem.,* **30**, 1008 (1938)

Index